BIOS INSTANT NOTES

Neuroscience

THIRD EDITION

BIOS INSTANT NOTES

Neuroscience

THIRD EDITION

Alan Longstaff
Associate Lecturer, Open University

GS Garland Science
Taylor & Francis Group
NEW YORK AND LONDON

Garland Science
Vice President: Denise Schanck
Editor: Elizabeth Owen
Editorial Assistant: Louise Dawnay
Production Editor: Ioana Moldovan
Copyeditor: Sally Huish
Typesetting and illustrations: Phoenix Photosetting, Chatham, Kent
Proofreader: Susan Wood
Printed by: MPG Books Limited

ISBN 978-0-4156-0769-8

Library of Congress Cataloging-in-Publication Data

Longstaff, Alan.
 Neuroscience / Alan Longstaff. -- 3rd ed.
    p. cm. -- (Bios instant notes)
 Includes index.
 ISBN 978-0-415-60769-8
 1. Neurosciences--Outlines, syllabi, etc. I. Title.
 RC343.6.L66 2011
 616.8--dc22

                        2011003160

Published by Garland Science, Taylor & Francis Group, LLC, an informa business,
270 Madison Avenue, New York NY 10016, USA, and 2 Park Square, Milton Park, Abingdon,
OX14 4RN, UK.

15 14 13 12 11 10 9 8 7 6 5 4 3 2 1

Garland Science
Taylor & Francis Group

Visit our web site at http://www.garlandscience.com

Preface

Neuroscience is multidisciplinary, having contributions from molecular biology, anatomy, physiology, pharmacology, psychology, and medicine to name the most obvious. Moreover, the neuroscience literature is huge. For these reasons textbooks of neuroscience tend to be large, making it hard for students to discriminate the conceptual heart of the subject from exemplars and enrichment material. This Third Edition of *Instant Notes in Neuroscience* is intended as a supplement to lectures that gives rapid and easy access to the core of the subject in an affordable and manageable-sized text.

When coming to a new subject, students commonly express three concerns; firstly, how to sort out the important ideas and facts from the wealth of detail; secondly, how to get to grips with the unfamiliar terminology; and thirdly, how to integrate their knowledge across the various disciplines, a necessary step for a good understanding of the subject. The *Instant Notes* format addresses these issues. Each topic is supported by a "Key Notes" panel which gives a concise summary of the crucial points. Whenever a term appears for the first time it is in bold and immediately followed by a definition or explanation. Extensive cross references are provided between topics so that students can forge the links that are important for integration.

Instant Notes in Neuroscience is a much slimmer volume than most neuroscience texts. Several features contribute to this. I have tried to minimize the amount of detail without compromising the need for students to have a database for autonomous learning. While many of the methods used by neuroscientists are included, *individual* experiments or items of evidence are included only where I thought it essential to illustrate a point. I have restricted examples to mammals, and the anatomy is largely human, even though there is a great deal of remarkable work in birds, fish, and invertebrates. This *third edition* is over one-fifth shorter than its predecessors. This has been achieved by removing two sections (Developmental Neurobiology and Brain Disorders), by simplifying the language of the text throughout, and by re-drafting many of the more complicated diagrams. However, I have not shied away from retaining conceptually challenging material where it is warranted.

The Third Edition of *Instant Notes in Neuroscience* has 15 sections. Section A sets the scene by introducing the cells of the nervous system, and looks at how the nervous system is organized by taking a broad view of neuroanatomy. The next three sections are essentially cellular neuroscience, concerned with electrophysiology, the properties of synapses, and an introduction to the principal neurotransmitters and their receptors. Elementary neural computing is introduced in Section E which looks at rate and temporal coding and how neurons are connected to make simple circuits. All the material thus far might reasonably be found in first year courses. The next six sections (F–K) form the core of systems neuroscience. These concentrate on sensory and motor neurobiology, but introduce visual attention and other cognitive aspects of brain function where needed. Section L tackles neuroendocrinology, how the brain controls metabolism, growth, and reproduction, and the autonomic nervous system.

The perspective of the next two sections is rather broader than what has come before. Behaviors, such as emotion, motivation, and sleep are explored in section M. Section N examines learning and memory, homing in on the episodic learning in the hippocampus and motor learning in the cerebellum as examples. Finally, Section O looks at the techniques used to explore the brain, neuroimaging, electrophysiology, and some of the remarkable methods being developed in the last few years such as brainbow and optogenetics.

As a student, how should you use this book? Restrict your reading only to the sections and topics covered by your current course. That said, sections A–E are likely to be part of any neuroscience program; you will probably need to work through these first. Later sections can be dipped into in any order. Read the main sections thoroughly first, making sure that you *understand* the ideas, and use the "Related topics" to make links. This is the stage to incorporate additional material from lectures, and other textbooks, in the gaps at the end of topics. For areas that particularly interest you, turn to "Further Reading" at the end of the book. Although by no means a comprehensive bibliography, following up the references in them will get you into the rest of the neuroscience literature. Studying *Instant Notes* "little but often" is a good strategy. The information density in the text is high, so many short, concentrated, bursts is the most effective way to study. The more times you work through a topic, the better your understanding will be, and the more likely you will remember it clearly. When it comes to revision, use the "Key Notes" as a prompt. In addition, you should aim to be able to write, from memory, a few sentences about each of the terms that appear in bold in the main text. Being able to reproduce the simpler diagrams is also an effective way of getting your point across in an exam.

I thank Elizabeth Owen and Louise Dawnay of Taylor & Francis for their hard work and encouragement; each helped shape the project in distinctive and important ways.

I hope you enjoy studying neuroscience as much as I do.

Contents

A1 Neurons

Key Notes

Cell body	The neuron cell body contains all the subcellular organelles found in a typical animal cell but it is specialized for high rates of protein synthesis.
Dendrites and axon	Dendrites are large extensions of the cell body and receive most of a neuron's synaptic inputs. A single thin axon arises from the axon hillock. Dendrites can synthesize proteins but axons cannot: axonal proteins come from the cell body. Both axons and dendrites have mitochondria.
Neuron classification	Neurons may be classified by their structure, connections, and neurotransmitters. Cells with one, two, or three or more neurites are classed as unipolar, bipolar, or multipolar respectively. The shape of the dendritic tree, the presence or absence of spines, and the length of the axon are used to categorize neurons. Connectivity distinguishes afferent neurons that provide input and efferent neurons that provide output. Neuron shape is often a good guide to the neurotransmitters it secretes, and so to its function.
Neuron numbers	It is estimated that the human brain has 86 billion neurons.
Related topics	(A3) Organization of the peripheral nervous system (A4) Organization of the central nervous system (B3) Action potential conduction (C1) Synapse structure and function

---

Cell body

The **cell body** (**perikaryon**) of a neuron (Figure 1) contains the nucleus, Golgi apparatus, ribosomes, and mitochondria, and maintains high levels of biosynthetic activity. The rough endoplasmic reticulum is densely packed with ribosomes, forming neuron-specific structures called **Nissl bodies** that allow very high rates of protein synthesis.

Neurons come in a great variety of shapes and sizes; cell bodies range from 5 to 120 μm across.

Dendrites and axon

Neurons are distinguished from other cells by **neurites**. These are long cylindrical structures: dendrites and axons. **Dendrites** are highly branched extensions of the cell body, up to 1 mm in length, and constitute up to 90% of the surface area of a neuron. A neuron may have one or many dendrites, arranged in a cell-typical pattern called a **dendritic tree**. The majority of synaptic inputs are with dendrites. Dendrites on **spiny neurons** are covered

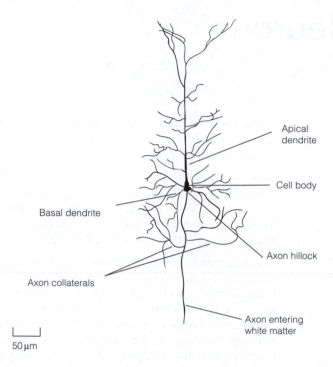

Figure 1. Key features of a neuron. A drawing of a pyramidal cell showing the distribution of neurites (dendrites and axon).

with hundreds of tiny projections termed **dendritic spines** on which synapses are made. Nerve cells lacking spines are called **aspiny neurons**.

Nerve cells generally have only one **axon**. It usually arises from the cell body but may emerge from a dendrite just where it leaves the cell body. In either case the site of origin is termed the **axon hillock**. Axons have diameters ranging from 0.2 to 20 μm in humans (but up to 1 mm in invertebrates) and vary in length from a few μm to over a meter. They may be encapsulated in a myelin sheath. Axons usually have branches, referred to as **axon collaterals**. The ends of an axon are swollen **terminals** (boutons) and usually contain mitochondria and vesicles. Some axons have swellings along their length called **varicosities**. Axon terminals and varicosities are the presynaptic components of chemical synapses. Axons are less highly branched than dendrites and unlike dendrites do not have protein synthetic machinery. Proteins made in the cell body are moved into and along the axon by **axoplasmic transport**. Both axons and dendrites have mitochondria. Axon terminals are rich in mitochondria.

Neuron classification

Nerve cells can be classified by their structure, connections, and neurotransmitters.

Structural classification is based on shape and size of the cell body, its dendritic tree, axon length, and the nature of the connections it makes. Neurons with one, two, or more than two neurites, are **unipolar**, **bipolar**, or **multipolar** respectively (Figure 2). Most neurons in vertebrate nervous systems are multipolar, but there are important exceptions. For example, bipolar neurons in the retina synapse with photoreceptors, and sensory

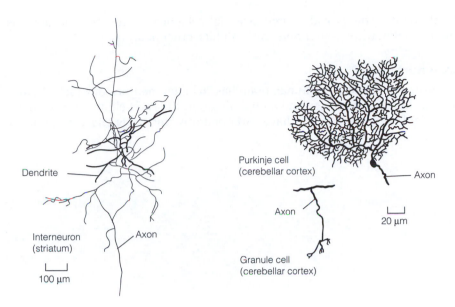

Figure 2. The morphologies of three common types of neuron. The full length of the axons is not shown. The bifurcating axon of the granule cell extends for several millimeters in each direction. Note how the axon of the interneuron branches extensively.

neurons in the dorsal root ganglion are described as **pseudounipolar** because they start life as bipolar cells but their two neurites fuse. Unipolar neurons predominate in invertebrates.

The shape of the dendritic tree, and the presence or absence of dendritic spines, helps determine the efficacy of its synaptic connections and hence the functionality of the cell. Pyramidal cells comprise some 60% of neurons in the cerebral cortex and have pyramidal-shaped cell bodies and dendritic trees. Other cortical cells are termed stellate cells because of the star-like appearance of their dendritic trees. Purkinje cells of the cerebellar cortex have the unique feature that their dendrites form a two-dimensional array.

Neurons can also be classified by axon length. **Projection** (principal, relay, or Golgi type I) **neurons** have long axons which extend into other regions of the nervous system. Pyramidal and Purkinje cells fall into this category. In contrast, **interneurons** (local circuit or Golgi type II) have short axons and produce direct effects only in their immediate neighborhood (e.g., stellate cells).

Neurons can be categorized by the connections they make, and this reflects their function. Any given region of the nervous system receives inputs from **afferent neurons** and projects by **efferent neurons** to other regions of the nervous system or an effector organ (such as a muscle or gland). Afferent neurons capable of responding directly to physiological stimuli are **sensory neurons**. Efferent projection neurons in motor pathways are called **motor neurons**.

Finally, neurons can be classed according to the neurotransmitters which they secrete. There is often a clear correlation between neuron structure and neurotransmitter. For example, pyramidal cells release glutamate whereas stellate and Purkinje cells secrete γ-aminobutyrate. This provides circumstantial evidence for function because usually glutamate excites, while γ-aminobutyrate inhibits, other neurons.

Neuron numbers

Recent estimates suggest the human brain has 86 billion neurons of which about 16 billion are in the cerebral cortex and 69 billion are in the cerebellum. Smaller mammals have smaller brains because they have fewer neurons (and fewer glial cells), not because their neurons are smaller.

A2 Glial cells

Key Notes

Types of glial cell	Glial cells are about as numerous as neurons overall and perform functions which support them.
Astrocytes	Astrocytes are large, star-shaped glia that surround neurons. They have elongated processes tipped with endfeet which form contacts with capillary endothelial cells and with the pia mater to form the glial membrane. The functions of astrocytes include homeostatic regulation of the extracellular K^+ concentration, the synthesis of the transmitters glutamate and γ-aminobutyrate, removal of neurotransmitters from the synaptic cleft, storing glycogen, and supplying lactate to neurons.
Oligodendrocytes and Schwann cells	Oligodendrocytes in the central nervous system (CNS) and Schwann cells in the peripheral nervous system are responsible for forming the insulating myelin sheath that surrounds many axons.
Microglia	Microglia are small immune cells derived from monocytes. In their macrophage guise they are key players in the inflammatory processes that accompany repair of nervous system injury.
Related topics	(A6) Blood–brain barrier (B3) Action potential conduction — (C4) Neurotransmitter inactivation

---

Types of glial cell

Glial cells outnumber neurons almost four-fold in the cerebral cortex, but in other regions (e.g., cerebellum) glia are heavily outnumbered. Overall the glia/neuron ratio in the human brain is thought to be close to one. They mediate functions which support neurons. Several distinct populations of glia are recognized: astrocytes, oligodendrocytes (including Schwann cells), and microglia.

Astrocytes

Astrocytes are irregularly shaped cells and many have long processes which superficially resemble the dendrites of neurons. Astrocytes can be distinguished from neurons by their lack of Nissl bodies and by the presence of specific astrocyte marker molecules. Astrocytes invest neurons and synapses leaving a gap just 20 nm across. They extend processes, forming **endfeet** which butt onto capillaries or onto the pia mater to produce a layer covering the surface of peripheral nerves and CNS called the **glial membrane**. Astrocytes are coupled to each other via electrical synapses.

Astrocytes have a wide variety of functions:

- Controlling local potassium concentration by removing K^+ that accumulates in the extracellular space as a result of neural activity and dumping it, via their endfeet, into capillaries.

- Uptake or synthesis of the precursors for the neurotransmitters glutamate and γ-aminobutyrate.

- Terminating the actions of small transmitter molecules by removing them from the synaptic cleft.

- Providing neurons with metabolic energy. Astrocytes take up glucose from blood and either store it as glycogen or convert it to lactate which is exported to neurons.

- Detoxification of ammonia (via the ornithine-arginine cycle), and of free radicals.

- Regulating the blood–brain barrier.

- Regulation of synapse formation in the developing brain and the production of new neurons in the adult brain.

- As radial glial cells they guide neurons to their proper destinations in the developing brain.

- **Gliotransmission**: astrocytes release transmitters under some circumstances (e.g. glutamate, adenosine) and may have a role in information processing, but are not excitable.

Oligodendrocytes and Schwann cells

Oligodendrocytes in the CNS and **Schwann cells** in the peripheral nervous system have the common function of providing the myelin sheath, an electrically insulating covering around many axons. Those axons with a myelin sheath are said to be myelinated, those without are termed unmyelinated.

Microglia

The smallest of the glial cells, microglia are immune system cells. Derived from bone marrow monocytes, they migrate into the nervous system during development where they secrete growth factors, guide axons, stimulate differentiation of other glial cells and the formation of blood vessels. They can transform into macrophages which phagocytose debris generated by developmental programmed cell death. In adults they cease being motile unless nervous system damage occurs (infections, trauma, and tumors) when they proliferate, revert to their macrophage lifestyle, and are responsible for inflammatory processes.

A3 Organization of the peripheral nervous system

Key Notes

Principal divisions of the nervous system

The brain and spinal cord comprise the central nervous system, while the peripheral nervous system, divided into somatic, autonomic, and enteric parts, is everything else.

Somatic nervous system

Thirty-one pairs of spinal nerves originating from the spinal cord and 12 pairs of cranial nerves arising from the brain make up the somatic nervous system. Almost all spinal nerves but only four cranial nerves have both sensory and motor fibers. Every spinal segment has a pair of spinal nerves, each with a dorsal root containing sensory fibers and a ventral root with motor fibers. The cell bodies of the sensory neurons lie outside the spinal cord in the dorsal root ganglia.

Peripheral nerves

Peripheral nerves consist of nerve fibers (axons surrounded by their associated Schwann cells) organized into bundles (fasciculi) and invested with connective tissue. Peripheral nerve fibers are classified by their diameters and conduction velocities.

Autonomic nervous system (ANS)

The visceral motor system originates with cell bodies in the CNS that give rise to preganglionic myelinated axons that secrete acetylcholine. They synapse with postganglionic unmyelinated axons in autonomic ganglia. The ANS has two divisions, sympathetic and parasympathetic. The sympathetic system arises from thoracic and lumbar spinal segments, has autonomic ganglia close to the cord. Its postganglionic axons usually secrete noradrenaline (norepinephrine). The parasympathetic system originates from the brainstem and sacral spinal cord. Its autonomic ganglia are located on or near the innervated organ and its postganglionic axons secrete acetylcholine.

Enteric nervous system

The nervous system of the gut is two interconnected cylindrical sheets of neurons embedded in the gut wall. The enteric nervous system coordinates gut motility and secretion autonomously although its activity is modified by the ANS.

Related topics

(J1) Nerve–muscle synapse	(L5) Autonomic nervous system (ANS) function

Principal divisions of the nervous system

The nervous system is comprised of the **central nervous system** (**CNS**) and **peripheral nervous system** (**PNS**). These divisions are contiguous both anatomically and functionally. The CNS includes the brain and spinal cord. The peripheral nervous system is everything else; namely nerve trunks going between the CNS and the periphery, and the networks of nerve cells with supporting glia in organs throughout the body. The PNS has three subdivisions, the somatic, autonomic, and enteric nervous system.

Somatic nervous system

In humans the **somatic nervous system** consists of 31 pairs of **spinal nerves**, each pair arising from a single segment of the spinal cord, and 12 pairs of cranial nerves which come from the brain. Afferent axons in spinal and cranial nerves entering the CNS carry sensory information from skin, muscles, joints, and viscera. The majority are wired to mechanoreceptors which inform about mechanical forces, others are nociceptors which signal tissue damage, and some (restricted to skin) are connected to temperature-sensitive thermoreceptors. Efferent fibers leaving the CNS are axons of motor neurons supplying skeletal muscles.

Almost all spinal nerves are mixed, containing both sensory and motor fibers. The exceptions are C1 which is motor and Cx 1 which is sensory. Of the cranial nerves only four are mixed (Table 1).

Each spinal nerve is formed from a **dorsal root** housing sensory axons and a **ventral root** carrying motor axons. The cell bodies of the primary afferent neurons lie within the **dorsal root ganglia** (**DRG**) just outside the spinal cord. There are a pair of DRG for each spinal segment. Efferent neuron cell bodies lie within the spinal cord (Figure 1).

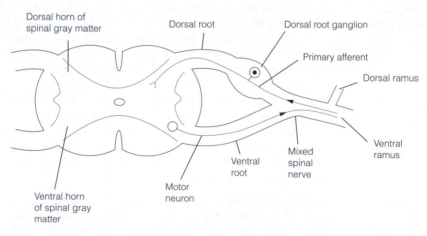

Figure 1. Origin of a spinal nerve from a spinal cord segment.

Peripheral nerves

All **peripheral nerves** have a common basic structure. A nerve fiber consists of an axon together with accompanying Schwann cells. Several unmyelinated axons are invested by a single glial cell. Individual nerve fibers supported by connective tissue, the endoneurium, are collected into bundles, **fasciculi**, surrounded by a connective tissue sheath, the

Table 1. Peripheral nerves

Nerve	Type	Region of origin or destination in CNS	Function
Cranial nerves			
I Olfactory	Sensory	Olfactory bulb	Smell
II Optic	Sensory	Forebrain LGN (thalamus)	Vision
		Midbrain superior colliculus tectum	Visual reflexes
III Oculomotor	Motor[a]	Midbrain	Motor to extrinsic eye muscles except superior oblique and lateral rectus, autonomic to intrinsic eye muscles
IV Trochlear	Motor	Midbrain	Motor to superior oblique extrinsic eye muscles
V Trigeminal	Mixed	Midbrain and hindbrain	Sensory from head and face, motor to jaw muscles
VI Abducens	Motor	Hindbrain	Motor to lateral rectus extrinsic eye muscles
VII Facial	Mixed[a]	Ventral lateral thalamus (sensory)	Sensory from tongue (taste) and palate
		Hindbrain (motor)	Motor to face, parasympathetic secretomotor to submandibular, submaxillary salivary glands and lachrymal glands
VIII Vestibulocochlear	Sensory	MGN (auditory division) Hindbrain (vestibular division)	Sensory from inner ear (hearing and balance)
IX Glossopharyngeal	Mixed[a]	Hindbrain	Sensory from tongue (taste) Motor to pharyngeal muscles Parasympathetic secretomotor to parotid salivary glands
X Vagus	Mixed[b]	Hindbrain	Sensory from viscera Somatic motor to pharyngeal and laryngeal muscles Parasympathetic to viscera
XI Accessory	Motor	Medulla, spinal cord C1–C5	Motor to palate and some neck muscles
XII Hypoglossal	Motor	Medulla	Motor to tongue
Spinal nerves			
C1–8	Mixed		
T1–12	Mixed (including sympathetic autonomic T1–12)		
L1–5	Mixed (including sympathetic autonomic L1, 2)		
S1–5	Mixed (including parasympathetic autonomic S2, 3)		
Cx 1	Mixed		

[a] Including autonomic.
[b] Large autonomic component.
LGN, lateral geniculate nucleus; MGN, medial geniculate nucleus.

perineurium. A nerve may be one or several fasciculi all encapsulated by a connective tissue epineurium.

Two systems for the classification of PNS nerve axons are in common use. They are based on axon diameter and conduction velocity and are summarized in Table 2. The Erlanger and Gasser system is used to classify both afferents and efferents. The Lloyd and Hunt scheme is used exclusively to define afferent axons.

Table 2. Classification of peripheral nerve fibers

Fiber (type/group)	Mean diameter (μm)	Mean conduction speed (m s^{-1})	Functions (example)
Erlanger/Gasser classification (type)			
Aα	15	100	Motor neurons
Aβ	8	50	Skin touch afferents
Aγ	5	20	Motor to muscle spindles
Aδ	4	15	Skin temperature afferents
B	3	7	Unmyelinated pain afferents
C	1	1	Autonomic postganglionic neurons
Lloyd/Hunt classification (group)			
I	13	75	Primary muscle spindle afferents
II	9	55	Skin touch afferents
III	3	11	Muscle pressure afferents
IV	1	1	Unmyelinated pain afferents

Autonomic nervous system (ANS)

The ANS is the visceral motor nervous system. By definition it includes no sensory components. However, the activities of the ANS are modified by sensory input that travels by way of the somatic nervous system, and by the CNS. The target tissues of the ANS are smooth muscle, cardiac muscle, endocrine and exocrine glands, liver, kidney, and adipose tissue. The synapses of autonomic neurons with their target cells are called neuroeffector junctions.

The **preganglionic neurons** of the ANS have their cell bodies in motor nuclei of cranial nerves, or the intermediolateral horn of the thoracic and lumbar spinal cord. Their axons are myelinated B fibers which secrete acetylcholine. The preganglionic axons synapse with postganglionic neurons in **autonomic ganglia**. The axons of the **postganglionic neurons** are unmyelinated C fibers. The ANS has two divisions, the **sympathetic** and **parasympathetic**, the main distinguishing features of which are summarized in Table 3.

In general the preganglionic axons of the sympathetic division are short and the postganglionic axons are long because the sympathetic ganglia lie close to the spinal cord in one of two locations:

- In paired paravertebral chains that run parallel to the vertebral column in the neck and down the posterior wall of the thorax and abdomen

Table 3. Divisions of the autonomic nervous system

Anatomy	Physiology	Postganglionic cell neurotransmitters

Craniosacral
    Preganglionic axons in cranial
    nerves III, VII, IX, X and spinal
    nerves S2, S3
	Parasympathetic	Acetylcholine
		Vasoactive intestinal peptide

Thoracolumbar	Sympathetic	Norepinephrine (but acetylcholine at selected neuroeffective junctions)
Preganglionic axons in spinal nerves T1–T12, L1, L2		Neuropeptide Y
		Adenosine 5′-triphosphate

● In subsidiary ganglia of autonomic plexuses adjacent to major blood vessels

The pathway taken by sympathetic axons is illustrated in Figure 2.

Preganglionic axons may synapse in the nearest ganglion, traverse the paravertebral chain to synapse in subsidiary ganglia, or ascend or descend the chain to synapse in a ganglion at a different level. Preganglionic sympathetic axons can modify the actions of up to 100 postganglionic cells. Most, but not all, postganglionic sympathetic axons secrete noradrenaline (norepinephrine). The **adrenal medulla** secretes adrenaline (epinephrine) directly into the circulation in response to activity in the preganglionic sympathetic fibers which supply it. The adrenal medulla is therefore regarded as part of the sympathetic system.

Parasympathetic autonomic ganglia are all subsidiary ganglia located close to the target organ. For this reason, in the parasympathetic division the preganglionic axons are long,

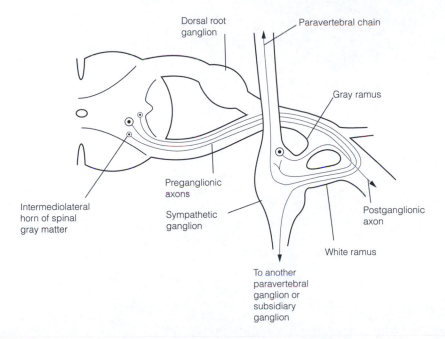

Figure 2. Sympathetic pathway from the spinal cord.

and the postganglionic ones are short. All postganglionic parasympathetic axons secrete acetylcholine. Although all the major organs, except the liver, have a parasympathetic supply, this division is far less extensive than the sympathetic. This is partly because only a few specialized blood vessels have a parasympathetic innervation, whereas all blood vessel smooth muscle receives a sympathetic supply.

Enteric nervous system

An interconnected network of about 10^8 neurons makes up the nervous system of the gut. It is organized into two thin cylindrical sheets that run along the length of the gut. The **myenteric** (**Auerbach's**) plexus lies between the longitudinal and circular smooth muscle layers and extends the whole length of the gut. The **submucosal** (**Meissner's**) plexus lies in the submucosa and extends from the pylorus of the stomach to the anus. There are extensive interconnections between these two plexuses. The enteric nervous system acts autonomously to coordinate gut motility and secretion. Its activity is modified by both divisions of the ANS.

A4 Organization of the central nervous system

Key Notes

Spinal cord

The human spinal cord contains about one hundred million neurons. Transverse sections reveal peripheral white matter (tracts of axons ascending/descending the cord) and central gray matter (neuron cell bodies). Sensory neuron fibers enter the dorsal horn of the gray matter in an ordered fashion. Motor neuron cell bodies lie in the ventral horn of the gray matter. The spinal gray is divided on morphological grounds into 10 columns which, on transverse section, are called Rexed laminae, each with distinctive connections and functions.

Brain

White matter consists of fiber tracts or pathways. Embedded in this are nuclei which are clusters of neuron cell bodies. Two large brain structures, the cerebrum and cerebellum are covered by cortex. The brain has three principal anatomical divisions; hindbrain, midbrain, and forebrain. The hindbrain consists of medulla, pons, and cerebellum. Together hindbrain (minus cerebellum) and midbrain are the brainstem from which emerge most of the cranial nerves. The brainstem is mostly concerned with cardiovascular control, breathing, arousal, sleep/wakefulness, and motivation. The cerebellum is involved in motor coordination. The forebrain consists of diencephalon and cerebrum. The diencephalon contains a dorsal thalamus, serving sensory functions among others, and a ventral hypothalamus, implicated in temperature and endocrine regulation, and appetitive behaviors. The cerebrum has two cerebral hemispheres and its cortex mediates motor, perceptual, and cognitive functions. The core of the cerebrum is occupied by nuclei which form two neural systems: the extrapyramidal motor system and the limbic system (which includes cortex) that is concerned with emotion and learning.

Related topics

(O1) Neuroimaging

(O4) Neuroanatomy imaging technologies

Spinal cord

The human **spinal cord** has about 10^8 neurons. A transverse section through the spinal cord shows a butterfly-shaped central **gray matter** which contains neuron cell bodies. The **white matter** surrounding the gray is largely axons in ascending and descending tracts and gets its color from the high content of myelin. In the middle is the central canal which contains cerebrospinal fluid (CSF), though in adults it is usually closed.

Sensory axons enter the spinal cord via the dorsal roots to synapse largely with cells in the **dorsal horns** of the spinal gray matter. Larger diameter fibers enter more medially and extend more deeply into the dorsal horn than smaller ones. Motor neuron cell bodies lie in the **ventral horns** of the spinal gray and their axons exit via the ventral roots.

In the spinal gray matter, 10 columns running through the cord can be distinguished on the basis of cell size. On transverse section these columns appear as **Rexed laminae** (Figure 1).

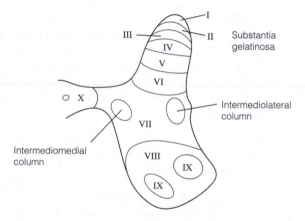

Figure 1. Rexed laminae. Lamina VI is only present in spinal segments supplying the limbs.

Each lamina has distinctive input–output relations which reflect functional specialization. Thus, while nociceptor afferents synapse on cells in lamina II, cutaneous mechanoreceptor afferents terminate in deeper layers of the dorsal horn, and lamina VII and lamina IX house preganglionic autonomic neurons and motor neurons respectively.

The white matter is organized into tracts specified by origin and destination. For example, the descending tract from the cerebral cortex is termed the corticospinal tract whereas the ascending pathway which terminates in the thalamus is the spinothalamic tract (Figure 2).

Brain

There are three main structural components to the brain:

- Tracts or **pathways** enter the neuraxis at various levels, ascend or descend, and these, together with internal tracts which go from one part of the brain to another, constitute the white matter.

- **Nuclei** embedded in the white matter are clusters of neuron cell bodies. Some neural structures are composed of groups of nuclei. The thalamus, for example, consists of some 30 nuclei.

- Two brain structures, the cerebrum and the cerebellum, are covered by **cortex**, a thin rind with a very high density of neuron cell bodies. In wiring terms cortex appears to be a simple circuit between just a few neuron types, repeated millions of times.

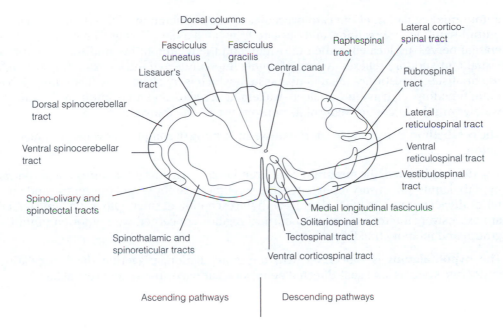

Figure 2. Pathways in the spinal cord white matter.

Together, the nuclei and cortex are the gray matter of the brain. Neural systems are comprised of interconnected nuclei and cortical regions that serve a common function. The visual system, for example, consists of the retinas, the lateral geniculate nuclei of the thalamus, the visual cortex and the pathways between them.

The most fundamental division of the brain into hindbrain, midbrain, and forebrain can be discerned in the human embryo by the end of the 4th week when the CNS is a hollow neural tube (Figure 3).

The **hindbrain** consists of **medulla**, **pons**, and cerebellum, although the hindbrain (minus the cerebellum) and midbrain together are often referred to as the **brainstem**. Much of the brainstem is occupied with vital (life-support) functions; for example,

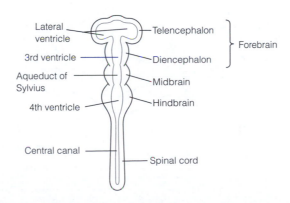

Figure 3. The human embryo neural tube, at 28 days gestation.

autonomic regulation of the cardiovascular system, generation of the rhythmic neural output required for breathing, and basic reflexes such as swallowing. The nuclei of most cranial nerves are located in the brainstem, and the tectum of the **midbrain** organizes visual and auditory reflexes. A core of highly interconnected nuclei extending through the brainstem constitutes the reticular system which is involved in orchestrating global brain functions such as arousal, sleep–wakefulness cycles, and motivation, and connects widely with the forebrain. Many of its neurons use amine transmitters.

The **cerebellum** is involved in coordination and timing of complex movements, and cognition.

The **diencephalon** of the **forebrain** is differentiated into a dorsal **thalamus** and a ventral hypothalamus. All sensory input enters the cerebral cortex by way of the ventral and posterior nuclear groups of the thalamus, with the exception of smell. Other thalamic nuclei are extensively interconnected with cortical regions concerned with emotion (anterior group) and memory (medial group).

The **hypothalamus** is concerned with thermoregulation, triggering sleep, regulating endocrine systems, and goal-directed behaviors (eating, drinking, and sexual behavior).

The smallest part of the diencephalon, the **pineal gland**, gets visual input and regulates circadian rhythms on the basis of the hours of light and dark.

The dominant part of the telencephalon is the **cerebrum**, two **cerebral hemispheres** linked across the midline by about 10^6 axons that constitute the **corpus callosum**. Each hemisphere is divided into four lobes named for the bones which overlie them (Figure 4). The surface is covered by cortex and is highly convoluted giving it a high surface area/volume ratio. The folds are called **gyri** (sing., **gyrus**), and the creases between them **sulci** (sing., **sulcus**). Most of the cerebral cortex is **neocortex** (new cortex) which has six layers. Cortical regions are mapped into **Brodmann areas** on the basis of differences in cellular

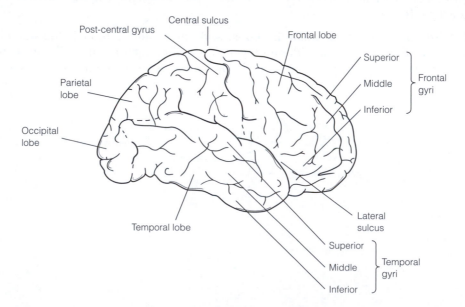

Figure 4. Lateral surface of the human right cerebral hemisphere.

makeup and relative thickness of the layers. The significance of this is that the Brodmann map corresponds quite well to how functions are localized in the cortex.

The layers of the cerebral cortex are numbered from I, nearest to the pial surface through to VI which is the deepest (Figure 5). The layers contain different proportions of two types of neurons, pyramidal cells which are output cells, and stellate cells that are interneurons. The relative thickness of the layers differs with the function of the cortical region. For example, the sensory cortex has a thick layer IV because of its large number of thalamic inputs, whereas motor cortex has a thick layer V because this is the location of motor neurons projecting to the brainstem and spinal cord.

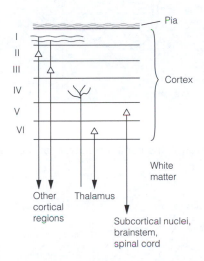

Figure 5. Inputs and outputs of the cerebral cortex.

The cerebral cortex is implicated in most brain activities, but is most often associated with the planning and execution of intentional movement, sensory perception, and cognitive (problem solving) functions. Those regions which are not specifically devoted to sensory or motor activities are called association cortex.

Within the core of each hemisphere lie clusters of nuclei that form components of two neural systems, the extrapyramidal system and the limbic system (Figure 6). The major component of the **extrapyramidal system**, responsible for organizing stereotyped patterns of movement, is the **basal ganglia**, consisting of the **striatum**, which lies in the forebrain, and two midbrain nuclei, the **subthalamus** and the **substantia nigra**. The striatum is subdivided into the **neostriatum**, itself composed of two nuclei, the **caudate** and **putamen**, and **paleostriatum** or **globus pallidus**. Anatomically the putamen and globus pallidus together form the **lentiform nucleus**.

The **limbic system** comprises nuclei (amygdala, septal nucleus, and mammillary bodies) plus regions of cerebral cortex which form a ring around the diencephalon, including the **cingulate cortex** and **hippocampus** (Figure 7). The hippocampus is **archaecortex** (ancient cortex) and has only three layers. The hippocampus and **amygdala** are concerned with certain types of learning, and the limbic system in general is implicated in emotion.

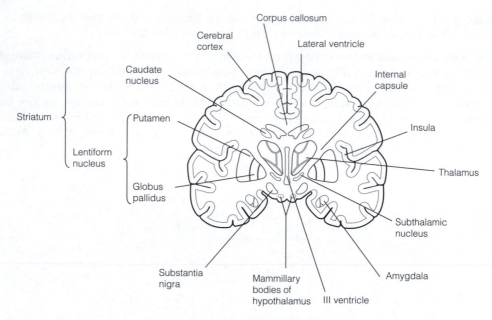

Figure 6. Coronal section through the human cerebrum at the level of the posterior hypothalamus.

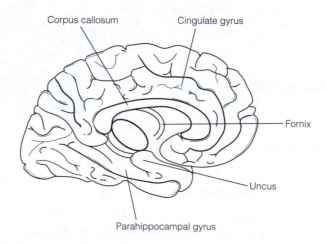

Figure 7. Medial surface of the human left cerebral hemisphere.

A5 Meninges and cerebrospinal fluid

Key Notes

Meninges

The brain and spinal cord are invested by three connective tissue layers, the meninges. Directly covering the brain is the pia mater, above which is the arachnoid mater. Between these layers lies the subarachnoid space which is filled with cerebrospinal fluid (CSF) and through which run blood vessels, branches of which enter the brain. Passive exchange of water and solutes across the pia mater keeps brain extracellular fluid and CSF in equilibrium. The tough outer layer is the dura mater which contains venous sinuses into which arachnoid villi project. Here tiny one-way valves allow the bulk flow of CSF from subarachnoid space into the venous circulation.

Cerebrospinal fluid (CSF) circulation

CSF is actively secreted by the choroid plexuses located in the ventricles. The direction of CSF flow is from lateral to 3rd to 4th ventricles, from where it enters the subarachnoid space. Finally it drains into the venous sinuses.

CSF secretion

About 500 cm^3 of CSF is secreted per day into a volume of between 100 and 150 cm^3. Choroid plexus epithelium contains a variety of active transport mechanisms. This results in secretion into the CSF of sodium, chloride, and bicarbonate but resorption of potassium, glucose, urea, and a number of neurotransmitter metabolites. The protein concentration of CSF is very much lower than that of blood plasma but its osmolality is the same.

CSF and meningeal functions

The CSF acts as a sink for metabolites that eventually are transported into the blood via arachnoid villi or choroid plexuses. Mechanical protective functions of CSF and meninges are to reduce the weight of the brain in the skull, to accommodate changes in intracranial pressure from altered brain blood flow and cushion the brain during head movements.

Related topics

(A4) Organization of the central nervous system

Meninges

The brain and spinal cord are surrounded by three connective tissue membranes, the meninges (Figure 1).

The **subarachnoid space**, filled with cerebrospinal fluid, separates the **pia mater** and **arachnoid mater**. Superficial cerebral blood vessels, invested by arachnoid mater, run

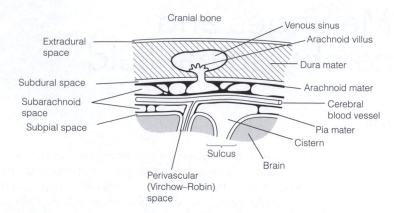

Figure 1. The meninges.

through the subarachnoid space and their branches penetrate the brain becoming surrounded by a cuff of pia mater that extends as far as the capillaries. The **perivascular** (**Virchow–Robin**) **space** between the vessel wall and the pia mater is continuous with the subarachnoid space. Here passive exchange of water and solutes across the pia mater keeps the CSF in equilibrium with brain extracellular fluid. At the cerebral capillaries the pia mater is lost and the single layer of capillary endothelial cells, with their basement membrane, are covered by glial cells. Expanded regions of the subarachnoid space are cisterns.

The **dura mater** is a thick tough outer layer with venous sinuses running through it. Small herniations of arachnoid mater called **arachnoid villi** protrude through the dura into the venous sinuses. Here bulk flow of CSF into blood occurs via mesothelial tubes in the arachnoid villi that act as valves, closing when the pressure in the venous sinus exceeds that of subarachnoid space to prevent reflux of blood into the CSF.

The **subdural space** is a potential space between the dura mater and the arachnoid mater. It is traversed by cerebral veins entering the venous sinuses in the dura.

Cerebrospinal fluid circulation

Cerebrospinal fluid (**CSF**) is actively secreted by choroid plexuses situated in the lateral, third, and fourth ventricles (Figure 2). Flow of CSF is from the **lateral ventricles** through the foramen of Munro into the **third ventricle**, and then through the aqueduct of Sylvius into the **fourth ventricle**. From here it drains via three orifices into the subarachnoid space. Here it equilibrates with extracellular fluid in the perivascular spaces. Finally it is dumped into the venous sinuses via the arachnoid villi.

CSF secretion

Choroid plexus consists of a cuboidal epithelium covering a core of highly vascular pia mater. In adult humans CSF is secreted at about 500 cm^3 day^{-1} into a steady state volume of 100–150 cm^3. Of this, about 30 cm^3 is in the ventricles and the rest in the subarachnoid space. Cerebrospinal fluid is turned over about every 5–7 hours.

The choroid plexus secretes some substances and absorbs others, most by active transport mechanisms. In this way it acts as a selective interface between blood and CSF, the blood–CSF barrier. The result is that by comparison with blood plasma CSF has somewhat

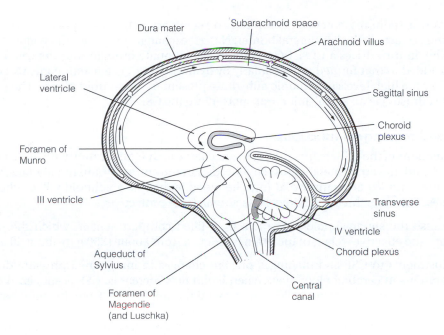

Figure 2. Cerebrospinal fluid circulation (arrow shows the direction of bulk flow).

higher Na^+, Cl^-, and HCO_3^- concentrations but lower K^+, urea, glucose, and amino acid concentrations. Although the protein concentration of CSF is about 1000-fold lower than blood plasma, its higher ionic concentration gives the two fluids the same osmolality.

Aspects of ion transport across the blood–CSF barrier are shown in Figure 3. Na^+, K^+-ATPase on the apical border of the epithelial cell pumps sodium into the CSF. This

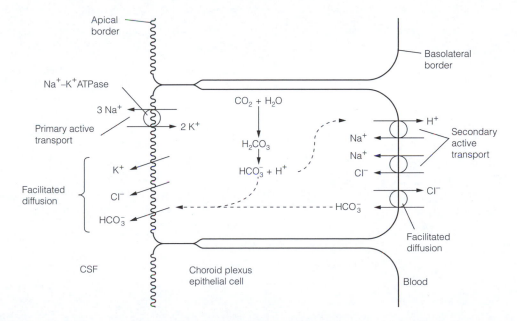

Figure 3. Aspects of ion transport across the choroid plexus.

generates a sodium gradient that drives two secondary active transport mechanisms bringing Na^+ across the basolateral border; $Na^+–H^+$ exchange and a $Na^+–Cl^-$ symport. The Cl^- influx in turn drives a $Cl^-–HCO_3^-$ antiport. Bicarbonate brought into the cell in this way is added to that formed intracellularly by hydration of CO_2, a reaction greatly accelerated by the high levels of carbonic anhydrase present in the choroid plexus. The bicarbonate diffuses via an apical anion transporter into the CSF.

CSF and meningeal functions

The functions of the CSF are metabolic and mechanical. By equilibrating with brain extracellular fluid unwanted metabolites (choline, dopamine and serotonin metabolites, urea, creatinine, and K^+) are transported into the blood, either via arachnoid villi or choroid plexuses. There are three mechanical effects which are protective:

- Because the subarachnoid space is a fluid-filled compartment in which the brain floats, the effective weight of the brain is reduced from about 1350 g to about 50 g.

- Adjustments to CSF and meninges prevent changes in intracranial pressure due to alterations in cerebral blood flow. When blood flow increases, CSF is squeezed from ventricles into the subarachnoid space around the spinal cord. Here the dura mater is more elastic and stretches to accommodate the rise in volume. Longer term increases in intracranial pressure can be offset by a rise in CSF flow into the venous sinuses through the arachnoid villi.

- The meninges (particularly the dura) support the brain and the CSF reduces the force with which the brain impacts the inside of the cranium when the head moves.

A6 Blood–brain barrier

Key Notes

Structure of the blood–brain barrier	The blood–brain barrier is formed by capillary endothelial cells which are coupled by tight junctions. A few regions of the brain, the circumventricular organs, are on the blood side of the blood–brain barrier and are able to secrete substances directly into the blood, or monitor the concentrations of materials in the blood.
Functions of the blood–brain barrier	The blood–brain barrier is a highly selective permeability barrier which allows the passage of water, some gases, and lipid-soluble molecules by passive diffusion, and contains specific carrier-mediated transporters for the selective transport of molecules crucial to neural function (such as glucose and amino acids). It prevents the entry of circulating neuroactive compounds and is able to exclude lipophilic, potential neurotoxins via P-glycoprotein. Cerebral edema is the accumulation of excess water in the extracellular space of the brain, and results when hypoxia causes the blood–brain barrier to open.
Related topics	(A2) Glial cells (A4) Organization of the central nervous system (L2) Posterior pituitary function

Structure of the blood–brain barrier

The **blood–brain barrier** governs what crosses from the blood into the brain extracellular fluid. The anatomical barrier is provided by brain capillary endothelial cells which are coupled to each other by **tight junctions** that are one hundred-fold tighter than is typical for other capillaries. Even small ions will not permeate between endothelial cells in brain capillaries. Moreover, brain capillary endothelial cells exhibit little pinocytosis (bulk transfer of fluid) or receptor-mediated endocytosis. Brain capillaries are entirely covered by the astrocyte endfeet which secrete growth factors (e.g., angiopoietin 1) that promote the efficacy of tight junctions (Figure 1).

The **circumventricular organs** (**CVOs**), for example, the posterior pituitary, lie on the blood side of the blood–brain barrier. They are isolated from the rest of the brain by ependymal cells (epithelial cells lining the ventricles) that are coupled together by tight junctions. The lack of a blood–brain barrier at the posterior pituitary permits oxytocin and vasopressin to be secreted directly into the systemic circulation. Other CVOs allow the brain to monitor osmolality, or the concentrations of specific ions or molecules for homeostatic functions.

Functions of the blood–brain barrier

Water, gases which are water or lipid soluble (e.g., O_2 or volatile general anesthetics respectively), and lipophilic molecules (e.g., steroids) passively diffuse across the plasma

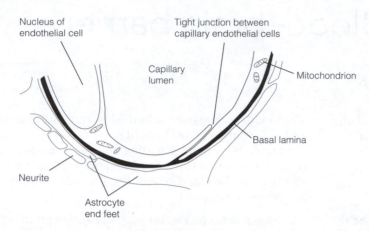

Figure 1. Structural features of the blood–brain barrier. The barrier is made by the tight junctions between the endothelial cells.

membranes of endothelial cells. The transport of ions or all but the smallest polar molecules must be carrier-mediated via ion channels and transporters.

The selective permeability of the blood–brain barrier ensures that crucial molecules, such as glucose—brain relies almost exclusively on glucose as a metabolic energy substrate (plus ketones in infancy or starvation)—and amino acids, are taken up by the brain. It protects neurons from the actions of neuroactive molecules in the blood, such as circulating catecholamines or glutamate. The blood–brain barrier is also able to actively exclude a wide range of potentially neurotoxic lipophilic compounds, many part of a natural diet. This is achieved by a transport protein in the plasma membrane of endothelial cells, **P-glycoprotein**. Lipophilic toxins which diffuse into the endothelial cell are rapidly pumped back out into the blood. Many brain tumor cells also express P-glycoprotein and so are able to exclude chemotherapeutic agents. Consequently chemotherapy is not generally successful in treating brain tumors.

B1 Membrane potentials

Key Notes

Excitable cells

Excitable cells are able to produce action potentials, brief reversals of the electrical potential across their plasma membrane. Excitable cells include neurons and muscle cells.

Resting potentials

The resting potential is the voltage across the plasma membrane of an unstimulated excitable cell. All membrane potentials are expressed as inside relative to outside. Resting potentials are inside negative, and range from about −60 mV to −90 mV in neurons. The resting potential is caused largely by the tendency for potassium ions to leak out of the cell, down their concentration gradient, so unmasking a tiny excess of negative charge on the inside of the cell membrane. Other ions (e.g., sodium) make a small contribution to the resting potential.

Action potentials

An action potential (nerve impulse) is a short-lived reversal of the membrane potential triggered by a stimulus that causes the potential to fall (depolarization). In neurons the spike lasts less than 1 ms and peaks at about +30 mV. The after-hyperpolarization that follows lasts a few milliseconds.

Action potential properties

Action potentials are triggered at the axon hillock and propagated along the axon. They obey the all-or-none rule; a stimulus must be sufficiently large to depolarize a neuron beyond a threshold voltage before it will fire and all action potentials in a given cell are the same size. The latent period is the short delay between the onset of the stimulus and the action potential. Neurons become inexcitable to further stimulation during the spike and harder to excite during the after-hyperpolarization. These constitute the absolute and relative refractory periods respectively. Refractory periods limit the maximum rate at which neurons can fire, and mean that action potentials are propagated in only one direction.

Related topics

(B2) Voltage-dependent ion channels
(B3) Action potential conduction

(O3) Classical electrophysiology

Excitable cells

A small potential difference exists across the plasma membrane that surrounds all cells. When sufficiently stimulated **excitable cells** generate **action potentials**, transient reversals in this membrane potential that are actively propagated over the cell surface. Excitable cells include neurons, skeletal, cardiac, and smooth muscle cells, and some endocrine cells. The membrane potential of an unstimulated excitable cell is called the resting potential.

Resting potentials

A **resting potential** (V_m) arises because there is a difference in the concentrations of ions between the inside and outside of the cell (Table 1) and because the cell membrane has different permeabilities for these ions. Extracellular fluid is an aqueous solution of sodium chloride. By contrast the intracellular fluid is an aqueous solution of potassium ions balanced by a variety of anions—including organic acids, sulfates, phosphates, some amino acids, and some proteins—to which the cell membrane is completely imperme-able. The cell membrane is permeable to K+ and because there is a concentration gradi-ent for K+ across the membrane there is a *diffusional force* acting to drive the K+ from the inside to the outside of the cell (Figure 1). However, the cell membrane is completely impermeable to the much larger anions which therefore remain inside the cell. As the potassium diffuses out a potential difference forms across the membrane because some of the intracellular anions are no longer neutralized by K+. The potential difference now means that an attractive *electrostatic force* is generated which acts to prevent potassium ions diffusing out. At some point the diffusional force driving K+ out is exactly balanced by the electrostatic force preventing K+ leaving.

Table 1. Ionic concentrations across mammalian membranes (mmol⁻¹)

Ion	Extracellular fluid	Axoplasm
K+	2.5	115
Na+	145	14
Cl⁻	90	6

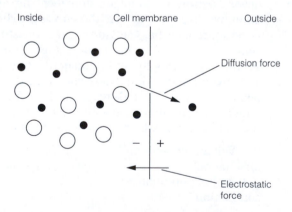

Figure 1. Illustration of how a potassium equilibrium potential is formed. A small potential exists across the membrane when the diffusional force equals the electrostatic force. Small filled circles represent K+ ions, large open circles represent anions.

At this equilibrium a small potential difference exists, termed an **equilibrium poten-tial** because at this potential there is no *net* flow of K+ ions across the membrane. In the case where the potential arises as a result of the distribution of diffusible K+ it is called a potassium equilibrium potential (E_K). Typically nerve cells have potassium equilibrium potentials around −90 mV. Three important points should be noted:

- Transmembrane potentials are always quoted as inside relative to the outside, which is taken to be zero. So, $E_K = -90$ mV means that the inside of the cell is negative with respect to the outside.

- The number of ions which migrate across the membrane to establish an equilibrium potential is extremely small.

- The potential difference exists only at the plasma membrane, which by storing charge acts as a **capacitor**.

Equilibrium potentials can be calculated using the **Nernst equation**:

$$E = (RT/zF) \ln C_e/C_i$$

R is the universal gas constant, T is temperature in Kelvin, z is the oxidation state of the ion, F is Faraday's number and C_e, C_i are extracellular and intracellular concentrations of the ion respectively.

That the potassium equilibrium potential is close to the resting potential (V_m) for excitable cells suggests that V_m arises largely as a result of the distribution of potassium ions across the cell membrane. Neuron resting potentials range between −60 mV and −90 mV. The discrepancy between E_K and V_m arises because ions other than potassium also make a contribution by virtue of their equilibrium potentials. The most important is sodium (E_{Na} = +55 mV), but since the permeability of Na^+ is low compared to K^+ it exerts only a modest influence. The effect of sodium is to drag the resting potential away from E_K towards E_{Na} by an amount that reflects the relative permeabilities of the two ions. The difference between the resting potential and the equilibrium potential for any ion, $V_m - E_{ion}$, is termed the ionic driving force and is a measure of the electrochemical force tending to move the ion across the cell membrane. At rest the driving force for K^+ is quite small but that for Na^+ is high. The reason that Na^+ does not flow into cells at rest is because the permeability for sodium is very low.

In most excitable cells the ionic driving force for chloride ions is close to zero (that is E_{Cl} = V_m). This is because Cl^- ions are passively distributed across the membrane according to the resting potential set up by the combined effects of E_K and E_{Na}. Chloride is passively distributed whilst K^+ and Na^+ directly determine the resting potential because the resting concentration gradients for potassium and sodium are actively maintained by the actions of Na^+/K^+-ATPase (cation pump), whereas there are no active transport mechanisms to maintain a fixed chloride concentration gradient.

Given the extracellular and intracellular concentrations of K^+, Na^+, and Cl^-, and the relative permeabilities of these ions, it is easy to calculate resting potentials.

The permeability of cell membranes to ions is conferred by protein ion channels termed **leak channels**. A family of 15 potassium leak channels (K_{2p}) has been identified.

Action potentials

Neurons are naturally excited either by the cascade of synaptic inputs onto their dendrites and cell body from other neurons or by receptor potentials generated by sensory organs. These stimuli correspond to the injection of an inward current into the neuron and cause its membrane potential to drop, that is, to become less inside-negative. This is termed **depolarization**. If a neuron is depolarized enough it will fire an action potential (nerve impulse). Intracellular recording of a nerve impulse (Figure 2) shows that the membrane potential rapidly depolarizes to zero, overshoots to about +30 mV then repolarizes back towards V_m all in less than 1 ms. This constitutes the **spike** of the action potential. Immediately after the spike the neuron membrane potential becomes larger (more inside negative) than the normal resting potential. This is called **hyperpolarization**. The **after-hyperpolarization** following an action potential spike decays over a few milliseconds, and the potential returns to its resting value.

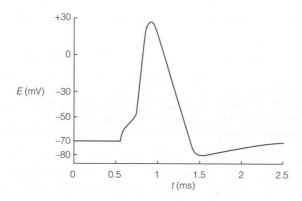

Figure 2. An intracellular recording of a neuron action potential. The resting membrane potential is −70 mV.

Action potential properties

Key properties of action potentials are:

- A minimum **threshold stimulus** is required to produce an action potential and it works by depolarizing the cell to a critical **threshold voltage**, typically some 15 mV less than V_m. A threshold stimulus is defined as the current that will cause a neuron to fire on 50% of occasions, and its amplitude depends on the size of the neuron because this affects the electrical resistance of the cell.

- Physiologically, action potentials are triggered at the axon hillock, because this region of the neuron has the lowest threshold, and are propagated along the axon towards its terminals.

- All action potentials are about the same size (in a given cell) *regardless* of the amplitude of the stimulus. The combined effects of this and property 1 are often paraphrased as the **all-or-none rule;** a neuron either fires completely or not at all.

- There is a short delay between the onset of the stimulus and the start of the action potential. This is called the **latency** (or **latent period**). The latency gets shorter as the strength of the stimulus increases.

- During the spike a neuron becomes completely inexcitable. This is the **absolute refractory period**, during which time a nerve cell will not fire again no matter how large the stimulus. After the spike, while the neuron remains hyperpolarized, the neuron can be excited only by suprathreshold stimuli. This period is the **relative refractory period** and is explained by the fact that when the cell is hyperpolarized a bigger current is needed to drive the membrane potential to the threshold voltage. That neurons are rendered temporarily refractory has two consequences:

- – It imposes an upper limit on firing frequency.
- – Physiologically, action potentials are propagated in only one direction (see property 2), because the segment of axon where the action potential has just been is inexcitable.

All of the above properties can be explained by the behavior of the ion channels responsible for the action potential.

B2 Voltage-dependent ion channels

Key Notes	
Voltage-dependent ion channels	Voltage-dependent ion channels are transmembrane proteins that are ion selective and voltage sensitive. They are named for the ion species to which they are most permeable. They can exist in at least two interchangeable states, open or closed, depending on the membrane potential across them.
Voltage-dependent sodium channels	Sodium channels are transmembrane glycoproteins found in most excitable cells. Normally closed, they are opened by depolarization of the membrane potential beyond threshold allowing sodium ions to permeate into the cell. This causes the depolarization phase of the action potential. The channels then flip into an inactivated state rendering them impermeable to sodium. This, together with the low ionic driving force for sodium at positive membrane potentials, curtails the spike amplitude. Sodium channel inactivation accounts for the absolute refractory period.
Voltage-dependent potassium channels	A delayed outward rectifying potassium channel is responsible for the downstroke of the action potential spike and the subsequent after-hyperpolarization. These transmembrane channels are activated by depolarization permitting potassium ions to flow out of the cell, carrying the membrane potential to increasingly negative values. They do not inactivate over the time course of the action potential.
Channel molecular biology	Molecular biology techniques have been used to sequence ion channels and deduce how they work. Voltage-dependent ion channels consist of four subunits, each with six α-helical transmembrane segments (S1–S6) arranged around a central pore. Segments S5 and S6 and an intercalated H5 loop form the pore. The positively charged S4 segment acts as voltage sensor. Three distinct mechanisms can close potassium channels.
Local anesthetics	Local anesthetics act by stabilizing the inactivated state of voltage-dependent sodium channels, blocking action potential propagation. Their preferential action on pain sensation when injected locally is because nociceptor afferents have a small diameter.
Related topics	(B1) Membrane potentials (O3) Classical electrophysiology

Voltage-dependent ion channels

Neurons are excitable because they have voltage-dependent ion channels in their plasma membrane. These transmembrane proteins have two properties, ion selectivity and voltage sensitivity. Voltage-dependent channels are classified according to the ion (Na^+, K^+ or Ca^{2+}) to which they are preferentially permeable. A large number have been identified to date. They are multi-subunit complexes. The ion channel α subunit(s) are accompanied by a variety of β subunits which modify the channel properties.

Voltage-dependent ion channels can exist in at least two interchangeable states: **open** (**activated**), when they allow ions to flow through them; or **closed**, when they do not. Whether they are open or closed depends on the voltage across them.

Voltage-dependent sodium channels

Voltage-dependent sodium channels (Na_vs) are large glycoproteins that span the full thickness of the plasma membrane of most excitable cells. Nine distinct Na_v α subunits have been identified, most are expressed in neurons. At the resting potential they are closed. If their region of membrane is depolarized by only a few millivolts they remain closed (Figure 1a). However, if the membrane is depolarized to the threshold voltage or beyond, Na_vs change shape so that the channels open, permitting Na^+ ions to flow down their electrochemical gradient into the neuron.

Channel opening, **activation** (Figure 1b), is an extremely fast event (~10 µs). During an action potential in a neuron a given sodium channel will remain open for about 0.5–1 ms. In this time about 6000 Na^+ ions will flow through the channel. The combined effect of sodium influx through a few hundred Na_vs will produce the upstroke of the spike that is the depolarizing phase of the action potential. Only a few voltage-dependent sodium channels need be driven beyond threshold initially to trigger an action potential because the localized influx of Na^+ causes a depolarization which drives other channels to open, driving a **positive feedback** explosive rise in Na^+ permeability.

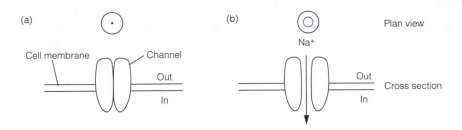

Figure 1. Behavior of a voltage-dependent sodium channel (a) at rest when it is in the closed state and (b) during the spike of an action potential, when it is activated.

At the top of the spike the self-regenerative rise in sodium permeability is halted for three reasons:

- All of the available Na_vs in the active region of membrane have opened.
- The ionic driving force for sodium gets less as the membrane depolarizes towards the sodium equilibrium potential.

- The voltage-dependent sodium channels flip into an **inactivated** state. During this state the channel does not permit the flow of any ions (Note that this is *not* the same as the closed state because the inactivated state *cannot* be made to open). In addition to curtailing the spike, it is this inactivation of Na_vs that is responsible for the absolute refractory period. The inactivation wears off after a few milliseconds as the channel relaxes into the closed state, from which it can be reactivated by subsequent depolarization.

Voltage-dependent potassium channels

Excitable cells have a variety of different types of **voltage-dependent potassium channel** (K_v), each with their own properties. One type, the delayed outward rectifier is involved in the repolarization phase of the action potential. These channels are transmembrane glycoproteins with a molecular structure related to that of voltage-dependent sodium channels. Like them, they are opened by depolarization. This allows potassium ions to flow down their electrochemical gradient out of the cell. The inside of the neuron becomes less positive, that is, it repolarizes. This accounts for the downstroke of the spike of the action potential.

At the bottom of the downstroke most of the Na_vs are inactivated so there is no flow of sodium into the cell. Delayed outward rectifying K_vs either do not inactivate or inactivate much more slowly than Na_vs (depending on channel type) so immediately after the spike the neuron remains highly permeable to K^+ whilst being impermeable to Na^+. The consequence is that for a few milliseconds after the spike, potassium ions continue to leave the cell carrying the membrane potential more negative than V_m. This is the **after-hyperpolarization** phase of the action potential. It accounts for the relative refractory period, because during this time, for a stimulus to excite the cell to threshold it must make the neuron depolarize from a more inside-negative state. Finally the membrane potential returns to the resting potential as the potassium channels flip into their closed or inactivated state in a time-dependent fashion. The changes in ion conductance during the action potential are summarized in Figure 2.

A couple of points follow from above:

- The size of the after-hyperpolarization is determined by the potassium equilibrium potential (E_K). When the K^+ efflux is sufficient to bring the membrane potential to E_K the ionic driving force on the potassium ions will be zero. No more K^+ will leave.

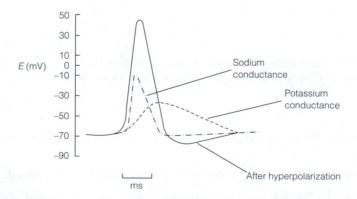

Figure 2. Changes in ion conductance during the action potential.

- Since depolarization of the neuron to threshold causes the opening of Na$_v$s (allowing Na$^+$ in) *and* opening of K$_v$s (allowing K$^+$ out) how does an action potential arise? The reason is that the sodium channels respond to the depolarization earlier than the potassium channels (hence the name *delayed*) so the increase in sodium permeability precedes the increase in potassium permeability.

Channel molecular biology

By cloning and sequencing the DNA that codes for a protein it is possible to deduce its primary amino acid sequence. This provides clues as to the secondary structure of the protein such as the presence of α-helical or β-pleated sheet regions.

Hydropathicity plots, which represent how hydrophobic or hydrophilic particular amino acid stretches are, indicate which regions of the protein might be located in the membrane. Recognizing consensus sequences for glycosylation or phosphorylation allow identification of extracellular and intracellular regions of the molecule respectively, providing evidence for how a protein is orientated in the membrane.

Site-directed mutagenesis, in which the DNA coding for a protein is altered in precise ways using genetic engineering techniques, and the mutated protein subsequently expressed in some convenient way, can be used to investigate the functions of particular amino acids in a channel.

The three-dimensional structure of ion channels in both closed and open states can be deduced by **X-ray crystallography** from which details of how the channel operates can be worked out. A major difficulty here is being able to crystallize the proteins.

There are numerous types of voltage-dependent ion channel, structures for many of which have been deduced using the technologies outlined above. The sodium channels and several potassium channels, including the delayed outward rectifier responsible for the downstroke of the action potential, share a common structure. The basic subunit has six highly hydrophobic segments (S1–S6), thought to be α-helices spanning the membrane (Figure 3). In the case of potassium channels each subunit is a separate molecule and a functioning channel is a tetrameric homo-oligomer, that is, it consists of four

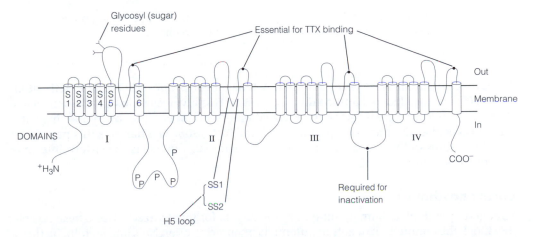

Figure 3. A cartoon depicting the secondary structure of an Na$_v$. Segments S1–S6 are labeled only in domain I. P, consensus sequences for phosphorylation.

similar subunits arranged around a central pore. In sodium channels the four subunits are linked by cytoplasmic loops to form a single huge molecule, but its tertiary structure is thought to resemble a potassium channel.

Segments 5 and 6 and the H5 loop between them contribute to the pore. This H5 loop is thought to line the channel pore because it contains amino acids that are crucial for conferring ion selectivity. The extracellular throat is lined with oxygen atoms that are crucial for ion permeation.

One of the most striking features of voltage-dependent ion channels is the S4 segment (Figure 4). It is highly conserved between channels, suggesting it has hardly altered during evolution and so serves a critical function. Although mostly hydrophobic, its N-terminal end has several positively charged lysine or arginine residues. Site-directed mutagenesis shows that S4 is part of the voltage sensor required for activation of the channel. X-ray crystallography implies that in response to depolarization, the S4 segment acts like a paddle swinging through the membrane towards the extracellular side, as its positive charges are repelled by the voltage change. This displacement of the paddle pulls an S4–S5 linker region, opening the intracellular throat of the pore.

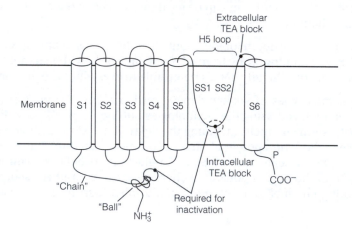

Figure 4. A cartoon depicting the secondary structure of an outward rectifying potassium channel subunit. Four subunits assemble to form a functioning channel.

Three mechanisms to close potassium channels have been proposed. One, **N-inactivation** occurs by the so called **ball and chain** mechanism. This involves a cluster of amino acids (the "ball") swinging up to block the internal mouth of the channel by interacting with an amino acid at the internal tip of the H5 loop. In *Drosophila* channels this peptide is the N-terminus of the standard (α) subunit. In mammals the ball is the N-terminus of a separate β-subunit.

Local anesthetics

Voltage-dependent sodium channels are the targets for **local anesthetics**. These reversibly block the conduction of action potentials by entering open Na_vs and stabilizing them in the inactivated state. In effect it can take up to 1000 times longer for inactivation to wear off in the presence of the drug than normally.

Most local anesthetics consist of a lipophilic group linked via an ester or amide to an ionizable amine, and are weak bases with pK_a around 8–9. Hence, at the pH 6.8–7.4 encountered in the tissues, the protonated form predominates. This is unable to penetrate the hydrophobic walls of the channel, so access to the site of action in the channel pore is denied until the channel has been opened. Moreover, the open and inactivated states of the Na_vs have a greater affinity for local anesthetics than the closed state. For these reasons local anesthetics demonstrate **use-dependent blockade**. The ester and amide linkages of local anesthetics are hydrolyzed by plasma cholinesterases or liver P450 respectively.

When injected locally pain sensation is preferentially blocked. Small diameter nociceptor afferents are more susceptible to local anesthetics than large diameter afferents because:

- The distance that local circuit currents can passively spread to propagate an action potential is shorter the smaller the diameter.

- Smaller diameter fibers fire at higher frequencies and have longer action potentials and so are more likely to suffer use-dependent blockade.

Myelinated fibers are more sensitive than nonmyelinated fibers of the same diameter because the nodes of Ranvier have fewer barriers to drug access.

B3 Action potential conduction

Key Notes

Propagation of action potentials

Action potentials are generated at the axon hillock (spike initiation zone) and propagate actively, at constant velocity, and without loss of amplitude, down the axon. Because the active zone, the region of the axon at which the action potential sits at a given instant, bears different charges to the axon at rest ahead of it, local circuit currents flow which depolarize the adjacent upstream membrane and so the action potential advances. Local circuit currents also spread backwards but do not allow the action potential to propagate in this direction because the membrane there is refractory.

Conduction velocity in nonmyelinated axons

In nonmyelinated axons the speed of conduction is between 0.5 and 2 m s^{-1}. The velocity is proportional to the square root of the axon diameter.

Myelination

Oligodendrocytes in the CNS and Schwann cells in the peripheral nervous system provide the myelin sheath, an electrically insulating covering around many axons. In the peripheral nervous system individual Schwann cells myelinate short sections of axons with a series of concentric layers of plasma membrane. Tiny gaps, nodes of Ranvier, expose the axon membrane to the extracellular fluid. Myelination in the CNS is similar except that each oligodendrocyte contributes to the myelination of several adjacent axons, saving space.

Conduction velocity in myelinated axons

Myelination produces dramatic increases in conduction speed for only modest increases in the overall diameter of axons. Myelinated axons conduct faster because local circuit currents flow around the electrically insulating myelin sheath so that only the axon membrane at the node of Ranvier needs to be depolarized to generate an action potential. The action potential appears to jump from one node to the next. The conduction velocity is proportional to axon diameter and varies from 7 to 100 m s^{-1}.

Related topics

(A2) Glial cells

(B1) Membrane potentials

Propagation of action potentials

In a neuron, action potentials are initiated at the axon hillock because this region has the greatest density of Na$_v$s and so the lowest threshold. Hence the axon hillock is also called the **spike initiation zone**. Once generated, action potentials are actively propagated (conducted) with constant velocity down the axon without loss of amplitude. Thus, action potentials are undiminished in size even when conducted along peripheral axons that in humans may be up to one meter long. This is one of the features that makes action potentials reliable signals for information transmission. The details of conduction are a little different depending on whether the neuron is myelinated or not.

In nonmyelinated neurons conduction works as follows (Figure 1). The region of an axon invaded by an action potential at a given time is called the **active zone**. It is a few cm long. The part of the active zone occupied by the overshoot of the spike will be inside positive. Far away from the active zone, ahead of the oncoming action potential or behind it, the membrane potential will be inside negative. The consequence of this is that a potential difference exists between different segments of axon membrane and hence **local circuit currents** flow passively between these axon segments.

Just ahead of the axon potential the currents drain positive charge from the external surface of the axon and simultaneously dump positive charge on the inside of the axon membrane. The net effect is to depolarize the axon immediately in front of the action potential. When this depolarization becomes suprathreshold Na_vs in this region activate and the action potential advances. Of course, local circuit currents flow in the same way along the axon behind the action potential but this region is refractory and so the

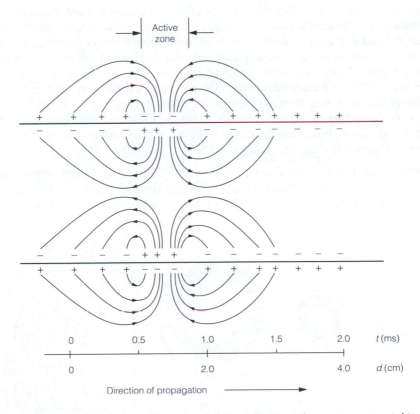

Figure 1. Local circuit currents in action potential conduction. The action potential is traveling from left to right and the leading edge of the spike (active zone) is 2 cm from the origin (lower scale) after 1 ms (upper scale). t, time; d, distance. By convention current direction is the flow of positive charge.

currents do not excite here. Thus action potentials propagate physiologically only in one direction.

Conduction velocity in nonmyelinated axons

The conduction velocity (θ), the speed with which nerve impulses are propagated, is quite slow in nonmyelinated axons. It varies between 0.5 and 2 m s^{-1} depending on the diameter of the axon. Small axons offer a higher resistance to the flow of currents through their cores than large ones, just as thin wires have a higher electrical resistance than thick ones. So, local circuit currents in the axoplasm of small axons spread less well than in larger axons and this is the reason for the slower speed. Very roughly the relationship is:

$$\theta = ka^{1/2}$$

where a is the axon diameter and k is a constant which depends on the internal resistance of the axon and its membrane capacitance.

Myelination

Oligodendrocytes in the CNS and **Schwann cells** in the peripheral nervous system provide the **myelin sheath**, an electrically insulating covering around many axons.

In the peripheral nervous system Schwann cells line up along the axon, surrounding it with a pseudopodium-like structure, the **mesaxon**. For unmyelinated axons the process stops at this point. For myelinated axons the mesaxon spirals around the axon some 8–12 times, ensheathing it with a series of concentric double layers of plasma membrane, because the cytoplasm gets left behind (Figure 2). Each Schwann cell myelinates between 0.15 and 1.5 mm of axon. Between adjacent ensheathed regions is a tiny (0.5 µm) gap of naked axon called the **node of Ranvier**. Here the axon membrane is directly exposed to the extracellular space. Because peripheral nerve axons can be long, hundreds of Schwann cells may be needed to myelinate them. Myelinated nerve fiber diameters are between 3 and 15 µm, but across this range the proportion of the diameter contributed by the myelin sheath is roughly constant.

Myelination proceeds in a similar way in the CNS except that each oligodendrocyte extends several processes so that it can contribute to the myelination of several adjacent axons. Hence fewer glial cells are needed for CNS myelination, saving space.

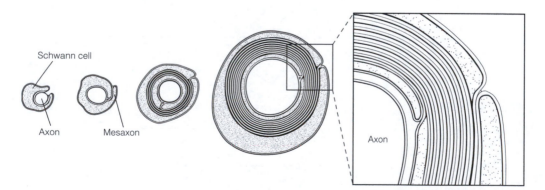

Figure 2. Myelination of a peripheral axon. The myelin sheath is generated by the growth of the mesaxon which wraps itself around the axon. From Susan Standring (ed) (2008) *Gray's Anatomy: The Anatomical Basis of Clinical Practice*, 40th ed. With permission from Elsevier.

Conduction velocity in myelinated axons

The function of the myelin sheath is to increase the conduction velocity substantially with relatively little increase in fiber diameter. The evolution of myelination has enabled vertebrates to have a large number of rapidly conducting axons without taking up too much cable space.

Because the myelin sheath consists largely of plasma membrane it has a large phospho-lipid content, giving it a high electrical resistance. Local circuit currents are forced to take paths of lesser resistance through the electrolyte solution around the sheath. The effect is that local circuits are established, not between adjacent regions of membrane as they are in nonmyelinated axons, but between adjacent nodes of Ranvier, which are relatively far apart. Local circuit currents ahead of an action potential arriving at the next down-stream node cause it to depolarize beyond the threshold and trigger an action potential. In this manner action potentials appear to jump from node to node, a mechanism called **saltatory conduction**. The density of Na_vs is about 100-fold greater at nodes than in non-myelinated axon membrane and the node threshold is consequently much lower. This greatly reduces the risk of nodes not firing in response to local circuit currents weakened by the long distances they must spread.

Conduction velocity (θ) is faster in myelinated than nonmyelinated axons for two reasons:

- The presence of a myelin sheath is functionally equivalent to increasing the thickness of the axonal membrane about 100-fold. This greatly reduces the amount of charge stored across the membrane, which means that much less time is taken to depolarize it. More technically put, myelination reduces the membrane capacitance (c_m), because the capacitance of a parallel plate conductor is inversely proportional to the thickness of the insulator, and since $\theta \propto 1/c_m$, the lower the capacitance the greater the speed of action potential propagation.

- The time taken up for Na_vs to respond to depolarization is less, because only channels at nodes have to be activated. In a nonmyelinated axon each little region of membrane has to be depolarized and respond in succession. For a myelinated axon, however, only the node membrane needs to be excited.

Conduction velocities of myelinated axons vary from about 7 to 100 m s^{-1}. As with non-myelinated axons, velocity depends on diameter, a, but the relationship is even simpler:

$$\theta = ka.$$

C1 Synapse structure and function

Key Notes	
Chemical synapses	Chemical synapses are where signaling between nerve cells occurs. Pre- and postsynaptic membranes are just 20–500 nm apart. Synapses can be classified according to where they are located on the receiving neuron. Axodendritic synapses are made on dendrites, axosomatic on the cell body, and axoaxonal on the axon.
Morphology of chemical synapses	The axon terminal contains small clear synaptic vesicles which store transmitter and both the presynaptic and postsynaptic membranes are thickened. Cortical axodendritic synapses are usually excitatory and asymmetrical, with thicker post- than presynaptic membranes. Axosomatic synapses are symmetrical and usually inhibitory. Synapses that secrete catecholamines or peptides have large dense-core vesicles and may have wide synaptic clefts. Many synapses contain both small clear vesicles and large dense-core vesicles and secrete more than one transmitter.
Chemical transmission	Neurotransmitter release from the nerve terminal following the arrival of an action potential is triggered by the influx of calcium through voltage-dependent calcium channels. After crossing the cleft, transmitter binds to postsynaptic receptors. These are either ligand-gated ion channels or G-protein-coupled receptors. Receptor activation either increases or decreases the chance that the postsynaptic cell will fire, responses described as excitatory or inhibitory respectively. Transmission mediated by ligand-gated ion channels is fast, whereas that by G-protein-coupled receptors is slow. A transmitter can act on both types of receptor. A given synapse can release more than one transmitter. Several mechanisms terminate transmitter action at synapses.
Electrical transmission	Electrical signaling occurs at electrical synapses formed by ion channels called connexons at gap junctions. Here small ions flow between cells so that they are electrically coupled. Via gap junctions action potentials can spread rapidly between cells without distortion.
Related topics	(C2) Neurotransmitter release (C3) Postsynaptic responses (C5) Neural integration

Chemical synapses

Most signaling between nerve cells occurs via chemical synapses. These are formed by the close association of an axon terminal of the presynaptic cell with some part of the postsynaptic cell. The gap between the cells, the **synaptic cleft**, is typically 20 nm across though in some synapses can be much wider. Synapses are classified by where on the receiving cell they are located. Most are **axodendritic** synapses made on dendrites; on spiny dendrites synapses are formed on spines. Particularly powerful are **axosomatic** synapses made on the cell body (named for an alternative term for the nerve cell body, the soma). Synapses between axon terminals and *axons* of postsynaptic neurons are **axoaxonal** synapses.

Morphology of chemical synapses

Electron microscopy has revealed numerous morphologically distinct types of synapse, but all share common features (Figure 1). Spherical, clear, **small synaptic vesicles** (**SSVs**), 50 nm across, store neurotransmitter, and are scattered throughout the terminal in association with microtubules that transport them to the presynaptic membrane. The presynaptic membrane is thickened and may have **dense projections** which are involved in the docking of synaptic vesicles at the **active zone**, the region from which transmitter release occurs. The dendrite cell membrane at the synapse is thickened as the **postsynaptic density**, formed by the proteins that constitute the postsynaptic machinery.

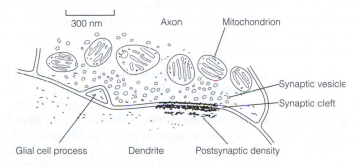

Figure 1. The structure of a chemical (axodendritic) synapse.

Cortical axodendritic synapses are asymmetrical in that they have an extremely well-developed postsynaptic density, they are usually excitatory. Axosomatic synapses are symmetrical in having pre- and postsynaptic densities of comparable thickness, they are usually inhibitory. The cerebral cortex may have as many as 10^{13} synapses.

Some synapses lack obvious specialized contact zones on both pre- and postsynaptic sides and have extremely wide (100–500 nm) synaptic clefts. These often secrete a catecholamine and have **large dense-core vesicles** (**LDCVs**) 40–120 nm across. LDCVs are also found in peptide-secreting neurons.

Many synapses contain both SSVs and LDCVs (Figure 2). This structural evidence supports physiological studies showing that many neurons secrete more than one transmitter.

Chemical transmission

Most synapses are chemical. At a typical CNS synapse the arrival of an action potential at the axon terminal may, or may not, trigger the release of neurotransmitter from a single

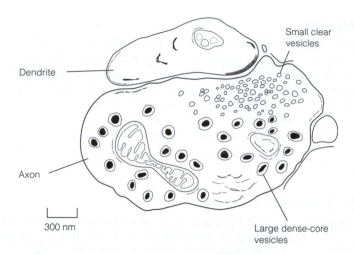

Figure 2. A type I synapse containing both small clear vesicles and large dense-core vesicles. From Revest PA & Longstaff A (1998) *Molecular Neuroscience*. BIOS Scientific Publishers.

presynaptic vesicle. Transmitter release requires a rise in intracellular Ca^{2+} brought about by calcium entry into the axon terminal via **voltage-dependent calcium channels**. The transmitter diffuses across the synaptic cleft in about 5 μs. At the postsynaptic membrane it binds to specific receptors, causing a change in the receptor conformation. What happens next depends on the receptor, but the overall result is to change the postsynaptic membrane permeability to specific ions.

Neurotransmitter receptors come in two **superfamilies**. The **ligand-gated ion channels** (**LGICs**), also referred to as **ionotropic receptors**, have ion-selective channels as part of the receptor. Binding of the transmitter to the receptor opens the channel, directly increasing its permeability. The second superfamily is the **G-protein-coupled receptors** (**GPCRs**). These are the largest group of **metabotropic receptors**, so called because they can modulate metabolic activities. Binding of transmitter to these receptors activates their associated G proteins that are capable of diverse and remote effects on membrane permeability, excitability, and metabolism. G proteins can influence permeability either by binding ion channels directly or by modifying the activity of second messenger system enzymes which phosphorylate ion channels, thereby altering their permeability.

The changes in membrane properties brought about by a transmitter produce **postsynaptic potentials**, small shifts in the membrane potential that can have essentially one of two effects. It may increase the probability that the postsynaptic neuron fires action potentials, in which case the response is **excitatory**. If the effect is to decrease the probability that the postsynaptic cell might fire, the response is **inhibitory**. It is commonly the case that a transmitter is excitatory at one of its receptors, but inhibitory at another, so it is a synapse that is excitatory or inhibitory rather than the transmitter *per se*.

Numerous molecules have been identified as neurotransmitters. The classical transmitters are small molecules, amino acids or amines. Quantitatively by far the most important are glutamate, which is almost invariably excitatory, and γ-aminobutyrate (GABA) which is usually inhibitory. This group also includes acetylcholine, the catecholamines such as dopamine and noradrenaline (norepinephrine), and the indoleamine, serotonin. A second, larger, group is an eclectic mix of peptides which includes the opioids (such as endorphins) and the tachykinins (e.g., substance P). See Table 1 for a far from exhaustive list.

Table 1. Key central nervous system neurotransmitters

Classical	Amino acids	Glutamate	
		Aspartate	
		γ-aminobutyrate	
		Glycine	
	(Mono)amines	Acetylcholine	
		Dopamine	⎫
		Norepinephrine	⎬ catecholamines
		Epinephrine	⎭
		Serotonin (5-hydroxytryptamine) indolamine	
Peptides	Opioids	Dynorphins	
		Endorphins	
		Enkephalins	
	Tachykinins	Substance P	
	Hormones	Cholecystokinin	
		Somatostatin	

Neurotransmission falls into two broad categories based on its time course. **Fast transmission** occurs whenever a neurotransmitter acts via LGICs, whereas **slow transmission** occurs when transmitters act at GPCRs. Glutamate, GABA, and acetylcholine (ACh) are together responsible for most fast transmission. However, each of these molecules also mediates slow transmission by activating their corresponding GPCR. Indeed, a transmitter can mediate both fast and slow transmission at the same synapse by activating multiple receptor populations. For example, ACh released from preganglionic cells acts on both nicotinic and muscarinic receptors on postganglionic cells in autonomic ganglia, mediating fast and slow effects of ACh respectively. Catecholamine and peptide transmission are invariably slow.

It is extremely common for a given synapse to release more than one transmitter. This **cotransmission** usually involves the release of a classical transmitter, coupled with the co-release of one or more peptides at higher firing frequencies.

Transmitters are rapidly cleared from the synaptic cleft after release by passive diffusion away from the cleft, reuptake into surrounding neurons or glia, or enzyme degradation.

Electrical transmission

Electrical transmission is mediated by electrical synapses, which are **gap junctions** between adjacent neurons. Gap junctions are arrays of paired hexameric ion channels called **connexons** (Figure 3a). The channel pores are 2–3 nm in diameter, allowing ions and small molecules to permeate between neighboring neurons. By electrically coupling neurons, gap junctions allow any potentials, for example action potentials, to spread between cells. Key features of electrical transmission are that it is extremely rapid, signals are transmitted with no distortion, and it works in both directions. Gap junctions between cells can close. Each connexon is made up of six subunits called **connexins**, and in response to a rise in intracellular Ca^{2+} concentration, the connexins rotate to close the central pore (Figure 3b).

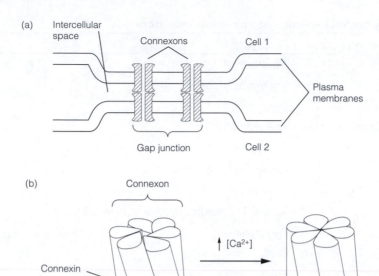

Figure 3. (a) Gap junction. (b) Change in configuration of the connexons to close a gap junction.

C2 Neurotransmitter release

Key Notes

Vesicular release

Neurotransmitter release occurs by calcium-dependent exocytosis from vesicles, in response to excitation of the axon terminal by action potentials.

Release is quantal

Transmitter is released in discrete packets, quanta, that correspond to exocytosis from a single vesicle. The spontaneous, random release of a single quantum causes miniature postsynaptic potentials at CNS synapses. Postsynaptic potentials arise from the release of several quanta simultaneously. At central synapses action potentials trigger neurotransmitter release in only a proportion of occasions.

The role of calcium

Following excitation of the nerve terminal calcium influx is restricted to a small region, but the local concentration reaches 200 µM, sufficient to trigger the exocytosis mechanism for small synaptic vesicles very rapidly.

Exocytosis from large dense-core vesicles

Amines and peptides are released by high-frequency stimulation, only after an appreciable delay, because large dense-core vesicles are situated away from the active zone.

Biochemistry of exocytosis

Several linked steps are involved in exocytosis, most require calcium. Recruitment shifts vesicles from a reserve pool into a releasable pool. Binding of vesicle-associated proteins and plasma membrane proteins permits the vesicles to be docked at the active zone in close proximity to voltage-dependent calcium channels. Partial fusion of the vesicle is achieved by priming. The final rapid stage of exocytosis occurs when excitation triggers Ca^{2+} influx.

Heuser–Reese cycle endocytosis

This recycles vesicles. Vesicle membrane is coated with clathrin so that it invaginates. Fission of coated vesicle is then triggered by hydrolysis of GTP bound to dynamin. Once in the cytoplasm the vesicle loses its clathrin coat.

Kiss-and-run cycle

In central synapses a fast kiss-and-run cycle allows high levels of release to be maintained by a small pool of vesicles.

Refilling

Classical transmitters are imported into vesicles driven by the efflux of H^+ via specific transporters. The proton gradient is generated by a vesicular proton ATPase. Peptides are packaged in the Golgi apparatus from which vesicles bud to be transported to the axon terminal.

Autoreceptors

Autoreceptors respond to the transmitter released by the neuron in which they are located. They occur at the presynaptic terminal, the soma, and dendrites. They regulate

Calcium channels	Calcium channels are responsible for neurotransmitter release, dendritic action potentials, and excitation–contraction coupling in muscles. There are several types of calcium channel that can be characterized by their electrophysiology, pharmacology, distribution, and functions. They have a close resemblance to sodium channels.
Related topics	(B2) Voltage-dependent ion channels (C1) Synapse structure and function (J1) Nerve–muscle synapse (M5) Sleep

(top of box:) neurotransmitter release, synthesis, and neuron firing rate, usually homeostatically.

Vesicular release

Most neurotransmitter release occurs by transmitter-loaded synaptic vesicles fusing with the presynaptic membrane and discharging their contents into the synaptic cleft. This is **exocytosis**. It is triggered by the arrival at the nerve terminal of an action potential which causes a transient and highly localized influx of Ca^{2+}. After release the vesicle membrane is recycled from the presynaptic membrane to form new vesicles by **endocytosis**. The vesicles are subsequently loaded with transmitter via active transporters localized in the vesicle membrane.

Release is quantal

Neurotransmitter is secreted in discrete packets or **quanta**. Each quantum represents the release of the contents of a single vesicle, about 4000 molecules of transmitter. At CNS synapses this causes a **miniature postsynaptic potential** (mpsp), either excitatory or inhibitory, due to the release of transmitter from the vesicle acting on 30–100 postsynaptic receptors beneath the active zone. Excitatory and inhibitory postsynaptic potentials represent the summation of multiple mpsps generated by an action potential invading several active zones simultaneously. This happens either because axons branch to form several discrete terminals or because some terminals have more than one active zone.

The active zone of many CNS synapses appears to have only one release site. This is known as the **one vesicle** or **one quantum** hypothesis. However, transmitter release does not happen every time an action potential arrives at the presynaptic terminal. Individual active zones behave in an all-or-none fashion because an action potential will either trigger the release of the single quantum or not. The proportion of successes will reflect the probability of release. At central synapses the probability of release varies between different sites and at least at some synapses it also depends on the recent history of the synapse.

The role of calcium

The arrival of an action potential at a nerve terminal causes voltage-dependent calcium channels at the active zone to open after ~300µs. The driving force for calcium entry is extremely high because of the large concentration gradient; the free Ca^{2+} concentration at rest in a terminal is 100 nM whilst the external concentration is about 1 mM. Despite this the presence of diffusion barriers and calcium buffers in the terminal restrict the rise in calcium concentration to within 50 nm of the channel mouth. The [Ca^{2+}] within 10 nm of the channel mouth rises to between 100 and 200 μM, which matches the half-maximal concentration of Ca^{2+} for glutamate release.

Exocytosis from large dense-core vesicles

In contrast to small clear synaptic vesicles (SSVs), the mechanism for release from large dense-core vesicles (LDCVs) takes longer and has a higher affinity for calcium (half-maximal release occurs at about 0.4 μM), because only a small amount of Ca^{2+} manages to diffuse to the LDCVs, which are some distance from the active zone. Hence exocytosis of amines and peptides occurs with a delay of about 50 ms and only in response to high frequency firing of the neuron which causes high levels of calcium influx.

Biochemistry of exocytosis

Exocytosis from SSVs involves several linked steps, most of which need calcium. Nerve terminals contain two pools of SSVs. The **releasable pool** is located at the active zone and can take part in repeated cycles of exocytosis and endocytosis at low firing frequencies. The **reserve pool** consists of vesicles tethered to cytoskeletal proteins, and can be recruited by repetitive stimulation to join the releasable pool. Liberation of a vesicle from the cytoskeleton requires Ca^{2+}-dependent phosphorylation of **synapsin I**, a protein that anchors vesicles to actin filaments in the terminal.

Vesicles are aligned at specific sites in the active zone by a process termed **docking**, which involves **SNARE** proteins (Figure 1). A vesicle-associated protein, **synaptobrevin** (v-SNARE, VAMP) binds with high affinity to a presynaptic membrane protein, **syntaxin** (t-SNARE). Syntaxin is closely associated with voltage-dependent calcium channels, ensuring that the release machinery is optimally placed to receive the Ca^{2+} signal. Synaptobrevin and syntaxin, together with a third protein crucial for docking, **SNAP-25**, are targets for **botulinum** and **tetanus toxins** that are powerful inhibitors of neurotransmitter secretion.

After docking comes another calcium-dependent step, **priming**, in which a number of soluble cytoplasmic proteins form a transient complex with the SNAREs, resulting in partial fusion of vesicle and presynaptic membranes. This step involves the hydrolysis of ATP.

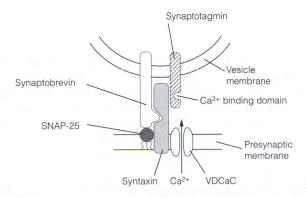

Figure 1. Proteins involved in the docking of neurotransmitter vesicles. VDCaC, voltage-dependent calcium channel.

Primed vesicles are poised for exocytosis, requiring only a large pulse of Ca^{2+} to permit complete fusion of the vesicle and presynaptic membranes and opening of the **fusion pore** through which exocytosis occurs. A calcium-binding protein located in the vesicle membrane, **synaptotagmin**, is the Ca^{2+} sensor in the exocytotic machinery. In the absence of calcium it prevents complete fusion, but when it binds Ca^{2+} it undergoes a conformational change which allows fusion to proceed.

Heuser–Reese cycle endocytosis

Following exocytosis, synaptic vesicles are recycled within 30–60 s by endocytosis. Firstly, vesicle membrane acquires a **clathrin** coat, distorting it so that it invaginates into the terminal. Next a guanosine triphosphate (GTP)-binding protein, **dynamin**, forms a collar around the neck of the invagination. Hydrolysis of bound GTP triggers the fission of the coated vesicle from the presynaptic membrane. The GTP-bound form of dynamin requires calcium, so the same rise in nerve terminal Ca^{2+} concentration responsible for exocytosis also enables endocytosis. Once free in the terminal the vesicle loses its clathrin coat (Figure 2).

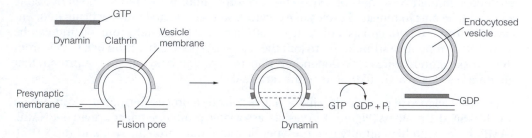

Figure 2. Vesicle endocytosis. From Revest PA & Longstaff A (1998) *Molecular Neuroscience.* BIOS Scientific Publishers.

Kiss-and-run cycle

A much faster **kiss-and-run cycle** also operates at central synapses. Here the vesicle membrane fuses with the presynaptic membrane to open a pore through which transmitter discharges, after which the pore closes and the vesicle disengages. There is no complicated endocytosis. Because this mechanism has a cycle time of only 1 s it is able to support extended periods of high release with only 35–40 vesicles in the recycling pool.

Refilling

Vesicles are reloaded with neurotransmitter in the nerve terminals. The vesicles are acidified by the action of a proton ATPase. The transport of transmitter into vesicles is then driven by secondary active transport with H^+ efflux providing the energy (Figure 3). Nerve terminals are incapable of protein synthesis. Peptide transmitters are synthesized in the cell body, secreted into the lumen of the rough endoplasmic reticulum (RER), and packaged for export by the Golgi apparatus, from which the loaded vesicles are budded. These are then moved to the terminal by fast axoplasmic transport.

Autoreceptors

Receptors are not confined to the postsynaptic membrane but also exist in the presynaptic membrane, where they are termed **presynaptic receptors**, and over the cell body

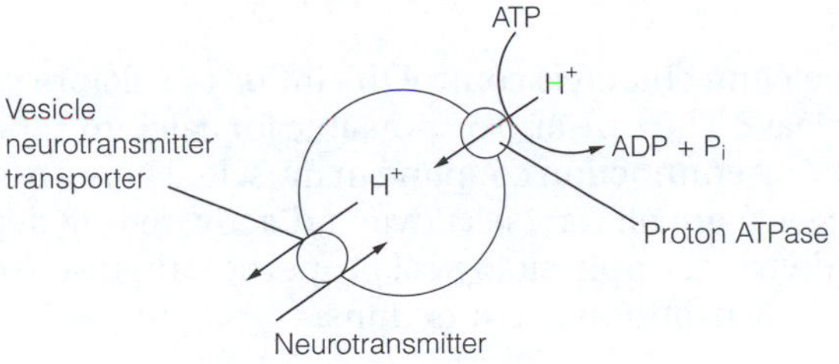

Figure 3. Vesicle refilling. From Revest PA & Longstaff A (1998) *Molecular Neuroscience.* BIOS Scientific Publishers.

and dendrites. If these are receptors for the transmitter released by the neuron in which they are located, they are **autoreceptors**. Autoreceptors, which are invariably metabotropic receptors, have several functions that are normally homeostatic. Those on the presynaptic membrane are involved in regulating neurotransmitter release. Most (but not all) presynaptic autoreceptors *decrease* the release of neurotransmitter by reducing calcium influx into the presynaptic terminal. This is a negative feedback mechanism either to avoid excessive excitation, or to curtail postsynaptic receptor desensitization, which would reduce the sensitivity of the synapse.

Autoreceptors can also decrease the synthesis of transmitter (e.g., of catecholamines and serotonin by their respective neurons) and some dopaminergic neurons have dopamine receptors on their dendrites and cell body which regulate neuron firing rate.

Heteroceptors are presynaptic receptors for transmitters *not* secreted by the neuron in which they are situated. These also regulate transmitter release. For example, $GABA_B$ receptors exist presynaptically at glutamatergic synapses where they reduce glutamate release. They are activated by GABA that has diffused from neighboring synapses.

Table 1. Calcium channel types

Type	Named for	Electrophysiology	Location
L	Long-lasting	HVA (−20 mV)	Pyramidal cells
		Slowly inactivating	Skeletal, cardiac, and smooth muscle
			Endocrine cells
T	Transient	LVA (−65 mV)	Cardiac muscle
		Rapidly inactivating	Neurons (e.g., thalamic)
			Endocrine cells
N	Neuronal	HVA (−20 mV)	Neurons
		Moderate inactivation	
P	Purkinje cell	HVA (−50 mV)	Cerebellar Purkinje cells
		Noninactivating	Mammalian neuromuscular junction
Q	Q after P	HVA	Cerebellar granule cells
R	Remaining	HVA and LVA	

HVA, high voltage-activated; LVA, low voltage-activated.

Calcium channels

Voltage-dependent calcium channels control the influx of calcium which couples excitation to transmitter release. They are also responsible for calcium action potentials in dendrites, and for excitation–contraction coupling in muscle. There are several distinct types of calcium channel. They are all Ca^{2+} selective and activated by depolarization but can be differentiated by their electrophysiological properties, their sensitivity to blockade by drugs and toxins, their distribution and functions.

Calcium channel types are summarized in Table 1. The N-, P-, and Q-type high voltage-activated (HVA) channels that need large depolarization to activate them are involved in transmitter release. L-type channels are located in proximal dendrites of pyramidal neurons and contribute to their excitability. They are the major calcium channels mediating excitation–contraction coupling. The rapidly inactivating low voltage-activated (LVA) T-type channels generate burst firing and are important in thalamic neurons.

Calcium channels share a high homology with voltage-dependent sodium channels.

C3 Postsynaptic responses

Key Notes

Fast transmission	Opening of ligand-gated ion channels is responsible for fast transmission. The flow of ions through the open channels causes postsynaptic potentials to be generated on the postsynaptic membrane. These decay with time and distance as they spread passively over the neuron surface. Postsynaptic potentials are either excitatory or inhibitory.
Excitatory postsynaptic potentials	Activation of ionotropic receptors permeable to Na^+, K^+, and Ca^{2+} results in depolarizing excitatory postsynaptic potentials (epsps). When recorded from the cell body they are caused by the activation of several synapses. They are graded in size from about 0.5 to 8 mV depending on the number of inputs stimulated, and decay exponentially after 10–20 ms. Most fast excitatory transmission is mediated by glutamate and acetylcholine.
Inhibitory postsynaptic potentials	Fast inhibitory postsynaptic potentials (ipsps) are caused by activating ionotropic receptors permeable to chloride ions. Inhibitory postsynaptic potentials have similar properties to epsps except for their inhibitory nature. An increase in Cl⁻ permeability will always be inhibitory because it will tend to stabilize the membrane potential at the chloride equilibrium potential, E_{Cl}. This is true even if the resting potential is greater than E_{Cl} and the ipsp is depolarizing. Most fast inhibitory transmission is mediated by GABA and glycine.
Slow transmission	Metabotropic receptor actions are intrinsically slower and longer lasting than those of ionotropic receptors. Slow transmitters produce slow epsps and ipsps, but have other diverse effects on cell behavior because of the capacity of second messenger systems to open or close a large variety of ion channels throughout a neuron. Slow transmitters modulate transmitter release, the action of fast transmitters, neuronal excitability, and gene expression.
Related topics	(B1) Membrane potentials (D2) G-protein-coupled (C5) Neural integration receptors (D1) Ligand-gated ion channel receptors

---

Fast transmission

The action of neurotransmitters on ligand-gated ion channels is rapid. Changes in membrane potential occur within a millisecond or so and return to resting potential in tens of milliseconds. This is **fast neurotransmission**. The binding of transmitter to ligand-gated ion channels increases the permeability of the postsynaptic membrane to whatever ions the channel conducts. This alters the potential of the postsynaptic membrane, generating a **postsynaptic potential**. These potentials are typically just a few millivolts in amplitude. They spread passively over the plasma membrane of the nerve cell, diminishing in size as they move away from the postsynaptic membrane, and with time since their generation. Postsynaptic potentials come in two flavors, excitatory and inhibitory.

Excitatory postsynaptic potentials

Activation of ionotropic receptors that are nonselective cation conductances (i.e., those permeable to Na^+ plus K^+ and possibly Ca^{2+} also) depolarizes the synaptic membrane. This is because the reversal potential for the current through these receptors is close to zero. Because this brings the membrane potential of the neuron *closer* to the threshold voltage at which action potentials are triggered, it *increases* the probability that the cell might fire. Hence depolarizing synaptic potentials are termed **excitatory postsynaptic potentials** (**epsps**). An *individual* epsp has no chance of making a central neuron fire, it is far too small to have a significant effect on the postsynaptic cell. Only when many epsps are generated on a neuron within tens of milliseconds of each other is there any prospect of driving the neuron across threshold. The major fast excitatory transmitters are glutamate and acetylcholine. Glutamate transmission was first studied in the spinal cord where sensory nerve axons from muscles form axodendritic synapses directly on motor neurons (Figure 1).

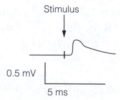

Figure 1. Excitatory postsynaptic potentials in spinal motor neurons in response to stimulating a single sensory axon.

There are several points to note about excitatory postsynaptic potentials:

- It is technically hard to record individual epsps at vertebrate synapses, so generally the epsp resulting from the activation of *several* synapses is recorded at the cell body.

- They are small and graded in size, ranging from fractions of a millivolt to about 8 mV. Stimulating more axons activates more synapses, which makes the epsp larger.

- There is a short delay of 0.5–1 ms between stimulating the afferents and the generation of an epsp. This is called the **synaptic delay**.

- They typically last for about 10–20 ms whilst decaying exponentially.

Inhibitory postsynaptic potentials

Activation of ionotropic receptors that conduct chloride ions channels *reduces* the probability of neuron firing. The most important fast inhibitory transmitters are γ-aminobutyrate (GABA) and glycine. The effect of GABA can be seen in the experiment illustrated in Figure 2. Pyramidal cells in the cerebral cortex have many GABAergic

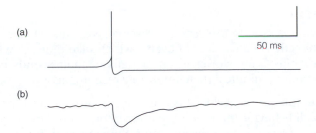

Figure 2. Inhibitory postsynaptic potential in a pyramidal cell produced by GABA release from an inhibitory (basket) neuron: (a) presynaptic action potential in basket cell, vertical scale bar 25 mV; (b) postsynaptic potential in pyramidal cell, vertical scale bar 0.5 mV.

axosomatic synapses impinging on them from interneurons. Activating the interneuron produces a modest hyperpolarization, an **inhibitory postsynaptic potential (ipsp)**, because it carries the membrane potential away from the threshold for firing action potentials. Inhibitory postsynaptic potentials have very similar properties to epsps.

Figure 3 shows what happens to the ipsp caused by GABA at different pyramidal cell membrane potentials. At −70 mV GABA produces no change in potential. This is the equilibrium potential for chloride, E_{Cl}. At this point there is no net flow of Cl⁻ through the activated GABA receptor. There are two other cases to consider. First, when the membrane potential of the neuron is more positive than E_{Cl} Cl⁻ enters the cell, making it more negative inside—the cell hyperpolarizes (this is also seen in Figure 2). Second, when the membrane potential of the neuron is initially more negative than E_{Cl} Cl⁻ leaves the cell so it becomes less negative inside—the cell depolarizes. This is still inhibitory. In both cases the effect of increasing Cl⁻ permeability is to force the membrane potential towards E_{Cl}. Whenever the membrane potential is not the same as E_{Cl} there will be an ionic driving force causing chloride ions to either leave or enter the cell. The increased Cl⁻ permeability therefore clamps the potential close to −70 mV, preventing it from being driven towards threshold by concurrent excitatory inputs. Because this inhibition effectively short circuits epsps it is referred to as **shunting inhibition**.

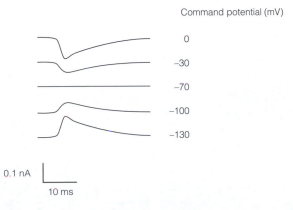

Figure 3. Reversal potential of fast GABA ipsps, found by voltage clamping. The reversal potential is −70 mV.

Slow transmission

Metabotropic receptors couple to second messenger systems. Hence, they affect ion channels only indirectly via a cascade of events which take time to switch on and off. Consequently their effects are slower in onset (tens of milliseconds to seconds) and longer lasting (seconds to minutes) than ionotropic receptor action.

Metabotropic receptors produce slow epsps and ipsps but can also have more diverse effects on nerve cell behavior than ionotropic receptors. This can be attributed to the widespread actions of second messenger systems. Being freely diffusible they can spread through dendrites into the cell body or even the axon. This brings them into contact with channels throughout a nerve cell. Second messenger systems target a great variety of ion channels (opening or closing them) including: voltage-dependent K^+ channels responsible for setting the overall excitability of the cell and voltage-dependent channels that generate action potentials (rendering cells more or less excitable), Ca^{2+} channels required for transmitter release (increasing or decreasing release), and ligand-gated ion channels (altering the efficacy of fast transmitters).

Metabotropic receptors also exert long-lasting effects on nerve cells, unrelated to their transient influence on cell permeability and electrical excitability, because second messenger systems act in the nucleus to alter gene transcription. This can alter the structure of neurons and how they connect to their neighbors. Such mechanisms underlie long-term memory.

C4 Neurotransmitter inactivation

Key Notes

Reason for inactivation	Inactivation of transmitter action occurs so that synapses can be modulated on a fast time scale, and to curtail receptor desensitization.
Enzyme degradation	Acetylcholine is hydrolyzed by acetylcholinesterase. The liberated choline is taken up into the nerve terminal via a high affinity Na^+-dependent cotransport system. ATP is also inactivated enzymically.
Transport	Reuptake from the synaptic cleft into neurons (and glia in the case of the amino acids) is a major mechanism for the inactivation of the classical transmitters. Two major families of transporters are involved. These molecules are unrelated to the vesicular transporters for transmitters. Cocaine exerts its effects by blocking the dopamine transporter. Serotonin transporters are thought to be the crucial targets for antidepressant drugs.
Diffusion	Diffusion away from the synaptic cleft is the major mode of inactivation of the peptides and is also probably important for glutamate and GABA. The large size of the peptides makes their diffusion slow and accounts for their protracted action.
Related topics	(C2) Neurotransmitter release (D7) Acetylcholine / (J1) Nerve–muscle synapse (M2) Motivation and addiction

Reason for inactivation

Inactivation allows synapses to respond to rapid changes in presynaptic neuron firing frequency. Without it the postsynaptic cell could not be updated on recent changes in incoming signals. In addition, by clearing transmitter from the synaptic cleft, inactivation limits receptor desensitization. There are three ways in which transmitter can be inactivated and they are not mutually exclusive: enzyme-catalyzed degradation, transport out of the synaptic cleft back into neurons or glia, or by passive diffusion away from the synapse.

Enzyme degradation

Many enzymes are involved in the catabolism of both classical and peptide transmitters and pharmacological inhibition of many of these can have consequences for synaptic transmission. However, only in a couple of cases is enzyme-catalyzed degradation important in the *physiological* inactivation of transmitter. Acetylcholine is hydrolyzed by

acetylcholinesterase (**AChE**), which cleaves the transmitter molecule into choline and acetate. Choline is taken back into the presynaptic nerve terminal by a Na^+-dependent transporter. AChE has an extremely high catalytic activity and at the neuromuscular junction can reduce the concentration of ACh from about 1 mM immediately after release to virtually zero in about 1 ms.

Adenosine 5′-triphosphate (**ATP**), a cotransmitter at some synapses, is the only other transmitter inactivated at the synapse by enzyme-catalyzed degradation.

Transport

Many classical transmitters are inactivated by their removal from the cleft via high affinity, saturable, secondary active transport. The amino acid transmitters may be transported into both neurons or glia whereas amines are transported only into neurons. Two families of transporter have been identified which serve this function:

- Na^+/K^+-cotransporter family constitutes the glutamate (and aspartate) transporters. Glutamate transport is electrogenic, that is, it results in a modest potential difference being set up across the membrane, inside positive (Figure 1). A consequence of this is that excessive depolarization of the membrane can reverse the direction of the transport causing glutamate *efflux* into the cleft. This can result in excitotoxicity. Glutamate transporters have been cloned and sequenced.

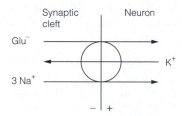

Figure 1. Glutamate transport by a Na^+/K^+ cotransporter.

- Na^+/Cl^--cotransporter family (Figure 2). This is a large family that includes transporters for GABA, glycine, noradrenaline/adrenaline (norepinephrine/epinephrine), dopamine, serotonin, and choline. The noradrenaline (norepinephrine) and serotonin transporters are the targets for the tricyclic antidepressant class of drugs. The serotonin transporters are targets of selective serotonin reuptake inhibitors (SSRIs), such as fluoxetine (Prozac), although questions are now being raised about their efficacy as antidepressants and their safety. The dopamine transporter is the target for cocaine.

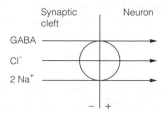

Figure 2. GABA transport by a Na^+/Cl^- cotransporter.

By inhibiting dopamine reuptake, cocaine deranges dopamine transmission in reward pathways in the brain, accounting partly for its addictive properties. Many members of this family have been sequenced. They are large glycoproteins with 12 transmembrane segments, but have no homology with vesicle transporters.

Diffusion

Despite the existence of transporters, diffusion out of the synapse may be important for the inactivation of glutamate and GABA at synapses in the cerebral cortex. There is no high affinity reuptake for peptides and although peptides may be internalized by neurons via receptor-mediated endocytosis and then degraded by nonspecific peptidases, this is probably not an important mechanism for their inactivation. Hence the major route for terminating the synaptic action of peptides is by diffusion. However, peptides are very much larger than the small classical transmitter molecules and there are significant barriers for free diffusion out of the cleft. This means that peptides clear only slowly from a synapse, which helps to explain why their actions can be so prolonged.

C5 Neural integration

Key Notes	
Neurons as decision-making devices	Small potentials (e.g., synaptic potentials) decay with time and distance. This behavior underlies how individual nerve cells treat their inputs and hence all information processing in the nervous system. Postsynaptic potentials (psps) generated on a neuron, both excitatory and inhibitory, add together (summate). If the result of this summation is that the axon hillock membrane potential is driven beyond threshold, the neuron will fire. So, whether or not a neuron will fire at any moment depends on how many excitatory and inhibitory synapses are active, and where they are located. It is by integrating synaptic inputs in this way that neurons act as computational devices.
Summation	The summation of psps generated at slightly different times is temporal summation. The summation of potentials arriving on different parts of the neuron is spatial summation. The geometry of a neuron determines the size and time course of synaptic potentials as they spread, and hence the extent of summation that occurs. If summation results in a sufficiently large depolarization of the axon trigger zone a nerve cell will fire.
Related topics	(C3) Postsynaptic responses

Neurons as decision-making devices

Individual synaptic potentials are too small to activate voltage-dependent sodium channels so they do not trigger action potentials. Instead they are passively conducted over the nerve cell membrane, getting smaller both with time and distance as they spread. This decay of small potentials is determined solely by the physics of the neuron. Generally, the smaller the diameter of a neuron, axon, or dendrite along which a potential is spreading, the shorter the distance over which it will decay, the faster this will happen, and the slower the potential is conducted. This is crucial in determining how neurons integrate their inputs and hence how information is processed in the nervous system. In addition it accounts for why action potentials, which do not decay in time and distance, are needed for long-distance transmission. Synaptic potentials decay to zero within a few millimeters in most neurites, so cannot carry information any great distance. However, some short interneurons (e.g., those in the retina) do not fire nerve impulses, but rely on synaptic potentials for transmission along their neurites.

Many thousands of synapses are formed on each neuron, both excitatory and inhibitory. At any given time a subset of these will be activated to generate epsps and ipsps. A special property of these graded potentials is that they **summate**, or add together. If a sufficient number of epsps are produced, in summing they will drive the axon hillock membrane potential across the threshold for triggering action potentials and the neuron will fire. The axon hillock is crucial because, being the region of a neuron with the highest density of voltage-dependent sodium channels, it has the lowest threshold. If at any instant insufficient excitatory synapses are activated, or a high level of excitatory synaptic input is more than offset by the generation of ipsps from inhibitory input, then the axon hillock will not be driven across the threshold and the cell will not fire. So, neurons are decision-making devices. The decision—to fire or not—is actually taken by the axon hillock on the basis of whether the sum total of epsps and ipsps causes its membrane potential to become more positive than the firing threshold. It is this operation that constitutes information processing by individual neurons. In engineering terms, a synapse converts digital signals (action potentials) into analog ones (postsynaptic potentials). The neuron then integrates all its analog signals over a short time and compares the result of that integration with a given threshold to decide whether to fire. When it does fire the output is digital.

Experiments on pyramidal cells show that about 100 excitatory synapses, on average, must be activated at the same time to trigger an action potential. However, the efficacy with which a synapse can influence firing depends on its position. Because postsynaptic potentials decay as they spread passively towards the axon hillock, a synapse far out on distal dendrite will have less effect than one closer to the cell body. In this context it is noteworthy that on pyramidal cells there are only about 250 inhibitory synapses on the cell body but 10000 or so excitatory axodendritic synapses. The relative strength of a synapse in contributing to a neuron's output is its **weighting**. This need not be a fixed property but may change with time.

Summation

If an afferent neuron fires a series of action potentials in quick succession, then the earliest psps generated in the postsynaptic cell will not have time to decay before the next psps arrive. Hence successive psps summate over time. This is referred to as **temporal summation**. If sufficient, temporal summation will cause the postsynaptic cell to reach firing threshold.

The summing of postsynaptic potentials generated at separate points on the neuron surface is called **spatial summation** (Figure 1). If a sufficient number of excitatory synapses are activated in relation to inhibitory ones the cell will fire.

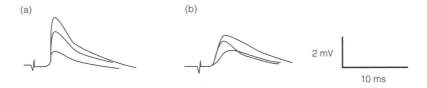

Figure 1. Spatial summation. In each case the upper trace is the summed response of the two lower epsps generated at synapses: (a) a long way apart, (b) close together.

Although temporal and spatial summation are described as separate processes, both occur together as a neuron is stimulated and it is their combined effect which dictates whether it will fire. The precise details of how summation works depends on a neuron's geometry because this determines exactly how all the synaptic potentials set up on the cell decay as they spread towards the axon hillock. The frequency with which a cell fires, and how long it fires, is determined by the amplitude and duration of the depolarization of the axon hillock membrane.

D1 Ligand-gated ion channel receptors

Key Notes

Ligand-gated ion channels

These receptors fall into two classes, the nicotinic receptor family and the ionotropic glutamate receptors. All consist of several subunits arranged around a central pore.

Cys-loop family

The nicotinic receptor superfamily members are pentamers assembled from several distinct subunits. Each subunit has four transmembrane segments that are either α-helices or β-pleated sheets, and extracellular N and C terminals. Both nicotinic cholinergic receptors (nAChR) and $GABA_A$ receptors show allostery (positive cooperativity) in ligand binding. The nAChR receptor ion channel is permeable to both Na^+ and K^+ while the $GABA_A$ receptor is a chloride permeable channel.

Nicotinic acetylcholine receptors

Nicotinic receptors are divided into muscle types and neuronal types. Each receptor binds two molecules of ACh and the binding of one molecule makes binding of the second easier, an example of allostery. Most nicotinic receptors are nonselective cation conductances allowing NA^+ influx and K^+ efflux, so binding of ACh produces depolarization.

$GABA_A$ receptors

The $GABA_A$ receptor has allosteric sites for binding benzodiazepines, barbiturates, and steroids. These drugs all enhance inhibition by increasing Cl^- conductance. Benzodiazepines have anti-anxiety and anticonvulsant profiles. Inverse agonists *decrease* Cl^- conductance and have an anxiety-producing, pro-convulsant profile. Endogenous steroids active in the brain, neurosteroids, are produced by neurons and glia. They potentiate $GABA_A$ receptors and are anxiolytic and anticonvulsant.

Ionotropic glutamate receptor family

Most fast transmission by glutamate is via AMPA/kainate receptors, which have a tetrameric structure. Their pore region resembles a potassium channel, but inside out in membrane orientation. Most native AMPA receptors are permeable to Na^+ and K^+ but some are also Ca^{2+} permeable. NMDA receptors are important because of their involvement in learning and memory, and neuropathologies, including epilepsy and strokes. They share only weak homologies with other ligand-gated ion channels. They are Ca^{2+} permeable and are blocked by Mg^{2+} except at depolarizing potentials. They have a requirement for glycine and serine as a co-agonists and binding sites for Zn^{2+}, Pb^{2+} and several drugs.

Related topics	(C1) Synapse structure and function	(J1) Nerve–muscle synapse
	(C3) Postsynaptic responses	(N4) Long-term potentiation
		(N5) Motor learning in the cerebellum: LTD

Ligand-gated ion channel receptors

Most ligand-gated ion channel (**ionotropic**) receptors are members of two families (Table 1), the cys-loop family (containing the nicotinic cholinergic, $GABA_A$, glycine, and 5-hydroxytryptamine ($5\text{-}HT_3$) receptors), and the ionotropic glutamate receptor family. The cys-loop receptors have a pentameric subunit structure; the glutamate receptors have four subunits. In both families the subunits are clustered around a central pore, like staves around a barrel. Because there are numerous subunit isoforms, receptors can be assembled from several subunit combinations. Each combination results in a receptor with distinct physiology and pharmacology, and is typically expressed in a specific population of cell types.

Table 1. Ligand-gated ion channels

Family	Receptor	
Cys-loop family	nAChR	
	$GABA_A$	
	$GABA_C$	
	glycine	
	$5\text{-}HT_3$	
Glutamate receptor family	$GluR_1$–$GluR_4$	(AMPA receptors)
	$GluR_6$	(kainate receptors)
	NMDAR	

Cys-loop family

All the **cys-loop family** of ligand-gated ion channels are responsible for fast neurotransmission. They are pentamers. Each subunit has extracellular N and C terminals and four transmembrane α-helical segments (M_1–M_4). All members have α subunits responsible for agonist binding. A disulfide bond between two cysteine residues at the N-terminus forms a loop required for binding (and is the reason for the "cys-loop" designation). The M2 domains of the α subunits form the ion pore. There are modest amino acid homologies between subunits within and across the cys-loop receptors.

Nicotinic acetylcholine receptors

Nicotinic receptors (nAChR) are divided into muscle types and neuronal types. The muscle types are assembled from α1, β1, δ, γ or ε subunits in a 2:1:1:1 stoichiometry, while the neuronal types are various combinations of 12 subunits, α2–α10 and β2–β4. (Figure 1).

Each of the α-subunits has a binding site for acetylcholine, thus each receptor binds two molecules of ACh. Binding of one molecule of ACh to a receptor makes binding of the second easier. This is an example of **allostery**. Most nAChR channels are nonselective cation conductances allowing Na^+ influx and K^+ efflux, though some configurations, for example the homomeric $(\alpha7)_5$, are calcium conductances. The net current is inward (i.e., depolarizing) so ACh is excitatory at nicotinic receptors.

(a)

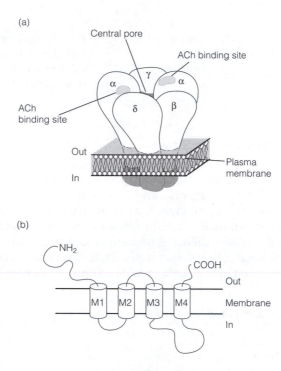

(b)

Figure 1. The Cys-loop family: (a) pentameric arrangement of subunits; (b) cartoon of subunit secondary structure.

GABA$_A$ receptors

GABA$_A$ receptors activation opens a channel selective for Cl$^-$. Although this normally results in an ipsp, in some neurons, especially in the embryo, the internal chloride concentration is so high that the chloride reversal potential is closer to zero than the resting potential. In these cases GABA produces an epsp; i.e. it is excitatory.

GABA$_A$ receptors are pentamers of various combinations of subunits designated α1–α6, β1–β3, γ1–γ3, δ, ε, θ and ρ1–ρ3 (Note that despite sharing the same nomenclature, these subunits are not the same as those in nAChR). Although a large number of receptors could in theory be assembled from these only 11 have been positively identified so far. The GABA-binding site is structurally similar to the acetylcholine-binding site, with two molecules of GABA acting allosterically to open the channel.

Various GABA$_A$ receptors contain binding sites for endogenous agents (endozepines and neurosteroids) and several major classes of drugs (benzodiazepines, barbiturates, ethanol, and volatile anesthetics). These compounds bind to a variety of sites distinct from the GABA-binding site and act by allosterically altering the binding of GABA and thereby modulating the chloride current.

Typical **benzodiazepines**, (e.g., diazepam) are agonists of the **benzodiazepine-binding site (BZ)**. On binding they increase the affinity of the receptor for GABA, which causes the chloride channel to open more frequently. The overall effect is to potentiate the inhibitory effect of GABA without prolonging it. That diazepam exerts its anti-anxiety and muscle relaxant effects via receptors containing α2 subunits, but its sedative and anti-convulsant effects through α1-containing receptors, shows the importance of subunit makeup.

Benzodiazepines have a place in the *short-term* treatment of anxiety (dependence is seen with long-term use) and (intravenously) as sedatives or basal anesthetics in minor surgery. Most benzodiazepines are too sedative to be used for maintenance therapy in epilepsy, but intravenously they can be life-saving in severe seizures.

Inverse agonists bind the BZ site and *decrease* channel opening. Not surprisingly they have the unpleasant pharmacological profile of being anxiogenic (anxiety-causing) and pro-convulsant. Endogenous inverse agonists at the BZ site have been identified. **Endozepines** are peptides (diazepam binding inhibitor and truncated versions of it) released by astrocytes. At $GABA_A$ receptors they increase anxiety. In addition they bind to G-protein-coupled receptors. These mediate anorexigenic effects of endozepines by acting on NPY/AgRP (neuropeptide Y/agouti-related peptide) and POMC/CART (proopiomelanocortin/cocaine- and amphetamine-related transcript) neurons in the hypothalamus.

Neurosteroids are endogenous steroids active in the brain and include pregnenolone and dehydroepiandrosterone and their sulfates. Neurosteroids are produced in the peripheral and central nervous system by glial cells and neurons, but supplemented by gonadal and adrenal steroids which easily cross the blood–brain barrier. They are produced during stress, rise during the luteal phase of the menstrual cycle, and reach high concentrations in pregnancy. Neurosteroids act on $GABA_A$ and NMDA (*N*-methyl-D-aspartate) receptors. At $GABA_A$ receptors they potentiate GABA-activated currents. Neurosteroids are anxiolytic, anticonvulsant, and possibly antidepressant.

Ionotropic glutamate receptor family

The **ionotropic glutamate receptors** (**iGluRs**) have a tetrameric quaternary structure and share only weak homologies with the cys-loop family. There are three populations of iGluRs, defined by selective agonists; **AMPA receptors** (named after α-amino-3-hydroxy-5methyl-4-isoxazole proprionic acid), **kainate receptors** and **NMDA receptors** (**NMDARs**). All are relatively nonselective cation channels, though NMDA receptors and some AMPA receptors favor calcium permeation.

The secondary structure for the AMPA and kainate receptor subunits (Figure 2) has four membrane-spanning segments (M), one of which (MII) is on a re-entrant loop which contributes to the pore. The MII-containing part of the molecule has a striking resemblance to the S5-H5-S6 region of potassium channels, albeit oriented in the membrane in the opposite sense. These receptors may thus have evolved from an ancient potassium channel.

Four subunits, $GluR_1$–$GluR_4$, contribute to AMPA receptors. Most are assembled from $GluR_2$ subunits together with either $GluR_1$ or $GluR_3$ subunits. $GluR_2$ subunits have several rather special roles in AMPA receptors, they:

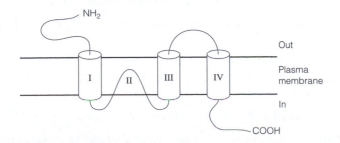

Figure 2. Ionotopic glutamate receptor subunit structure.

- Control the voltage gating of the receptor

- Come in two forms that differ in their MII region by just one amino acid (this changes the receptor from a Na^+ and K^+ channel to a Ca^{2+} channel)

- Control the transport of AMPA receptors to the postsynaptic membrane

- Stimulate the growth of dendritic spines on cortical pyramidal cells

The NMDA receptor is named after the selective agonist **N-methyl-D-aspartate**. Its natural agonists are glutamate and aspartate. The receptor is important because it is implicated in key aspects of brain function such as development, learning and memory, and in pathologies, for example strokes and epilepsy.

NMDA receptors are heterotetramers containing two NR1 and two NR2 subunits. There are eight NR1 isoforms generated by alternative splicing of a single gene and four NR2 isoforms. NR1 subunits are expressed ubiquitously in the brain whereas NR2 isoforms are expressed in a region-specific way. This gives rise to multiple NMDA receptors with distinct brain distributions and functional properties arising from the particular combination of subunits. There are also two NR3 isoforms that confer inhibitory properties on NMDA receptors. The secondary structure resembles that in Figure 2. The extracellular N-terminal domain of the NR2 subunits binds the neurotransmitter while the extracellular domain between MIII and MIV is a modulatory domain that binds the co-agonists glycine or serine. There are structural similarities between the agonist-binding domain of NR2 subunits and those of other ionotropic glutamate receptors and with ancient amino acid-binding proteins of bacteria. The extensive cytoplasmic domain has phosphorylation sites and regions for binding structural proteins.

NMDA receptors have some unusual properties that are summarized below:

- Glycine and D-serine act as allosteric **co-agonists** to potentiate the effects of glutamate. D-serine is synthesized and released by astrocytes co-localized with neurons containing NMDA receptors, may also be released by neurons, and might be the primary ligand at the co-agonist site.

- NMDA receptors are voltage-gated. At resting membrane potentials glutamate will bind to the receptor but the ion channel is blocked by Mg^{2+} ions. This blockade is lifted only by a large depolarization. In other words, the ion channel is only opened if glutamate binds and the receptor experiences depolarization at the same time. This behavior is critical to NMDA receptor functions.

- The ion channel is permeable to Ca^{2+} as well as Na^+ and K^+. Activation of NMDA receptors raises intracellular Ca^{2+} concentrations and this can activate several second messenger cascades, many of which alter AMPA receptor functions. This is critical for long-term potentiation (LTP) and long-term depression (LTD).

- Most NMDA receptor types are inhibited by H^+ and show partial inhibition at physiological pH.

- They contain a redox modulatory site where reductants enhance and oxidants depress NMDA channel activity. This site is presumably the target for endogenous redox agents such as glutathione.

- The channel is blocked by Zn^{2+} ions, and Pb^{2+} is a potent antagonist.

- NMDA receptors are a target for dissociative anesthetics (e.g., ketamine), nitrous oxide, several opiates, and for the psychotomimetic drug phencyclidine.

D2 G-protein-coupled receptors

Key Notes

G-protein-coupled receptors

Many neurotransmitter and hormone receptors, and sensory transduction molecules are G-protein-coupled receptors (GPCRs). They have seven transmembrane segments. A cytoplasmic loop interacts with G proteins to modulate intracellular events. The metabotropic glutamate receptors form a separate family from those GPCRs that have amine or peptide agonists.

G proteins

Metabotropic receptors are coupled to ion channels or second messenger enzymes via trimeric GTP-binding proteins, G proteins. There are several different G protein families with their own particular targets. Binding of ligand to receptor causes liberation of a GTP-bound form of G protein which activates its targets. The intrinsic GTPase activity of the G protein rapidly hydrolyzes the GTP, hence curtailing its own activity.

Activation of adenylyl cyclase

G_S proteins activate adenylyl cyclase, which converts ATP to cyclic adenosine monophosphate (cAMP). This second messenger molecule activates protein kinase A which phosphorylates its target proteins. The cAMP is subsequently degraded by phosphodiesterases. Depletion of cAMP, the action of phosphatases and receptor desensitization all act to curtail the effects of the cAMP second messenger system.

Inhibition of adenylyl cyclase

G_i proteins inhibit adenylyl cyclase. The activity of this enzyme and hence the concentration of cAMP in a cell at any time therefore depends on activation of receptors coupled to G_S relative to those coupled to G_i.

Phosphoinositide second messenger system

G_q proteins activate phospholipase C which cleaves a membrane phospholipid to generate two second messenger molecules. Diacylglycerol (DAG) activates protein kinase C. Inositol trisphosphate (IP_3) mobilizes calcium from internal stores to raise cytoplasmic Ca^{2+} concentration, which activates calcium-dependent protein kinases.

Related topics

(C1) Synapse structure and function
(G3) Photoreceptors

(I1) Olfactory receptor neurons
(N4) Long-term potentiation

G-protein-coupled receptors

G-protein-coupled receptors (**GPCR**s) form a huge superfamily of **metabotropic recep-tors**. It includes receptors for slow neurotransmitters, many hormones, and sensory transduction molecules important in vision, smell, and taste.

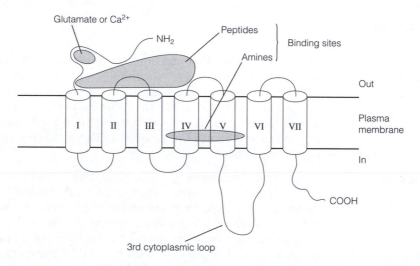

Figure 1. G-protein-coupled receptors: cartoon showing transmembrane segments and ligand binding sites; each of the Roman numerals designates a transmembrane segment.

The iconic feature of these receptors are the seven membrane-spanning segments (MI to MVII) (Figure 1). GPCRs bind small molecules (amines, glutamate) and a wide variety of peptides. The metabotropic glutamate receptors have little homology with the others. The third cytoplasmic loop between MV and MVI couples to G proteins that are critical for signal transduction.

G proteins

G proteins, trimers consisting of α-, β-, and γ-subunits, are so called because the α-subunit binds GTP. Binding of neurotransmitter to GPCRs activates their associated G proteins which may do one or both of the following:

• They may interact directly with ion channels causing them to open or close.

• They may interact with enzymes, for example, adenylyl cyclase and phospholipase C, to switch on or off second messenger cascades that regulate ion channels and other proteins by phosphorylation.

The cycle of events by which a G protein couples GPCR activation to second messenger modulation is shown in Figure 2.

Binding of the transmitter allows the receptor and G protein to couple. GDP leaves the α-subunit in exchange for GTP. In its GTP-bound form the G protein dissociates into separate α and β/γ-subunits. The α-subunit activates a second messenger enzyme. The α-subunit has an intrinsic GTPase activity that cleaves the terminal phosphodiester bond in the GTP converting it to GDP. In its GDP-bound form the α-subunit uncouples from the enzyme, which reverts to its basal activity. One consequence of this cycle is that it

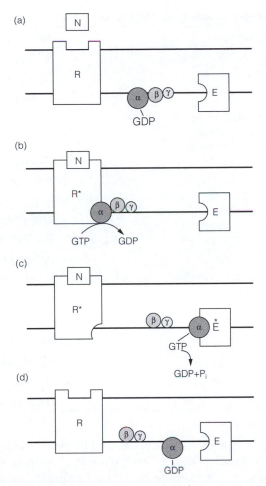

Figure 2. Coupling of metabotropic receptors to second messenger systems by G proteins. N, neurotransmitter; R, receptor; E, enzyme. Asterisks indicate activated states of receptor or enzyme.

amplifies the effect of a small signal. A single transmitter-binding event results in several cycles of G protein shuttling between receptor and enzyme. Furthermore the enzyme will have time to catalyze the synthesis of hundreds of second messenger molecules before it is switched off by the hydrolysis of the G-protein-bound GTP.

There are several distinct G proteins, differing largely in their α-subunits. G_S and G_I interact with adenylyl cyclase, G_q with phospholipase C. Despite this multiplicity, G proteins serve as a point for convergence of signals impinging on a neuron because many receptors talk to just a few second messenger systems. Table 1 lists some of the major G-protein-linked receptors for selected transmitters, together with the second messenger systems they are coupled to.

Activation of adenylyl cyclase

Adenylyl cyclase is activated by a specific family of G proteins, the **G_S proteins**, so called because their action on adenylyl cyclase is stimulatory. The enzyme catalyzes the conversion of ATP to **cyclic adenosine-3′,5′-monophosphate** (**cAMP**). This second messenger

Table 1. Second messenger coupling to selected neurotransmitter receptors

G protein	Second messenger	Receptor
G_s	Increased cAMP	β_1, β_2, β_3 adrenoceptors
		D_1, D_5 (dopamine)
		H_2 (histamine)
G_i	Decreased cAMP and/or opening of K^+ channels closing of Ca^{2+} channels	α_2 adrenoceptors
		D_2, D_3, D_4 (dopamine)
		$GABA_B$
		$5\text{-}HT_1$ (serotonin)
		mGlu, types II and III (glutamate)
		M_2, M_4 (muscarinic)
		μ, δ and κ opioid
G_q	Increased phosphoinositide metabolism	α_1 adrenoceptors
		CCK (cholecystokinin)
		mGlu, type I (glutamate)
		$5\text{-}HT_2$ (serotonin)
		M_1, M_3, M_5 (muscarinic)
		H_1 (histamine)
		NK (tachykinin)

molecule diffuses freely through the cytoplasm and binds to a kinase enzyme, **protein kinase A** (**PKA**), which is thereby switched on (Figure 3). The kinase then phosphorylates target proteins that have the appropriate amino acid sequence to recognize the kinase.

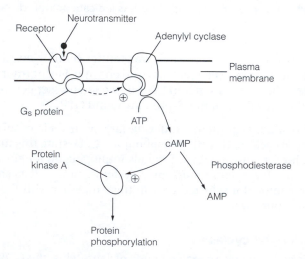

Figure 3. The adenylyl cyclase-cAMP second messenger system. The activated G_s protein uncouples from the receptor to switch on adenylyl cyclase.

Targets include ion channels (the phosphorylation state of a channel often determines whether it is open or closed), and transcription factors—allowing the cAMP second messenger system to modify gene expression.

A single activated PKA molecule is able to phosphorylate many target proteins, adding to the amplification. Second messenger cascades must rapidly turn off if their signals are to be modulated over a time course of tens or hundreds of milliseconds. For the cAMP system this occurs in three ways:

- Cyclic AMP is hydrolyzed to AMP by the action of a specific **phosphodiesterase** in the cytoplasm.

- There are specific **phosphatases** responsible for dephosphorylating the target proteins. Hence the phosphorylation state of a protein at a given time will depend on the balance of the activities of kinases and phosphatases.

- Prolonged occupation of the receptor by the transmitter causes it to **desensitize**. This involves phosphorylation by a specific **kinase** that recognizes the agonist-bound form of the receptor followed by the binding of an **arrestin** protein. The resulting complex is unable to recognize the G protein.

Inhibition of adenylyl cyclase

Some neurotransmitter receptors are negatively coupled to adenylyl cyclase. These receptors associate with G_I **proteins** that inhibit the activity of the enzyme. Both their α-subunit and β/γ-subunits independently block the isoform of adenylyl cyclase common in neurons. The outcome is that the activity of adenylyl cyclase, and hence the amount of cAMP in the cell at any given instant, will reflect the balance of activation of receptors coupled to G_S and those coupled to G_I.

Phosphoinositide second messenger system

Many receptors are coupled via the G_q **protein** to activation of **phospholipase C** (Figure 4). This enzyme cleaves a minor phospholipid in the inner leaflet of the plasma membrane, **phosphatidyl inositol-4,5-bisphosphate** (**PIP$_2$**), to give **diacylglycerol** (**DAG**) and **inositol-1,4,5-trisphosphate** (**IP$_3$**), both of which are second messengers. DAG, a

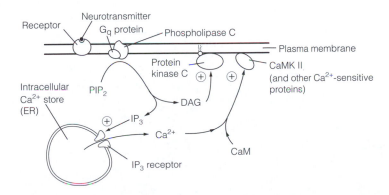

Figure 4. The phosphoinositide second messenger system. CaM, calmodulin; CaMKII, calcium-calmodulin-dependent protein kinase II; DAG, diacylglycerol; ER, endoplasmic reticulum; IP$_3$, inositol trisphosphate; PIP$_2$, phosphatidyl inositol bisphosphate.

hydrophobic molecule, diffuses within the lipid where it activates **protein kinase C** (**PKC**). In turn this kinase phosphorylates its protein targets, affecting metabolic, receptor, and ion channel functions.

IP_3 is water soluble and freely diffusible in the cytosol. Its target is the **IP_3 receptor**, a large IP_3-gated calcium channel located in the membrane of **smooth endoplasmic reticulum** (**SER**). The SER in neurons (and its equivalent, the **sarcoplasmic reticulum** in muscle cells) acts as an intracellular Ca^{2+} store. The binding of IP_3 to its receptors causes the calcium channels to open and Ca^{2+} flows out of the SER into the cytosol. A rise in intracellular calcium concentration has diverse and widespread effects that are cell typical. An obvious example is that by binding the protein **troponin** in striated muscle, calcium triggers the cascade of biochemical events that leads to muscle contraction. Neurons contain a calcium binding protein called **calmodulin** (**CaM**) which shares considerable homology with troponin. On binding Ca^{2+}, calmodulin activates a number of enzymes including **calcium–calmodulin-dependent protein kinase II** (**CaMKII**). CaMKII, and the many other calcium sensitive proteins, mediate the effects of raised intracellular calcium, such as changes in membrane permeability and gene expression.

D3 Amino acid transmitters

<table>
<tr><td colspan="2">Key Notes</td></tr>
<tr><td>Excitatory amino acids</td><td>Glutamate is the major mammalian CNS excitatory transmitter. In the glutamate–glutamine cycle, astrocytes remove glutamate from the cleft and convert it to glutamine for re-export to neurons which use it to synthesize transmitter glutamate. Glutamate acts on ionotropic (AMPA, kainate, and NMDA) receptors that are cation conductances, and on metabotropic receptors (GPCRs).</td></tr>
<tr><td>Inhibitory amino acids</td><td>GABA and glycine are the major inhibitory transmitters in the mammalian CNS. GABA is synthesized from glutamate by glutamic acid decarboxylase which is found only in GABAergic neurons. After reuptake from the cleft it is broken down by GABA transaminase. GABA acts on $GABA_A$ receptors that are ligand-gated chloride channels and $GABA_B$ receptors that are GPCRs. Glycine receptors are closely related to $GABA_A$ receptors.</td></tr>
<tr><td>Related topics</td><td>(C4) Neurotransmitter inactivation
(D1) Ligand-gated ion channel receptors
(N) Learning and memory</td></tr>
</table>

Excitatory amino acids

Glutamate and aspartate are the major CNS excitatory transmitters with glutamate by far the most predominant. Most of the major sensory pathways and some motor pathways are glutamatergic (Table 1). All pyramidal cells in the cerebral cortex and granule cells in the cerebellar cortex release glutamate.

Neurotransmitter glutamate is synthesized in neurons from glutamine, a reaction catalyzed by **glutaminase**. Glutamate is then pumped into vesicles. After release glutamate is removed from the synaptic cleft by glutamate transporters in neurons and glia. In neurons the glutamate is probably metabolized, although some may be recycled as a transmitter. In glia the glutamate is converted by **glutamine synthetase** to glutamine which is then liberated into the extracellular space for uptake by neurons. This closes the **glutamate–glutamine cycle** (Figure 1). It allows astrocytes to export transmitter glutamate to neurons in a form—glutamine—that cannot spuriously activate glutamate receptors.

Glutamate acts on ionotropic receptors—AMPA, kainate, and NMDA, named for selective antagonists—which are cation conductances, and on metabotropic receptors (GPCRs).

Inhibitory amino acids

Gamma-amino butyrate (**GABA**) and glycine are the major inhibitory amino acids in the CNS with glycine having higher abundance in brainstem and spinal cord than the forebrain. Many pathways involved in motor control are GABAergic, as are most of the interneurons in both cerebral and cerebellar cortices, and the Purkinje cells that provide the entire output of the cerebellar cortex (Table 1).

Table 1. Major glutamatergic and GABAergic neurons/pathways

Glutamate	GABA
Primary afferents of cranial and spinal nerves	Interneurons of cerebral cortex
Visual system; photoreceptors, bipolar cells, ganglion cells	Interneurons of cerebellar cortex
Dorsal column-medial lemniscus	Cerebellar Purkinje cells
Thalamocortical neurons	Efferents of caudate nucleus, putamen (dorsal striatum)
2nd order neurons of proprioceptor pathways	Efferents of nucleus accumbens (ventral striatum)
Cerebral cortical pyramidal cells; corticopontinecerebellar tract, corticospinal tract	Efferents of globus pallidus and substantia nigra pars reticulata
Hippocampal pyramidal cells	Interneurons of hippocampus
Granule cells of dentate gyrus	Inhibitory dorsal horn cells
Rubrospinal tract	
Lower motor neurons	
Propriospinal neurons	
Cerebellar granule cells	

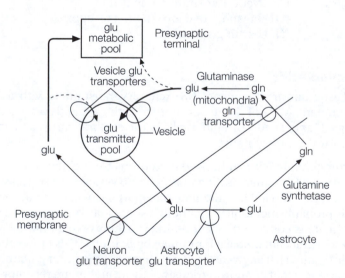

Figure 1. The glutamate (glu)–glutamine (gln) cycle.

GABA is synthesized from glutamate by **glutamic acid decarboxylase** (**GAD**), an enzyme virtually exclusive to GABAergic neurons. After release it is taken up by transporters into both neurons and glia. It is catabolized to succinic semi-aldehyde by the mitochondrial enzyme **GABA transaminase** (Figure 2).

GABA acts on GABA$_A$ receptors that are ligand-gated chloride channels and GABA$_B$ receptors that are GPCRs. Classically, responses of these two classes of receptor can be discriminated by the GABA$_A$ antagonist **bicuculline** and the GABA$_B$ agonist **baclofen**.

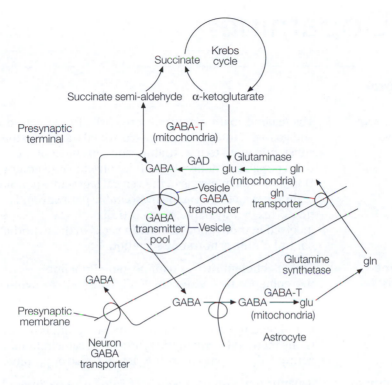

Figure 2. GABA shunt. GAD, glutamic acid decarboxylase; GABA-T, GABA-transaminase; glu, glutamate; gln, glutamine.

Neurotransmitter **glycine** is synthesized from serine by mitochondrial **serine transhy-droxymethylase**. Glycine transporters remove it from the synapse. In the spinal cord **Renshaw cells** express nAChR and are excited by collaterals of motor neurons. Renshaw cells use glycine as a transmitter and inhibit the motor neurons that excite them, among others. This is an example of recurrent inhibition. It serves to dampen the output of motor neurons. The **glycine receptor** resembles the GABA$_A$ receptors and is a Cl⁻ channel. It is blocked by **strychnine**. Moreover **tetanus toxin** blocks glycine release. Both of these agents are convulsants because they remove Renshaw cell inhibition.

D4 Dopamine

Key Notes

Dopaminergic pathways

The major dopaminergic pathways arise from the midbrain and go to the forebrain. The nigrostriatal tract from the substantia nigra to the striatum contains most of the brain dopamine neurons and is involved in movement. Dopaminergic neurons in the ventral tegmentum project to limbic structures via the mesolimbic pathway and to the cortex by way of the mesocortical pathway. These are involved in motivation. Dopamine cells in the hypothalamus control pituitary hormone secretion.

Dopamine synthesis

The catecholamines (dopamine, noradrenaline (norepinephrine), adrenaline (epinephrine)) are synthesized from tyrosine. The first, rate-limiting step which generates L-dopa is catalyzed by tyrosine hydroxylase. This enzyme is inhibited by catecholamines. This end point inhibition is one method by which the synthesis of catecholamines is controlled. L-dopa is decarboxylated to give dopamine.

Inactivation of dopamine

Synaptic dopamine is taken back into nerve terminals by a high affinity dopamine transporter. This process is inhibited by amphetamines and cocaine. Dopamine which escapes reuptake is catabolized to homovanillic acid by catechol-O-methyltransferase then monoamine oxidase (MAO). Dopamine free in the cytoplasm is converted to dihydroxyphenyl acetic acid by mitochondrial MAO.

Dopamine receptors

The five metabotropic receptors for dopamine fall into two families. The D1 receptor family (D_1 and D_5) increase cAMP concentrations, whereas the D2 receptor family (D_2, D_3, and D_4) decrease cAMP concentrations.

Related topics

(C2) Neurotransmitter release
(K6) Anatomy of the basal ganglia

(K7) Basal ganglia function
(M2) Motivation and addiction

Dopaminergic pathways

Dopamine neurons are widely distributed in the nervous system, being found in the retinal amacrine cells, olfactory bulb, and autonomic ganglia. Most, however, are confined to a few nuclei in the brainstem, sending their axons to many regions of the forebrain including the cerebral cortex. These major pathways are illustrated in Figure 1.

About 80% of dopamine neurons are in the zona compacta of the **substantia nigra** (**SNpc**), which constitutes the A9 group of catecholaminergic cells. (These groups range from A1 to A16, the higher the number the more rostrally they are located.) SNpc neurons

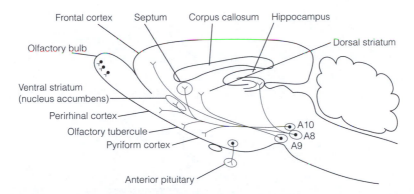

Figure 1. Major dopamine pathways in a sagittal section of the rat brain. The A8 and A10 group of dopamine neurons give rise to the mesolimbic and mesocortical tracts. The nigrostriatal tract originates in the substantia nigra (A9). A12 neuron axons run in the tuberoinfundibular pathway.

project to the striatum as the **nigrostriatal pathway**. These cells are involved in basal ganglia regulation of movement and their loss results in Parkinson's disease. Dopamine cell clusters (groups A8 and A10) in the ventral tegmentum of the midbrain project to limbic structures (e.g., nucleus accumbens) or to associated cortical areas (e.g., medial prefrontal and cingulate cortex), giving rise to the **mesolimbic** and **mesocortical** systems respectively. These are implicated in motivation, drug addiction, and in schizophrenia. Several small groups of dopaminergic cells in the hypothalamus project axons in the **tuberoinfundibular pathway** to the pituitary to inhibit the secretion of prolactin or growth hormone.

Dopaminergic neurons are small with a thin unmyelinated axon (which arises from one of the dendrites) which bears numerous varicosities along its length. Action potentials in dopamine neurons are long lasting (2–5 ms) and propagated very slowly (0.5 m s^{-1}).

Dopamine synthesis

The precursor for all catecholamine transmitters (dopamine, noradrenaline (norepinephrine), adrenaline (epinephrine)) is the amino acid, L-tyrosine. This is hydroxylated by **tyrosine hydroxylase** to give 3,4-dihydroxy phenylalanine (L-dopa) which is rapidly decarboxylated by the unspecific enzyme L-**aromatic amino acid decarboxylase** to give dopamine (Figure 2).

Tyrosine is actively transported into the brain, and the brain concentration of tyrosine is normally enough to saturate tyrosine hydroxylase (TH), so administration of tyrosine cannot alter the rate of dopamine synthesis. The hydroxylation of tyrosine is the rate-limiting step for catecholamine synthesis under basal conditions and TH is subject to regulation:

- Increased expression of TH genes, leading to *de novo* synthesis of the enzyme.

- Phosphorylation by protein kinases which increases its activity.

- Inhibition by catecholamines. This is an example of **end point inhibition**.

Dopamine is taken into vesicles by a **vesicular monoamine transporter** (**VMAT**) which actively transports catecholamines and serotonin using the co-transport of protons from

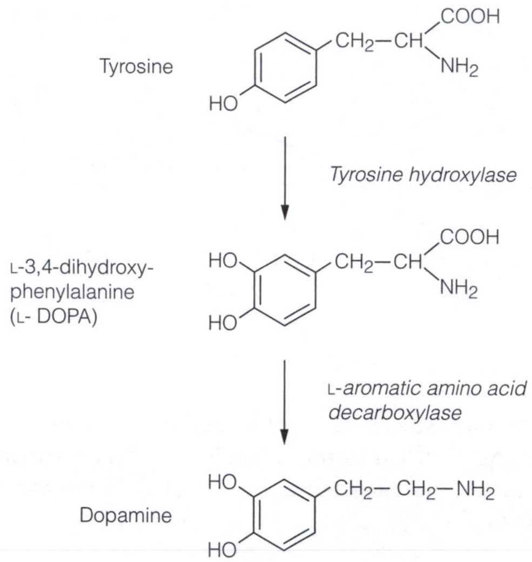

Figure 2. Synthesis of dopamine from the amino acid tyrosine.

the vesicle to provide the energy. VMATs are blocked by the drug **reserpine** which, by preventing vesicular storage, drastically impairs monoamine neurotransmission.

Inactivation of dopamine

Inactivation is by diffusion and reuptake into the nerve cell by a high affinity Na^+/Cl^--dependent **dopamine transporter**. Neurons that release their dopamine into the hypothalamic-pituitary portal system lack the dopamine transporter. The dopamine transporter is competitively inhibited by amphetamines and by cocaine, which thus potentiate the effects of dopamine at the synapse. This contributes to the powerful reinforcing properties of these drugs which make them addictive.

Two major enzymes are involved in catecholamine catabolism. The primary dopamine metabolites in the CNS are **homovanillic acid** (**HVA**) and **dihydroxyphenyl acetic acid** (**DOPAC**). In primates, the major catabolic route is via HVA and this is the fate of dopamine released into the cleft which escapes reuptake. It requires the sequential action of **catechol-O-methyl transferase** (**COMT**) and **monoamine oxidase** (**MAO**), both of which are present in neuronal membranes (Figure 3). Cytoplasmic dopamine that is not transported into vesicles remains free in the axon where it is catabolized by MAO located on the outer membrane of mitochondria, then by **aldehyde dehydrogenase**, a soluble cytosolic enzyme, to dihydroxyphenyl acetic acid.

Dopamine receptors

Five dopamine receptors have been identified (Table 1, Section D2), all are G-protein-coupled receptors and fall into two families. The D1 family (D_1 and D_5) are coupled to G_s and activate adenylyl cyclase to increase cAMP synthesis. The D2 family (D_2, D_3, and D_4) are coupled to G_i and inhibit adenylyl cyclase to reduce cAMP synthesis.

D1 receptors are postsynaptic and located mostly in the striatum and substantia nigra. There are two splice variants of D_2. The short variant (D_{2S}) is an autoreceptor on nigrostriatal and ventral tegmental neurons where it regulates dopamine synthesis by lowering

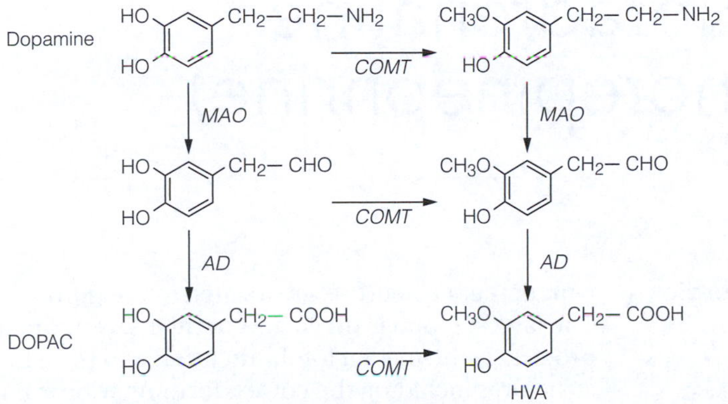

Figure 3. Metabolism of dopamine. DOPAC, dihydroxyphenyl acetic acid; HVA, homovanillic acid; COMT, catechol-*O*-methyl transferase; MAO, monoamine oxidase; AD, alcohol dehydrogenase.

cAMP concentrations and so phosphorylation of tyrosine hydroxylase. The long variant (D_{2L}) is postsynaptic in the striatum. D_3 receptors are presynaptic autoreceptors and reduce dopamine release by closing presynaptic Ca^{2+} channels. D_4 receptors are more abundant in the cortex than striatum. D_5 receptors are extremely widely distributed in the brain.

Bromocriptine has about 100-fold the affinity for D_2 than D_1 receptors and classic antipsychotic antagonists (e.g., **haloperidol**) block postsynaptic D_2 more strongly than D_1 receptors. **Clozapine** is a fairly selective D_4 receptor antagonist.

D5 Noradrenaline (norepinephrine)

Key Notes

Noradrenergic pathways	Noradrenergic neurons are located in the pons and medulla. The largest group is the locus coeruleus. Noradrenergic axons project via the medial forebrain bundle to most forebrain structures including the cortex, forming wide synapses which allow considerable diffusion of the transmitter. Noradrenergic pathways form an arousal system
Noradrenaline (norepinephrine) and adrenaline (epinephrine) synthesis	Dopamine-β-hydroxylase catalyzes the synthesis of noradrenaline (NA; norepinephrine) from dopamine. In adrenergic neurons in the brain and chromaffin cells of the adrenal medulla, NA is further metabolized to adrenaline (epinephrine).
Noradrenaline (norepinephrine) inactivation	A high affinity transporter is responsible for reuptake of NA from the synaptic cleft. The transporter is inhibited by tricyclic antidepressant drugs. The enzymes MAO and COMT are responsible for NA catabolism producing 3-methyl-4-hydroxyphenyl glycol, which is then excreted.
Adrenergic receptors	Adrenoceptors are GPCRs activated by NA and adrenaline (epinephrine). α_1 receptors are typically postsynaptic and coupled to the IP_3/DAG second messenger system. α_2 receptors are presynaptic and reduce cAMP. All β adrenoceptors are coupled to G_s proteins and raise cAMP levels.
Related topics	(F3) Pain (M5) Sleep (L5) Autonomic nervous system (ANS) function

Noradrenergic pathways

Cell bodies of noradrenergic neurons are located in the pons and medulla (cell groups A1–A6, except A3) (Figure 1). The most caudal groups A1 and A2 send their axons into the spinal cord where they form synapses with the terminals of primary afferents. The others project in two bundles, a dorsal and a ventral bundle, which unite to form the **medial forebrain bundle** (**MFB**) that ascends to supply the hypothalamus, amygdala, thalamus, limbic structures, hippocampus, and neocortex. The major noradrenergic cell group is the **locus coeruleus** (**LC**, group A6) which contributes most of the axons of the dorsal noradrenergic bundle and projects to the cerebellum. Noradrenergic neurons are small with fine, highly branched axons that ramify widely. The axons bear varicosities along their length, but they do not form close synaptic contacts, so noradrenaline (norepinephrine) is able to diffuse to reach widespread targets. This is termed **volume transmission**. Firing

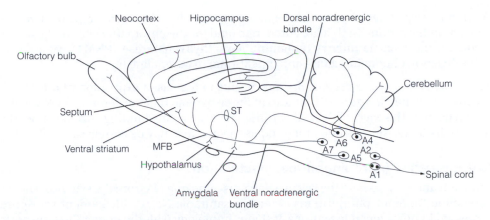

Figure 1. Major noradrenergic pathways in a sagittal section of the rat brain. A6 is the locus coeruleus. MFB, medial forebrain bundle; ST, stria terminalis.

of noradrenergic cells is low in sleeping animals and increases with arousal level. Hence noradrenaline (norepinephrine) is implicated in controlling sleep–waking cycles and in maintaining arousal. It increases the signal-to-noise ratio of cortical processing.

Noradrenaline (norepinephrine) and adrenaline (epinephrine) synthesis

The first steps in noradrenaline (NA; norepinephrine) synthesis require the synthesis of dopamine from tyrosine. **Dopamine-β-hydroxylase (DβH)**, an enzyme present in the synaptic vesicle membrane then catalyzes the synthesis of noradrenaline (norepinephrine) (Figure 2). NA is actively taken into synaptic vesicles by the vesicular monoamine transporter where it is stored together with ATP.

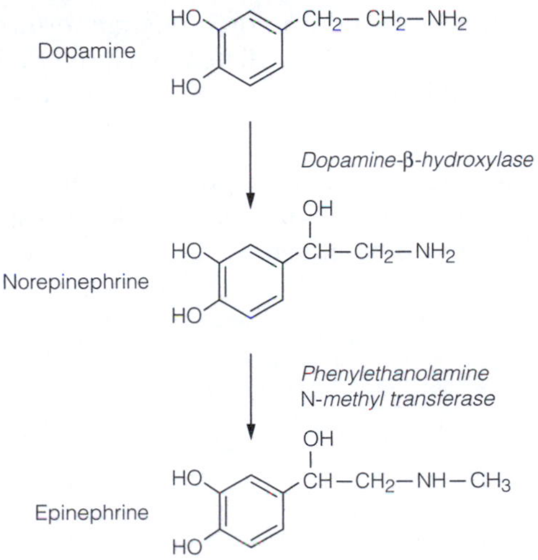

Dopamine

$$HO-\bigcirc-CH_2-CH_2-NH_2$$
HO

Dopamine-β-hydroxylase

↓

Norepinephrine

$$HO-\bigcirc-\overset{\overset{OH}{|}}{CH}-CH_2-NH_2$$
HO

Phenylethanolamine N-methyl transferase

↓

Epinephrine

$$HO-\bigcirc-\overset{\overset{OH}{|}}{CH}-CH_2-NH-CH_3$$
HO

Figure 2. Synthesis of norepinephrine and epinephrine. These catecholamines, like dopamine, are derived from tyrosine. Early synthetic steps are shown in Figure 2 in Section D4.

For noradrenergic neurons the reactions stop at this point. However, for the relatively few neurons in the hindbrain that are adrenergic (and for the chromaffin cells of the adrenal medulla), the enzyme **phenyletholamine *N*-methyltransferase** (**PNMT**) catalyses the *N*-methylation of noradrenaline (norepinephrine) to adrenaline (epinephrine).

High activity by locus coeruleus neurons results in increased expression of the tyrosine hydroxylase genes and *de novo* synthesis of the enzyme so that the demand for NA synthesis can be met. The effect of this is that DβH becomes the rate limiting enzyme rather than tyrosine hydroxylase; thus dopamine and its metabolites may be co-released with NA.

Noradrenaline (norepinephrine) inactivation

Diffusion and reuptake are the key mechanisms removing NA from the synapse. The noradrenaline (norepinephrine) transporter is a saturable Na^+/Cl^--dependent transporter expressed in noradrenergic neurons. It shares homology with the dopamine transporter. The noradrenaline (norepinephrine) transporter is inhibited by the tricyclic antidepressant group of drugs.

The metabolic degradation of NA is not important for its inactivation and occurs via different routes in the periphery and CNS. In the CNS, monoamine oxidase (MAO) catalyzes the formation of 3,4-dihydroxy phenylglycoaldehyde which is then reduced to the corresponding alcohol, **3,4-dihydroxy phenylglycol** (**DOPEG**). Finally this is methylated by COMT to give **3-methoxy,4-hydroxy phenylglycol** (**MOPEG**) which is excreted in the urine.

Adrenergic receptors

Adrenoceptors are GPCRs activated by both noradrenaline (norepinephrine) and adrenaline (epinephrine). α_1 receptors are typically postsynaptic and coupled via G_q to phospholipase C so are excitatory by increasing the concentrations of the second messengers inositol trisphosphate and diacylglycerol concentrations. All β adrenoceptors are coupled to G_s proteins and raise cAMP levels. In the CNS, α_2 receptors are presynaptic autoreceptors, reducing NA release by lowering cAMP-mediated phosphorylation of Ca^{2+} channels. Presynaptic β receptors are also found on noradrenergic terminals in the brain. These facilitate NA release by increasing cAMP-mediated phosphorylation and opening of Ca^{2+} channels. Both excitatory and inhibitory effects of NA release are seen postsynaptically in CNS neurons.

Classically **phenylephrine**, **clonidine**, and **isoproterenol** act as fairly selective agonists on α_1, α_2, and β receptors respectively. **Prazosin** blocks α responses. β-blockers (e.g., **propranolol**) are selective β receptor antagonists.

D6 Serotonin

Key Notes

Serotonergic pathways

Serotonin neurons are located in raphe nuclei which lie close to the midline throughout the brainstem. Some project into the spinal cord to inhibit nociceptor input. Most project in the medial forebrain bundle to most forebrain structures. Serotonin pathways are implicated in anxiety, mood, sleep, and the control of CSF secretion and cerebral blood flow.

Synthesis of serotonin

Serotonin (5-HT) is synthesized from tryptophan, the plasma concentration of which can affect serotonin levels in the brain. The rate limiting step in 5-HT synthesis is the hydroxylation of tryptophan catalyzed by tryptophan hydroxylase. The activity of this enzyme increases with neuron firing rate so that transmitter synthesis keeps pace with neural activity.

Inactivation of serotonin

Reuptake of serotonin by a transporter terminates its transmitter action. The transporter is inhibited by tricyclic antidepressants and selective serotonin reuptake inhibitors. 5-HT is catabolized by MAO to 5-hydroxyindoleacetic acid.

Serotonin receptors

Of the many subtypes of 5-HT receptor all but 5-HT_3 receptors are GPCRs. 5-HT_3 receptors are ligand-gated cation channels similar to nicotinic receptors and produce fast depolarization. Most 5-HT receptors are postsynaptic but the 5-HT_{1A} subtype is a presynaptic autoreceptor that inhibits serotonin release. Apart from 5-HT_1 subtypes, receptors for 5-HT are excitatory by stimulating phospholipase C or adenylyl cyclase.

Related topics

(F3) Pain
(M4) Brain biological clocks

(M5) Sleep

Serotonergic pathways

Clusters of serotonin neurons (designated B1–B9) are scattered throughout the brainstem, mostly towards the midline in the **raphe nuclei**. Projections into the spinal cord that terminate in the dorsal horn reduce nociceptor input into the spinothalamic tract. Forward projections run into the medial forebrain bundle to go to the hypothalamus, amygdala, striatum, thalamus, hippocampus, and neocortex (Figure 1). Some of these serotonergic neurons may be involved in the expression of anxiety. They are inhibited by GABAergic neurons, enhancing the activity of which reduces anxiety. Several potent anxiolytic drugs are 5-HT receptor ligands. Serotonin transmission is modulated by social status; that is, by how dominant an animal is within its social group. This may underlie the effect of serotonin on mood. Deficits are associated with depression and increased risk of suicide. Serotonin is also involved in sleep, satiety, and regulation of CSF secretion and cerebral blood flow.

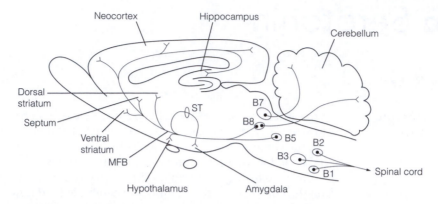

Figure 1. Major serotonin (5-HT) pathways in a sagittal section of the rat brain. The cell groups B1–B8 correspond to the 5-HT containing raphe nuclei (except B4 and B6). MFB, medial forebrain bundle; ST, stria terminalis.

Synthesis of serotonin

The precursor for serotonin is the amino acid tryptophan. The plasma concentration of tryptophan, which varies according to dietary intake, can alter brain serotonin levels. Serotonin is hydroxylated by **tryptophan hydroxylase** to give 5-hydroxytryptophan (5-HTP) and this reaction is the rate limiting step in serotonin synthesis. Decarboxylation

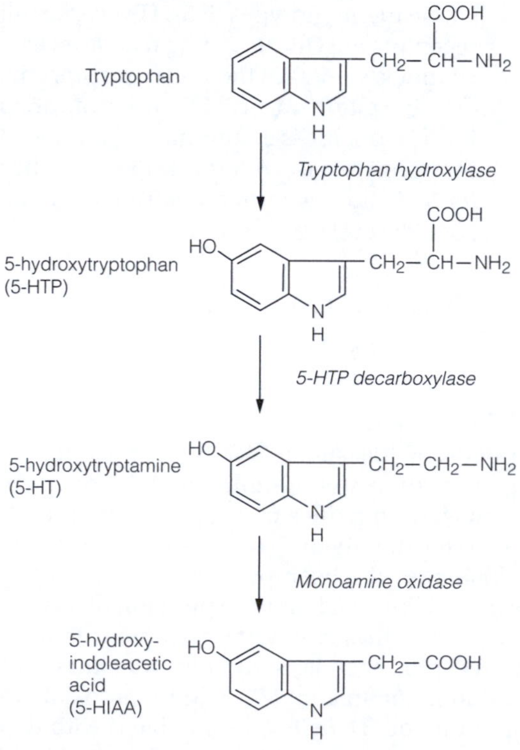

Figure 2. Synthesis of serotonin from the amino acid tryptophan.

of 5-HTP by L-aromatic amino acid decarboxylase (the same enzyme found in catechol-aminergic neurons) gives **serotonin**, also referred to as **5-hydroxytryptamine (5-HT)** (Figure 2).

Serotonin synthesis is matched to the firing frequency of the neuron. Higher firing rates allow increased Ca^{2+}-dependent phosphorylation of tryptophan hydroxylase, the activity of which goes up.

Inactivation of serotonin

Diffusion and reuptake via a saturable Na^+/Cl^--dependent transporter terminates the synaptic action of serotonin. The transporter is inhibited by tricyclic antidepressants and the **selective serotonin reuptake inhibitors (SSRIs)**. However, the action of at least one SSRI, fluoxetine (Prozac) may relate to its action in releasing neurosteroids. The psycho-stimulant ecstasy (3,4-methylene-dioxymethamphetamine) competes with serotonin for this reuptake system, which may partly explain the euphoria it produces. Oxidative deamination of serotonin by MAO yields its principle metabolite, **5-hydroxyindoleacetic acid (5-HIAA)**.

Serotonin receptors

There are many subtypes of serotonin receptor and all except the 5-HT_3 receptor are GPCRs. The 5-HT_1 subtypes decrease cAMP, while the 5-HT_4 and 5-HT_7 subtypes increase cAMP. The 5-HT_2 subtypes are coupled via G_q to the phospholipase C so are excitatory by increasing inositol trisphosphate and diacylglycerol. 5-ht_5 and 5-ht_6 subtypes are currently putative.

The 5-HT_3 receptors are ligand-gated ion channels similar to the nicotinic receptor and form pentamers. There are several isoforms. As cation conductances they mediate fast depolarization and are distributed widely in the nervous system. A number of drugs that bind to nicotinic receptors also bind to 5-HT_3 receptors but there are selective antagonists at some isoforms (e.g., **ondansetron**). These are anti-emetics because they block 5-HT_3 receptors in the area postrema, the chemosensitive cells of which trigger vomiting in response to toxins in the blood.

Most 5-HT_1, 5-HT_2, and 5-HT_3 receptor subtypes are located postsynaptically but 5-HT_{1A} autoreceptors inhibit the release of serotonin. Agents which reduce serotonin transmission—5-HT_2 and 5-HT_3 antagonists, and 5-HT_{1A} agonists—have proved to be potent anti-anxiety agents.

D7 Acetylcholine

Key Notes

Cholinergic pathways	Somatic and autonomic preganglionic motor neurons that project from the brainstem and spinal cord are cholinergic. Central cholinergic projections come from three principle sources. The pontine reticular formation send axons to spinal cord or forward to forebrain structures. These help regulate sleep and waking. Basal forebrain nuclei make massive connections with the cortex and the septum projects to the hippocampus. They produce cortical arousal and facilitate learning and memory.
Acetylcholine synthesis and inactivation	Acetylcholine (ACh) is produced from acetyl CoA and choline by acetylcholine transferase, a marker enzyme for cholinergic neurons.
	ACh is hydrolyzed to choline and acetate in the synaptic cleft by acetylcholinesterase which terminates its transmitter action. Choline is taken back into the nerve terminal by a Na^+-dependent choline transporter.
Acetylcholine receptors	Nicotinic receptors (nAChR) are ligand-gated cation channels and muscarinic receptors (mAChR) are GPCRs. In the brain nAChR are found in sensory cortex, the hippocampus, and in the ventral tegmentum where they are presumed to mediate nicotine addiction. Presynaptic nAChRs enhance transmitter release. Central mAChR are widely distributed with M1 receptors being postsynaptic and excitatory by activating phospholipase C and M2 receptors being presynaptic where they inhibit transmitter release by decreasing cAMP. In the periphery, nAChR allow fast transmission in autonomic ganglia and at the neuromuscular junction of skeletal muscle. Muscarinic receptors are present in smooth muscle, cardiac muscle, and glands and respond to ACh released from the autonomic nervous system.
Related topics	(J1) Nerve–muscle synapse (M5) Sleep (L5) Autonomic nervous (N3) Hippocampus and system (ANS) function episodic learning

---

Cholinergic pathways

Motor neurons in motor nuclei of the cranial nerves and ventral horn of the spinal cord are cholinergic, as are preganglionic autonomic neurons, and the axons of all these cells project into the peripheral nervous system. Cholinergic interneurons are present in the striatum and the nucleus accumbens.

Three regions within the brain contain cholinergic neurons that project centrally. Most caudal are those of the **pontine reticular formation** that send axons into the spinal cord or forward to the amygdala, thalamus, and basal forebrain. These are important in regulating sleep and wakefulness.

A second region, the basal forebrain, contains the **nucleus basalis of Meynert** (**nBM**) which projects extensively to the cerebral cortex. The third region, the **medial septum**, gives rise to the **septohippocampal pathway** (Figure 1). In primates, cholinergic neurons of the nBM show brief changes in firing rate during behavioral tasks, particularly when reinforcing stimuli are presented. Lesions of the nBM produce impairment in recall for tasks learnt before the surgery and in acquisition of new learning. ACh produces long-term facilitation of neurons in the neocortex and hippocampus by acting at muscarinic receptors to close potassium (K_M) channels, making the cells more likely to fire in response to excitatory inputs. Hence the forebrain cholinergic system seems to be a *selective* arousal system, activated by rewarding or salient events, which facilitates learning.

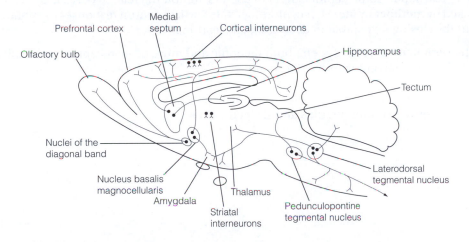

Figure 1. Major cholinergic pathways in a sagittal section of the rat brain. The nucleus basalis magnocellularis of the rat is known as the nucleus basalis of Meynert in primates.

Acetylcholine synthesis and inactivation

Synthesis of acetylcholine (ACh) from choline and acetyl CoA is catalyzed by **choline acetyl transferase** (**ChAT**), a cytoplasmic enzyme. Acetyl CoA is derived from glycolysis and must be transported out of the mitochondria of cholinergic neurons. This supply of acetyl CoA, rather than ChAT activity, is thought to be rate limiting for ACh synthesis. Cholinergic neurons express a Na^+-dependent choline transporter which is saturated at plasma choline concentrations and is responsible for the uptake of choline into neurons. ACh is loaded into vesicles by a transporter that is related to the vesicular monoamine transporters.

The synaptic action of ACh is terminated by its hydrolysis in the cleft to choline and acetate by **acetylcholinesterase** (**AChE**). The liberated choline is recovered by the Na^+-dependent choline transporter.

Acetylcholine receptors

Nicotinic receptors (**nAChR**) are ligand-gated ion channels found throughout the CNS. They mediate fast acetylcholine transmission from the septum to GABAergic inhibitory interneurons in the hippocampus. This is thought to help synchronize the rhythmic firing of hippocampal pyramidal cells. They also mediate fast ACh transmission from brainstem nuclei to the ventral tegmental area, stimulating the dopamine reward pathways. This may be the route by which nicotine is addictive. Presynaptic nicotinic receptors enhance transmitter release at several sites. Interestingly they potentiate glutamate transmission via NMDA but not AMPA receptors in the prefrontal cortex, suggesting a role in memory.

Slow cholinergic transmission by acetylcholine in the CNS is mediated by **muscarinic receptors (mAChR)**. All mAChR responses can be blocked by **atropine**. There are five mAChR subtypes, all GPCRs. M_1, M_3, and M_5 are all coupled via G_q to the phospholipase second messenger system so are excitatory by elevating inositol trisphosphate and diacylglycerol. M_2 and M_4 decrease cAMP, though the M_2 also has other effects.

Postsynaptic mAChR are commonly M_1 which can be selectively blocked by **pirenzepine**, while presynaptic autoreceptors are M_2 receptors, inhibit the release of ACh, and can be blocked by **methoctamine**. M_1 receptors facilitate cortical neuron responses to excitatory input (by closing K_M potassium channels), and learning.

In the peripheral nervous system, both nicotinic and muscarinic receptors are involved in cholinergic transmission in autonomic ganglia, whereas muscarinic receptors only are found at the neuroeffector junctions of the ANS; that is, on smooth muscle, cardiac muscle, and glands. Nicotinic receptors mediate fast transmission at the neuromuscular junction between motor neurons and skeletal muscle.

D8 Purines and peptides

Key Notes

Purines

The purine transmitters are ATP and adenosine. ATP acts on ionotropic receptors and is excitatory on smooth muscle and neurons. It is implicated in transmission in the hippocampus, by autonomic neurons, sensory neurons and is implicated in pain signaling. It is inactivated enzymically. Adenosine is not stored in vesicles or released in a calcium-dependent way. It is generated from ATP and ADP and acts on metabotropic receptors. It is probably the molecule responsible for physiological termination of seizure activity. It is inactivated by re-uptake.

Peptides

There are over 50 peptides transmitters. They are grouped by amino acid homology and by derivation from a common precursor. Peptides are packaged into vesicles, subject to posttranslational processing and moved by axoplasmic transport to nerve terminals for secretion.

Tachykinins

Tachykinins are a group of excitatory peptides. They act via receptors coupled to phospholipase C second messenger systems. Substance P is the transmitter of small diameter primary afferents and implicated in pain transmission and neurogenic inflammation.

Opioids

Opioid peptides are implicated in natural analgesic pathways, sexual, and aggressive/submissive behaviors. Falling into three families, the enkephalins, endorphin, and dynorphins, they act on metabotropic receptors that produce inhibitory responses by decreasing cAMP.

Related topics

(C3) Postsynaptic responses   (M2) Motivation and addiction
(F3) Pain

Purines

Both ATP and its catabolite adenosine are purine transmitters. ATP is stored in synaptic vesicles and co-released with classical transmitters from postganglionic autonomic fibers and central synapses.

There are two families of receptor for ATP (Table 1). P_{2X} **receptors** are ligand-gated ion channels distinct from the cys-loop or glutamate receptor families. They are permeable to Na^+, K^+, and Ca^{2+} and exert excitatory effects. P_{2Y} **receptors** are GPCRs that are excitatory by stimulating phospholipase C and/or adenylyl cyclase. ATP is synaptically inactivated by an ecto-5′-nucleotidase.

Examples of ATP transmission include:

- The fast phase of smooth muscle contraction in response to sympathetic stimulation

Table 1. Purine receptors

Receptor	Second messenger/effector	Endogenous ligand	Actions
A_1	↓ cAMP	Adenosine	Presynaptic, inhibition of transmitter release. Postsynaptic inhibition (terminate seizure activity, anxiolytic, hypnogenic)
A_2	↑ cAMP	Adenosine	Heart nociceptors (mediate ischemic pain)
P_{2X}	Ionotropic cation channel	ATP/ADP	Fast transmission in CNS and sympathetic terminals. Located on polymodal nociceptors
P_{2Y} [a]	↑ cAMP, ↑ IP_3/DAG	ATP/ADP	Excitatory

a There are multiple subtypes of P_{2Y} receptors that couple to different G proteins

- Excitation of dorsal horn cells and motor neurons in the spinal cord by ATP release from primary afferents
- In the CA3 region of the hippocampus
- Nociceptor signaling at a number of sites (e.g., urinary bladder)

Adenosine is an atypical transmitter in that it is not stored in vesicles or released in a Ca^{2+}-dependent way. It is generated locally by enzyme-catalyzed breakdown of released ATP and ADP. Adenosine receptors are GPCRs which modulate cAMP. Synaptic actions of adenosine are inactivated by a nucleoside transporter. Adenosine transmission:

- Terminates epileptic seizures
- Protects neurons from oxidative stress

Peptides

Over 50 small peptides are thought to be neurotransmitters. Some are also hormones or neuroendocrines. They can be grouped into families on the basis of:

- Similarities in their amino acid sequence.
- Being derived by cleavage of a common large precursor polypeptide encoded by a single mRNA molecule. Often the peptides generated from a common polypeptide have related functions. Different cells may process the same precursor or its mRNA in different ways. For example, different peptides are made from pro-opiomelanocortin by hypothalamic neurons and endocrine cells in the anterior pituitary.

Tachykinins

The first peptide transmitter to be discovered was **substance P** (**SP**). It is an excitatory transmitter in several brain regions including the cerebral cortex, striatum, and substantia nigra. It is released by both central and peripheral terminals of C fiber primary afferents. The central terminals synapse with dorsal horn cells to convey information about pain and temperature. Release from the peripheral terminals results in neurogenic inflammation. SP-containing terminals are found adjacent to cerebral blood vessels and abnormal release of SP may play a role in migraine and other headaches.

The gene which codes for SP also encodes other transmitters of the tachykinin family, such as **substance K** and **neurokinins A** and **B**.

The three **tachykinin receptors** (NK_1, NK_2, and NK_3) are G-protein-coupled receptors coupled to phospholipase C and (for NK_1 and NK_2) increase in cAMP. The tachykinins have different affinities for the receptors, with SP the preferred ligand of NK_1.

Opioids

The opioids are a group of neurotransmitters which act on opioid receptors, the targets for opiate drugs such as morphine. They are generally co-released with classical transmitters, typically GABA and serotonin, and are usually inhibitory. Opioid transmission is important in analgesia pathways in the CNS. Opioids are encoded by three precursor genes:

- The enkephalin precursor encodes **met-enkephalin** and **leu-enkephalin** (so called because they differ in just one amino acid) and is expressed mainly in short interneurons throughout the brain.

- Pro-opiomelanocortin encodes **β-endorphin** and is expressed in neurons of the hypothalamus which project to the thalamus or brainstem.

- The dynorphin precursor codes for leu-enkephalin and **dynorphins**.

There are three populations of opioid receptors, the properties of which are summarized in Table 2. They are GPCRs that allow direct coupling of G proteins to ion channels. By opening K^+ channels and closing Ca^{2+} channels they hyperpolarize neurons.

Two peptides with high affinity and specificity for μ receptors have recently been identified. The gene or precursor protein for these **endomorphins** has not yet been found.

Some opioids also interact with NMDA receptors and with σ receptors.

Table 2. Opioid receptor pharmacology[a]

		μ	δ	κ
Location:	supraspinal	+++[b]	–	–
	spinal	++	++	+
	peripheral	++	–	++
Endogenous ligands:	β-endorphin	+++	+++	+++
	enkephalins	+	+++	–
	dynorphin	++	+	+++
Agonists:	morphine, meperidine, fentanyl	+++	+	+/–
Weak agonists:	methadone	+++	–	–
Mixed partial agonists-antagonists:	buprenorphine	+++	–	××
	nalorphine	××	–	++
	pentazocine	×	+	++
Antagonists:	naloxone, naltrexone	×××	×	×××

[a] All opioid receptors ↓ cAMP and are inhibitory by opening K^+ channels, closing Ca^{2+} channels and inhibiting presynaptic transmitter release.
[b] +, agonist activity; ×, antagonist activity; –, inactive.

E1 Rate coding

Key Notes

Neural coding

The firing rate of a sensory neuron is determined by stimulus intensity and how this changes with time whilst that of a motor neuron helps determine the force of muscle contraction. How a sensory neuron is connected encodes the nature (modality) and location of a stimulus. Motor output is similarly encoded.

In some cases a sensation or movement can be specified by relatively few neurons (sparse coding) but more usually it requires the concerted activity of many cells (population coding). Temporal coding allows precise timing of events and may be important in perception, attention, and memory.

Static and dynamic rate coding

Stimulus intensity is encoded by the firing frequency of a neuron. This is rate coding. Slowly adapting receptors cause their afferents to fire at a rate that reflects the size of a constant stimulus, so-called static responses. Afferents of rapidly adapting receptors show reduced firing to application of a constant stimulus as the receptor adapts. These afferents respond to the rate of change of stimulus intensity, that is, dynamic responses.

Stimulus intensity versus firing frequency

The relationship between the intensity of a stimulus and the firing frequency of a sensory neuron may be linear or more complicated. For many mechanoreceptors and all photoreceptors, the firing frequency is proportional to the logarithm of the stimulus intensity. This allows a large range of intensities to be captured within the firing frequency range of the neuron.

Error protection

Action potentials are binary digital signals which are intrinsically less error prone than analog signals. Population coding confers redundancy on neural systems which consequently fail gracefully rather than catastrophically.

Disadvantages of rate coding

A neuron integrates inputs over time. Hence for rapid decisions, when there is time only for short chains of action potentials, accuracy of stimulus intensity coding is reduced. This might be partly offset by population coding.

Related topics

(B1) Membrane potentials   (F1) Sensory receptors
(C5) Neural integration   (F2) Touch
(E4) Temporal coding

Neural coding

Individual neurons encode information by virtue of two properties. Firstly, the **firing frequency** signals stimulus intensity and how it changes with time. In the same way motor neuron firing rate encodes the timing and force of contraction of a discrete population of muscle fibers. This is **rate coding**. Secondly, the address of an afferent neuron, that is, how it is connected via its inputs and outputs, encodes the **spatial location** of a stimulus, and the nature of the stimulus (i.e., touch, vision), a property termed **modality**. The address of a motor neuron contributes to the type of movement executed and its direction.

In both sensory and motor systems the accurate encoding of a given feature can sometimes be specified by few neurons (e.g., skin itch, whisker movements in rats). This is called **sparse coding** and has the advantage of being energy efficient. More often features are specified by activity in an ensemble of cells (e.g., skin temperature or the direction of a limb movement). This is referred to as **population coding**. It is thought to compensate for the fact that generally neurons are very noisy, although because the noise in one neuron in a population is not independent of noise in the others (the noise is said to be correlated) it is proving hard to fathom how population coding works.

Temporal coding refers to a variety of situations in which neurons fire at very precise times (e.g., only at a particular phase of a sound wave or when two inputs are coincident), which allows neural systems to time events much more precisely than is possible by rate coding. A temporal code in which neurons widely distributed in the brain are driven to fire synchronously is a key area of current neuroscience research because it could be important in perception (the binding problem), attention, and memory.

Static and dynamic rate coding

Stimulus intensity is encoded by the mean rate at which a sensory neuron fires. Afferents fall between two end-members depending on the nature of their sensory receptor. A **slowly adapting receptor** responds to a protracted stimulus for as long as the stimulus lasts, causing its sensory neuron to fire repetitively with a frequency that relates to the magnitude of the stimulus. These neurons exhibit **static** (**tonic**) responses to a constant stimulus. In contrast, **rapidly adapting receptors** respond only briefly to a constant stimulus because they soon become insensitive, or adapt, to it. These receptors respond best to *changes* in stimulus intensity. Their afferents show **dynamic** (**phasic**) responses. Many afferents display a mixture of dynamic and static responses. Examples of static and dynamic responses are shown in Figure 1 which compares three classes of afferent in the skin that are wired to different types of mechanoreceptor.

The beginning and end of a stimulus will be signaled by changes in the rates of firing of slowly adapting afferents, and by transient bursts of firing from rapidly adapting afferents. In this way **stimulus duration** is encoded.

Stimulus intensity versus firing frequency

The relationship between stimulus intensity and response for static sensory neurons can be linear, as for example in skin thermoreceptor afferents. Often, however, the relationship is more complicated. Commonly, for example, the firing rate rises with the *logarithm* (log_{10}) of the intensity. Many skin mechanoreceptors and all photoreceptors fall into this category. This allows a very wide range of stimulus intensities to be accommodated within the **dynamic range** of neurons, that is, the maximum difference in firing

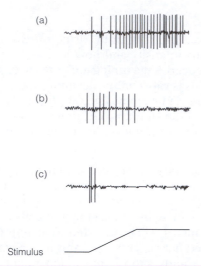

Figure 1. Static and dynamic afferent neuron responses to skin displacement: (a) static response of Ruffini organ afferent; (b) dynamic response of Meissner's corpuscle afferent to velocity of displacement; (c) dynamic response of Pacinian corpuscle afferent to acceleration.

frequency a neuron is capable of. It has the disadvantage that for high intensities the ability to discriminate intensity differences is reduced.

Error protection

The accuracy of sensory coding or motor output depends on the fidelity with which action potentials transmit information. This is facilitated in three ways:

• Because action potentials are all-or-none they are binary digital signals. Binary digital information coding is less prone to error (corruption of the signal by noise) than analog signaling because only two states need to be discriminated.

• With rate coding, the spurious absence or inclusion of occasional action potentials will not change the mean frequency of a train of action potentials much, unless the train is short.

• Most stimuli are sensed, or motor output generated, by populations of neurons operating in concert. This makes for a degree of **redundancy**. Firing errors in a few neurons will be swamped by proper firing of the majority. Even if there are a sizeable number of rogue cells the system will not fail catastrophically, all that happens is that the information conveyed will be less precise. Systems that fail in this way are said to show **graceful degradation**.

Disadvantages of rate coding

For rate coding what counts is firing frequency, so a *single* action potential (AP) carries no information. A minimum of two APs is necessary because instantaneous frequency can then be derived from the **interspike interval**, the time between two successive APs. Given noise and inherent variability of neurons, two or three APs may not accurately encode information about stimulus intensity. For accuracy sufficient time must elapse to sample a reasonable number of APs. Thus for neural systems making rapid decisions, that is, short integration times, the accuracy of stimulus intensity encoding is sacrificed. This problem is partly solved by population coding which briefly samples simultaneous output of many neurons carrying the same information.

In addition, individual neurons are temporal integrators so cannot convey information about the precise timing of events. Precise timing requires temporal coding in ensembles of neurons.

E2 Coding of modality and location

Key Notes	
Modality segregation	Modality refers to the qualitative nature of a stimulus. The sparse coding (labeled line) hypothesis is that everything about a sensation can be encoded in just a few neurons. The population (ensemble) hypothesis is that many neurons are required to specify a sensation. Although there are some exceptions, population coding seems to be more common. It explains how compound sensations (e.g., wetness) arise from the simultaneous activation of multiple types of receptor. Stimulus quality is conferred by the sense organ.
Receptive fields	The region of a sensory surface which when stimulated causes a neuron to respond is the cell's receptive field. In sensory systems, proximal neurons have larger receptive fields than distal neurons because of convergence onto proximal cells of inputs from several distal neurons, and more complex receptive fields because proximal cells can receive inputs from many sources. Many receptive fields show lateral inhibition, in which the cell responses are different in the centre and surround of the field. It enhances contrast at sensory boundaries.
Topographic mapping	Sensory pathways are organized anatomically so that information about the stimulus modality and location are preserved. In consequence many brain structures have ordered maps of the sensory space. Three broad categories of map exist. Discrete maps are anatomically accurate representations of a sensory surface, though area is usually distorted, and arise from local connections between neurons. Patchy maps have discontinuities which distort anatomical relations and represent interactions between distant parts of the body. Diffuse maps are not ordered by any property of the sensation.
Related topics	(F1) Sensory receptors (F2) Touch (F5) Balance (G3) Photoreceptors (G4) Retinal processing (G5) Early visual processing (H2) Anatomy and physiology of the ear (H4) Central auditory processing (I2) Olfactory pathways

Modality segregation

The qualitative nature of a sensation is termed its **modality** though it is poorly defined in the neuroscience literature (but see Table 1 for a version). There are two end-member hypotheses that account for how stimulus quality is encoded.

Table 1. Classification of sensory receptors

Type	Organ/receptor	Stimulus	Modality	Submodality
Exteroceptors				
Special	Retinal photoreceptors	Light photons	Vision	Gray scale brightness, color
	Cochlea hair cells	Sound	Hearing	Tone, loudness
	Olfactory chemoreceptors	Diverse molecules	Smell	No agreed primary qualities
	Gustatory chemoreceptors	Diverse molecules	Taste	Salt, sour, sweet, bitter, umami
	Vestibular hair cells	F_G[a], Head angular velocity and acceleration	Balance	
Somatosensory	Mechanoreceptors	F_M[b]	Touch	Light touch, pressure, vibration, flutter
	Thermoreceptors	Heat	Temperature	Warm, cold
	Nociceptors	F_M, heat, diverse molecules	Pain	Fast well-localized pain
				Slow diffuse pain
	Itch receptors	Histamine	Itch	
Proprioceptors	Muscle/joint mechanoreceptors	F_M	Body position and movement	
Interoceptors	Visceral mechanoreceptors	F_M	Visceral senses	
	Visceral nociceptors	F_M, diverse molecules	Visceral senses	

[a] gravitational force; [b] mechanical force

The **sparse coding** (**labeled line**) hypothesis is that a single class of sensory receptor and its afferent are necessary and sufficient to account for each type of sensation. The correspondence between receptor class and the nature of the sensation occurs because a sensory receptor responds only to a specific type of stimulus (e.g., sodium chloride, pressure). For example, a labeled line exists for skin itch since firing of a single population of afferents, and no other, causes itching.

The **population coding** (**ensemble, across-fiber**) hypothesis is that firing of several types of afferent is required to produce a given sensation. This is clearly the case for color vision in which three populations of wavelength-selective cones are necessary to see colors.

But this is a false dichotomy because most sensory systems lie somewhere along a spectrum between these two end-members. Even in a single class, the receptors and their afferents differ in properties such as threshold, and the range of the stimulus quality over which they operate. For example, auditory neurons vary in the range of sound pressure levels they are sensitive to. Neurons that fire over a wide range are said to be broadly tuned. Population coding is required to account for **compound sensations** which must involve simultaneous activation of several receptor types by a single stimulus. With processing this makes possible a rich variety of higher order sensory experience (e.g., texture, wetness).

Stimulus quality is determined by the sense organ. Surgically rerouting visual pathways to auditory cortex resulted in animals which behaved as if they interpreted input into the redirected pathway as light, not sound. This further suggests that sensory cortex may be a rather general purpose machine.

Receptive fields

The spatial location of a stimulus on a sensory surface (skin, retina, etc.) is given by which particular subset of neurons respond. The **receptive field** (**RF**) of a neuron is the region of a sensory surface which when stimulated causes a change in the firing rate of the neuron. Primary afferents generally have small RFs, the size of which is governed by the distribution of the cluster of sensory receptors which supply the afferent. Receptive fields of neighboring neurons responding to the same type of stimulus tend to overlap.

More proximal neurons in a sensory pathway have receptive fields that are composites of the RFs of more distal neurons. This gives rise to two features:

● In general, proximal neurons have larger RFs because of **convergence**; several afferents may synapse on a single more proximal (i.e., downstream) neuron. Low convergence is seen where high **spatial resolution** (the ability to sense stimuli that are close together as independent) is important, such as between cones and bipolar cells in the retina. In contrast, high convergence is required where it is necessary to integrate weak signals from a number of receptors to achieve high sensitivity. This is the case between rods and bipolar cells in the retina, where it permits vision in dim light.

● The more proximal a neuron the more complex its receptive field. This is because proximal neurons get inputs from a wider range of sources than distal neurons. This reflects the fact that extensive information processing occurs in sensory systems. Greater complexity of receptive fields also arises as a consequence of an extremely common characteristic of sensory pathways, **lateral** (**surround**) **inhibition**. At its simplest, this is where the RF of a neuron has two zones, a central area and a surround, from which opposite and antagonistic effects are produced in the cell when stimulated. It is seen in somatosensory, visual, and auditory pathways. Figure 1 shows the receptive field of a somatosensory cell. Stimulation of the center causes an increase in firing so the RF is said to have an excitatory center. Stimulation of the surround reduces firing and is brought about by inhibitory interactions. A cell behaving in this way is described as an **on-center cell**. **Off-center cells** are also common. For the on-center cell, maximum firing rate would be seen with a stimulus that just managed to stimulate the entire center. A larger stimulus that encroached upon the surround would be less effective, by causing some inhibition. In this way lateral inhibition sharpens spatial resolution and enhances contrast at boundaries between stimuli.

In skin mechanoreceptor afferents this improves **two-point discrimination**. By similar means, light–dark contrast at edges is enhanced in the retina, and tone discrimination sharpened by central auditory neurons. In general, lateral inhibition happens between

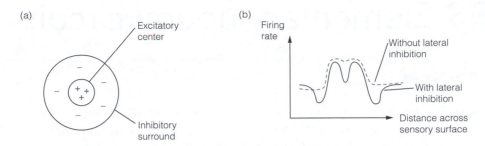

Figure 1. Lateral inhibition. (a) Receptive field of an on-center sensory neuron showing lateral inhibition; an off-center neuron would have an inhibitory center and an excitatory surround. (b) Contrast enhancement in the presence and absence of lateral inhibition.

neurons coding the same type of sensation. However, color vision depends on lateral inhibition between cells that respond to different wavelengths.

Topographic mapping

In most sensory pathways primary afferents are wired to specific subsets of more central neurons in a strictly ordered fashion so that nearest neighborhood relations are conserved. This means that information about stimulus location is not lost in more proximal parts of a pathway. This arrangement is called **topographic mapping**. Receptive fields are aligned to produce an ordered map across brain structures such as the thalamus or the cerebral cortex. These maps are neural representations of a sensory surface or some feature of a sensation. Key examples are: somatotopic maps which represent skin surface; retinotopic maps that reflect the visual fields; and tonotopic maps which represent the pitch of a sound. In addition are numerous motor maps, particularly in the cerebral and cerebellar cortices in which movements are represented in a systematic way. The motor mapping is preserved in descending pathways so that connections with motor neurons are precisely those needed to execute the mapped movement.

Three broad types of map are recognized, thought to be determined by the extent of the connections between the neurons involved in the mapping:

- **Discrete maps** such as somatotopic or retinotopic maps are anatomically accurate and complete representations of a sensory surface, though they are usually distorted, in that the area of the surface is not faithfully proportioned. Fingers and lips get far more than their fair share of the cortex in somatotopic maps. Discrete maps arise because neurons are connected mostly to their neighbors, allowing **local interactions** between cells. In other words, most of the comparisons the CNS needs to make of, say an image, are between adjacent pixels of retina.

- **Patchy maps** consist of several domains within each of which the body is accurately represented. However, adjacent domains map regions that are not anatomically close or which are disoriented. Cerebellar motor maps are of this kind and said to exhibit **fractured somatotopy**. Patchy maps arise because while some groups of neurons are locally connected others are wired to distant neurons allowing global interactions to take place. Serving a tennis ball requires the coordination of movements in distant parts of the body.

- **Diffuse maps** are those which have no underlying topography. Distinct smells are mapped to particular sites in the olfactory bulb but not in any orderly fashion. Smells are not arranged within the brain in any systematic way by property.

E3 Elementary neural circuits

Key Notes	
Neural networks	The operation of groups of interconnected neurons is not well understood. This is a problem because much of the nervous system consists of such neural networks.
Divergence	Neural pathways in which few neurons upstream connect to many neurons downstream exhibit divergence. It allows signals to be spread to many targets.
Convergence	Neural pathways in which many upstream neurons connect to few downstream ones exhibit convergence. It allows data compression and integration of weak signals.
Feedforward	In feedforward, signals flow downstream from lower order to higher order neurons; that is, in the input-to-output direction. Feedforward inhibition is responsible for the surround inhibition in sensory pathways and the reciprocal inhibition seen in motor reflexes.
Feedback	In feedback circuits signals flow upstream. They can be excitatory but are more usually inhibitory when their effect is negative feedback. Neurons that feedback onto themselves do so via recurrent axon collaterals.
Central pattern generators	Neural circuits that generate cyclical patterns of neural activity, such as respiration or locomotion are called central pattern generators.
Related topics	(E2) Coding of modality and location (J4) Spinal motor function	(J5) Brainstem postural reflexes (K5) Cerebellar function (L6) Control of vital functions

Neural networks

Neurons connect together to form networks. The operation of neural networks, even those containing just a few distinct types of nerve cell, is poorly understood in general. This is arguably the most serious problem for contemporary neuroscience because large regions of the nervous system (e.g., the cerebral cortex) apparently consists of the same circuit repeated millions of times. How the same circuit serves functions as diverse as those of the cerebral cortex is currently a mystery. The scale of the difficulty is illustrated by *Caenorhabditis elegans*, a small nematode worm. This animal has just 302 neurons and the circuit diagram of its nervous system is completely known virtually down to the last synapse. Although some aspects of its behavior are beginning to be understood in terms of its network operations (e.g., chemotaxis) most of its behavior cannot yet be modeled.

However, some patterns of neural organization are both common and comprehensible.

Divergence

Few cells connecting with many downstream is termed **divergence** (Figure 1a). It serves to disseminate information to a wide variety of targets. Examples include:

- Primary afferents which relay with many interneurons so that other inputs and motor output in the cord and brainstem can be modified.

- Small numbers of preganglionic autonomic neurons supply up to 100-fold greater numbers of postganglionic neurons.

Convergence

The funneling of connections from many cells to few is called **convergence** (Figure 1a). It is the means by which target cells are able to integrate information from several sources. Convergence must involve data compression. Examples include:

- The retina, which has 100 million photoreceptors but only one million output neurons.

- The spinal cord, where motor neurons are outnumbered by primary afferents about 10-fold.

Low convergence is seen where high **spatial resolution** (the ability to sense stimuli that are close together as independent) is important, such as between cones and bipolar cells in the retina. In contrast, high convergence is required where it is necessary to integrate

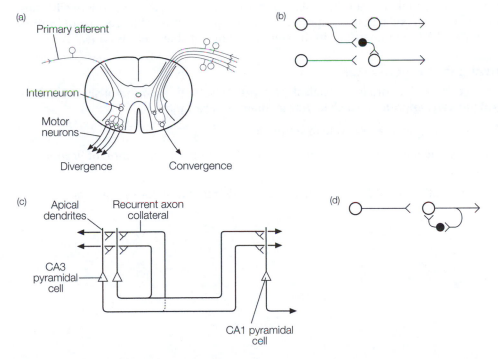

Figure 1. Basic neural circuits. (a) Divergence and convergence in the spinal cord. (b) Feedforward inhibition. (c) Recurrent excitation in the CA3 region of the hippocampus. (d) Feedback (recurrent) inhibition. Excitatory neurons are open circles, inhibitory neurons are filled circles.

weak signals from a number of receptors to achieve high sensitivity. This is the case between rods and bipolar cells in the retina, where it permits vision in dim light.

Feedforward

In feedforward circuits (Figure 1b), input neurons establish connections (either excitatory or inhibitory) with cells that are closer to the output (i.e., higher order) neurons than themselves. In **feedforward inhibition** lower order cells excite inhibitory interneurons which project forward to dampen the activity of neighboring higher order cells. This results in only the strongest signals being propagated further. Feedforward inhibition, in the form of GABAergic interneurons, is responsible for generating the surround inhibition seen in sensory pathways, and may also contribute to **selective attention**, the facility to attend to one stimulus in preference to others.

A special case of feedforward inhibition is the enhancing of a response by attenuating an opposing action, a mechanism known as **reciprocal inhibition**. This operates in spinal cord reflexes that time the activities of limb flexors and extensors. Here, inhibitory interneurons reduce the activity of extensor muscles during flexion and *vice versa*.

Feedback

In feedback circuits, higher order cells establish connections to lower order cells. The connections may be excitatory but are more usually made via inhibitory interneurons to cause feedback inhibition, the neural equivalent of negative feedback. Feedback inhibition allows motor systems to correct errors during the execution of a movement. A neuron may feedback on itself by making recurrent connections. **Recurrent excitation** (Figure 1c) by axon collaterals is important in the hippocampus, while **recurrent inhibition** (Figure 1d) of motor neurons in the spinal cord by Renshaw cells is crucial.

Central pattern generators

Neural networks that produce cyclical patterns of activity autonomously are called **central pattern generators** (**CPGs**). They mediate, for example:

- The inspiratory–expiratory cycle of ventilation
- Limb movements during locomotion that involves alternate activation of flexors and extensors

The basic operation of CPGs is modified or overridden by extraneous pathways.

E4 Temporal coding

Key Notes	
Precise timing by neurons	Neurons may precisely time events by acting as coincidence detectors. This means that only simultaneously active inputs cause the cell to fire. It requires that the neuron has a short time constant.
Brain oscillations	In several brain regions large populations of neurons show synchronized oscillations in membrane potential. These can arise from intrinsic properties of the neurons or via feedback from GABAergic inhibitory interneurons. Brain oscillations occur over several characteristic frequency ranges, each correlated with particular behavioral states. They organize neurons to fire in synchrony, a property thought to be critical for perception, attention, and learning.
Perceptual binding	In many sensory systems different features of a stimulus are processed in different neurons in widely dispersed brain regions. How these features are combined in a unified percept is termed the binding problem. One solution is for all neurons encoding a specific object to fire in synchrony and it is brain oscillations which do this.
Spike-timing-dependent plasticity	Learning is thought to be due to changes in the strength of synapses. Hebb's rule proposes that a synapse between two neurons is strengthened when the neurons are activated together. Long-term strengthening (LTP) or weakening (LTD) of synapses are produced by differences in activating pre- and postsynaptic neurons during a narrow time window.
Related topics	(G6) Parallel processing in the visual system (H4) Central auditory processing

Precise timing by neurons

Rate coding depends on mean firing frequency, so cannot be used to encode precise timing of events. Yet there is considerable evidence that precise timing is crucial for many neural functions and under these circumstances the brain uses **temporal coding**. This is where exact spike timing or high frequency firing fluctuations carry timing information. Studies of the cat visual system show that the resolution of temporal coding is on a millisecond timescale. There are several different temporal coding mechanisms and all rely on features other than mean firing frequency. Temporal coding is used for exact timing of events and as a way in which the firing of many neurons can be synchronized.

Precise timing of events uses **correlation detection** (Figure 1a). Correlated inputs, those occurring within a short time of each other, increase the probability that a neuron will fire. Uncorrelated inputs will reduce spike probability. Exactly how close in time the inputs must be depends on the extent of temporal summation of the inputs and this

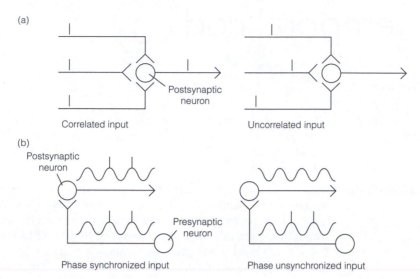

Figure 1. (a) Correlation detection; firing probability of postsynaptic cell depends on precise timing of inputs. (b) Brain oscillations modulate firing.

depends on the time constant of the neuron. Neurons with fine dendrites or being bombarded by numerous ipsps, which consequently have short time constants, will correlate inputs only if they are very close together. This allows for high precision timing.

Pyramidal cells in the cortex are thought to act as coincidence detectors in this way and modulate their behavior to inputs arriving just 5 ms apart. Coincidence detection underlies the way in which the brainstem allows tiny differences in arrival time of a sound in the two ears to be used by the auditory system to help localize the sound source.

Brain oscillations

In several brain regions, notably the cerebral cortex and thalamus, neurons exhibit regular oscillations in membrane potential. These can be large enough to cause rhythmic fluctuations in firing rate which drive changes in synaptic activity. The oscillations arise from intrinsic pacemaker properties of individual neurons, or arise by feedback through GABAergic inhibitory interneurons, described as **hub neurons** because of their widespread connectivity. The oscillations are synchronized over large populations of neurons and the changes in potential are consequently large enough to be recorded via scalp electrodes (electroencephalography) or by electrodes implanted into the cortex of behaving animals. The oscillations occur over several characteristic frequency ranges that tend to be associated with particular behavioral states (Table 1), although there are significant species and developmental differences. For example, theta frequencies, while prominent in rats engaged in moving and exploring are not recorded as regularly in primates; delta activity is normally only seen in deep sleep in adult humans, but is seen in young children when awake.

Neural oscillations are generated in a variety of ways. Those associated with wakefulness (e.g., alpha) are established by reciprocal connections between the thalamus and the cortex. With the onset of sleep, changes in thalamic neurons results in altered connectivity with cortical neurons which consequently switch into oscillations of a different frequency (delta). In rats theta rhythms are seen predominantly in the hippocampus

Table 1. Brain oscillations and behavior

Class[a]	Frequency band/Hz	Associated behaviors
δ	1–4	Slow wave sleep
θ	4–8	Spatial navigation learning (rats), REM sleep
α (also μ, κ)	8–12	Quiet wakefulness, somatosensory processing
β	13–30	Active wakefulness, motor control
γ	30–70	Spatial navigation learning, attention, perceptual binding

[a] Higher and lower frequency oscillations have also been documented.

where they are important in spatial learning. Hippocampal neurons require cholinergic excitatory input from the medial septum to show theta activity though the intrinsic theta pacemaker lies with hippocampal neurons.

A critical function of neural oscillations is to modulate the output of a neuron depending on the relative timing or synchrony between presynaptic and postsynaptic neuron. A neuron is more likely to fire if an input coincides with the depolarizing phase of its neural oscillation than its hyperpolarizing phase (Figure 1b). This mechanism is important in synchronizing the output of neurons and this is thought to be involved in perception, attention, and memory.

Perceptual binding

In many sensory systems different features of a stimulus (e.g., the color, shape, and motion of an object or the loudness, location, and pitch of a sound) are processed in different neurons in widely dispersed brain regions. How does the brain produce a unified percept from all the disparate bits of information it has about the object? This question is known as the **binding problem**. One possible solution is that all the segregated bits of information pertaining to a single stimulus are bound by synchronous firing of the neurons involved. This might be achieved by brain oscillations, perhaps of thalamic neurons which between them are reciprocally connected to all cortical areas. Synchronization of visual system neurons by the thalamus underlies visual attention.

Spike-timing-dependent plasticity

A core idea of neuroscience is that learning occurs by changes in the strength of synapses. A mechanism to account for how this might occur was proposed in 1949 and is called **Hebb's rule**. This states that all synapses between two neurons become stronger if both of the neurons are activated at the same time. Synapses which show this type of plasticity are said to be **Hebbian**, and can mediate associative learning because they act as coincidence detectors that associate firing of the presynaptic *and* postsynaptic cell. Hebb's rule is summarized by the aphorism; "what fires together, wires together."

Several mechanisms to bring about synaptic modifications are now known which either increase synaptic weighting, long-term potentiation (LTP); or decrease synaptic weighting, long-term depression (LTD). These are examples of **spike-timing-dependent plasticity** (**STDP**), alterations to synaptic weighting that are fashioned by precise timing of neural activity. LTP and LTD are both long lasting, but other more transient varieties of STDP exist.

Both LTP and LTD occur in the hippocampus, LTP is also seen in the neocortex, amygdala, and at other sites in the nervous system, while LTD also occurs in the cerebellum and spinal cord. Both require that activity in the pre- and postsynaptic cell occurs within a narrow time window of 10–20 ms. If the presynaptic spike occurs before the postsynaptic spike the result is LTP, but if the presynaptic spike follows the postsynaptic spike then LTD ensues. The critical conditions needed for synchrony to trigger STDPs are supplied by gamma and theta frequency oscillations and the phase relations between them.

LTP and LTD are regarded as cellular substrates of learning. Some neuroscientists have argued that most, if not all, excitatory synapses in the CNS are capable of STDPs. If so, then learning is a fundamental property of the simplest neural networks.

F1 Sensory receptors

Key Notes

Receptor potentials

Sensory receptors, except photoreceptors, produce a depolarizing receptor potential in response to a stimulus. Photoreceptors hyperpolarize. Receptor potentials are small amplitude, graded, passively conducted potentials that decay with time and distance (cf. synaptic potentials). Receptors adapt in that their response declines with time.

Mechanoreceptors

Skin mechanoreceptors respond to mechanical forces. They are classified as slowly or rapidly adapting and within each of these they fall into two types. Type I are superficial, have small receptive fields (RFs) with clear boundaries and are concerned with shape and texture sensation. Type II are deep, have large RFs with fuzzy edges. The density of receptors is variable, being highest in fingertips and lips.

Thermoreceptors

Thermoreceptors are slowly adapting. Warm receptors increase firing in response to skin warming, whereas cold receptors fire faster as skin cools. Temperature perception relies on comparing the responses of warm and cold receptors.

Nociceptors

Mechanical nociceptor afferents are Aδ fibers responsible for the sensation of sharp pricking pain. Thermal nociceptors respond to high or low skin temperature activating Aδ afferents. Polymodal nociceptor afferents are C fibers and respond to intense mechanical forces, heat, and a number of chemicals released during tissue damage. Itch afferents are C fibers that respond to histamine.

Related topics

(E1) Rate coding
(E2) Coding of modality and location

(F3) Pain

Receptor potentials

Sensory receptors generate a **receptor potential**, a change in their membrane potential, in response to appropriate stimulation. This process is called **transduction** and is different in different receptors. For somatosensory systems the sensory receptor is the modified ending of the primary afferent neuron, and is depolarized directly by the stimulus. In other sensory systems the sensory receptor is a specialized cell type which forms synaptic connections with the first afferent neuron. Here, alterations in membrane potential alters sensory cell neurotransmitter release, with effects on the primary afferent. In vertebrates, all sensory receptors, except photoreceptors, depolarize when stimulated. Photoreceptors are hyperpolarized by light.

Receptor potentials share many of the properties of synaptic potentials (Figure 1). They are small amplitude, graded in size depending on the stimulus strength, passively

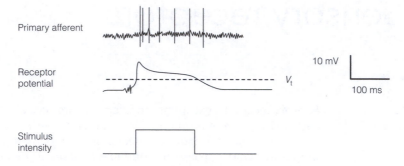

Figure 1. Receptor (generator) potential (middle trace) and discharge (upper trace) of a slowly adapting cutaneous mechanoreceptor afferent in response to 150 ms indention of skin (lower trace). V_t, threshold voltage.

conducted over the receptor cell surface or along neurites, decay with time and distance and can be summated. A receptor potential will trigger action potentials for as long as it remains beyond the firing threshold, the frequency of firing will be higher the greater its amplitude. Sensory receptors demonstrate adaptation, a decline in response over time to a constant stimulus.

Receptors are classified as mechanoreceptors, found in skin, muscles, joints, and viscera; thermoreceptors, confined to the skin; and nociceptors which are found almost everywhere except the brain.

Mechanoreceptors

Skin mechanoreceptors (Figure 2) are classified as slowly or rapidly adapting (SA or RA respectively) and, separately, as being of two types, type I and II, distinguished by their location and receptive fields (RFs).

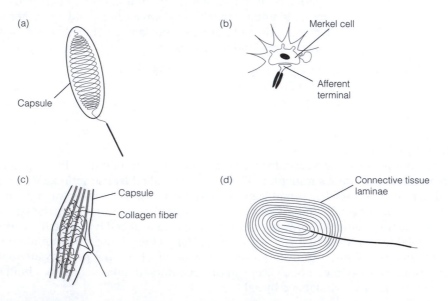

Figure 2. Morphology of glabrous (non-hairy) skin mechanoreceptors: (a) Meissner's corpuscle; (b) Merkel's disc; (c) Ruffini corpuscle; (d) Pacinian corpuscle.

The **Ruffini organ** is slowly adapting so its afferent has a frequency of firing that is directly proportional to the extent to which overlying skin is indented by a mechanical force. This receptor codes skin *position*. The **Meissner's corpuscle** is rapidly adapting and its afferent fires only when skin displacement is changing with time. It codes the *velocity* with which the skin is displaced. Finally, the **Pacinian corpuscle** adapts so rapidly that its afferents respond to skin *acceleration*. Hence the three afferents between them encode a wealth of dynamical information about the stimulus (Figure 1 in Section E1).

Type I mechanoreceptors are superficial, lying at the boundary of epidermis and dermis and have small RFs with well-defined boundaries. These include Meissner's corpuscles and **Merkel's discs**. Type II are deep in the dermis and have large RFs with poorly defined edges, and include Ruffini corpuscles and Pacinian corpuscles. Type I receptors are more directly concerned with form and texture perception than type II. The density of type I receptors varies across the body surface being highest in the fingertips, lips, and tongue and lowest in the trunk. Areas with higher density have proportionally greater representations in somatotopic maps. Receptor convergence varies with receptor; whereas each Merkel's disc afferent receives input from 2 to 7 receptors, a one-to-one ratio is the case for Pacinian corpuscles and their afferents.

Pacinian corpuscles can respond to skin indentation as little as 1 µm. The force is transmitted through the corpuscle to deform the neurite within. This causes the opening of stretch-sensitive Na^+ channels in the membrane (not to be confused with voltage-dependent Na^+ channels) resulting in a brief depolarization. Membrane potential returns to normal extremely fast because the receptor adapts by individual connective tissue layers of the corpuscle sliding over each other, which relieves the neurite deformation.

Table 1. Cutaneous receptors

Receptor	Adaptation		Fiber type	Sensation
Mechanoreceptors				
Meissner's corpuscle	RA1	velocity	Aβ	Touch, flutter, stretch
Pacinian corpuscle	RA2	acceleration	Aβ	Vibration
Merkel's disc	SA1	velocity and displacement	Aβ	Touch, pressure
Ruffini corpuscle	SA2	displacement	Aβ	Stretch
Lanceolate ending[a]	RA1	velocity	Aα	Hair movement
Pilo-Ruffini ending[a]	SA2	displacement.	Aβ	Hair movement
Hair follicle receptor[a]	RA1	displacement.	Aβ	Hair movement
Thermoreceptors				
Warm, bare nerve ending	SA		C	↑Skin temperature
Cold, bare nerve ending	SA		Aδ	↓Skin temperature
Nociceptors				
Mechano- bare nerve ending	Nonadapting		Aδ	Sharp pain
Polymodal bare nerve ending	Nonadapting		C	Burning pain

[a] Hairy skin only

Human skin is either **hairy** or **glabrous** (non-hairy). Innervation of hairy skin differs in having a lower density of Merkel's disks and in possessing two additional types of mechanoreceptor closely associated with hairs (Table 1).

Thermoreceptors

Skin thermoreceptors are the naked terminals of small diameter afferents. They are slowly adapting and tonically active. Thermoreceptor afferents each have just three to four terminals and have very small RFs, although infrared radiation is poorly localized.

There are two types of thermoreceptor, warm and cold, which fire over different temperature ranges (Figure 3). They do not respond to noxious temperatures. Skin temperature is perceived by comparing the relative activities of the warm and cold receptors. Thermoreceptors signal the *direction* in which temperature changes. Skin cooling briefly silences warm receptors and causes cold receptor firing rates to rise. Similarly, skin warming silences the cold receptors and boosts warm receptor firing.

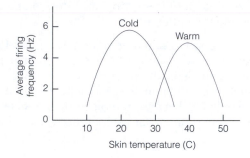

Figure 3. Frequency response of populations of cutaneous cold and warm thermoreceptors.

Nociceptors

The bare endings of small diameter afferents are **nociceptors**, receptors for noxious (tissue-damaging), pain-producing stimuli. Nociceptor afferents are described as high threshold because they require intense stimulation to excite them. Unlike thermoreceptors, normally they have no background firing. They are classified by what excites them:

- **Mechanical nociceptors** are stimulated by intense mechanical forces and those in skin give rise to sharp, pricking pain. Each nociceptor is the ending of one of 5–20 branches of an Aδ afferent with low conduction velocities. Mechanoreceptors in the visceral peritoneum that invests the gut respond to excessive distension.

- **Thermal nociceptors** fall into two groups, one excited by temperatures greater than 45°C, the other by temperatures less than 5°C. They also respond to intense mechanical stimuli. Their afferents are Aδ or C fibers.

- **Polymodal nociceptors** in skin respond to puncture, temperatures in excess of 48°C, and to a wide variety of molecules liberated as a result of tissue damage including: K^+, H^+, bradykinin, prostaglandins, serotonin, and histamine. Stimulation of these nociceptors causes burning or aching pain. Their afferents are C fibers which conduct at less than 1.0 m s^{-1}. Because C fiber conduction is so slow, the pain they produce arrives last after a blow. They are also responsible for visceral and muscle pain, and toothache.

- **Itch receptors** belong to a separate class of C fiber that respond to histamine released from mast cells.

F2 Touch

Key Notes

Dorsal column–medial lemniscal (DCML) pathway
Each dorsal root receives input from a skin dermatome. Low threshold mechanoreceptor and proprioceptor afferents enter the dorsal roots to synapse with interneurons involved in spinal reflexes in the dorsal horn. Afferent axons ascend in the dorsal columns to synapse in dorsal column nuclei (DCN) in the medulla. Axons of DCN neurons cross the midline and ascend in the medial lemniscus to the thalamus. From here neurons project to the primary somatosensory cortex (SI). Somatotopic mapping at each stage preserves stimulus location and type; both skin and proprioceptor input are represented. SI is concerned with tactile discrimination. The secondary somatosensory area (SII) gets input from both sides of the body and is involved in guiding movement in the light of somatosensory input.

Descending connections
Reciprocal connections between the somatosensory cortex and DCML system nuclei are formed which have the same mapping as the ascending pathway. These descending connections probably filter somatosensory inputs.

Related topics
(E2) Coding of modality and location
(F1) Sensory receptors
(F3) Pain

Dorsal column–medial lemniscal (DCML) pathway

The region of skin innervated by a dorsal root is a **dermatome**. These are numbered for the spinal cord segment served by the dorsal root. Cutaneous low threshold (easily stim-ulated) mechanoreceptor primary afferent axons relaying skin mechanoreceptor and

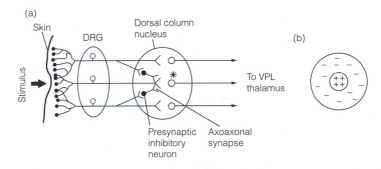

Figure 1. Lateral inhibition in the dorsal column nucleus: (a) circuitry, the dorsal column neuron marked by an asterisk has a receptive field with inhibitory surround because transmitter release from its primary afferents is reduced by presynaptic inhibitory neurons driven by afferents in the surrounding skin; (b) the receptive field of the indicated cell.

proprioceptor signals enter the dorsal roots to synapse with interneurons, **dorsal horn cells** (**DHCs**), in Rexed laminae III–VI. DHCs mediate or modify spinal reflexes. Each afferent sends a collateral up the **dorsal columns** to synapse with neurons in the **dorsal column nuclei** (**DCN**) in the medulla. The **cuneate nucleus** receives input from C1–8 and T1–6, whereas the **gracile nucleus** gets its inputs from T7–12, lumbar and sacral spinal segments. Lateral inhibition in DCN shapes this input (Figure 1).

Axons of DCN neurons cross the midline to ascend on the opposite side of the cord as the **medial lemniscus**, terminating in the **ventroposterolateral** (**VPL**) division of the ventro-basal thalamus (Figure 2). VPL neurons give rise to thalamo-cortical axons, which project to the **primary somatosensory cortex** SI (Brodmann's areas 1, 2, 3a, and 3b), situated over the postcentral gyrus. SI neurons in turn project to SII (Figure 3a).

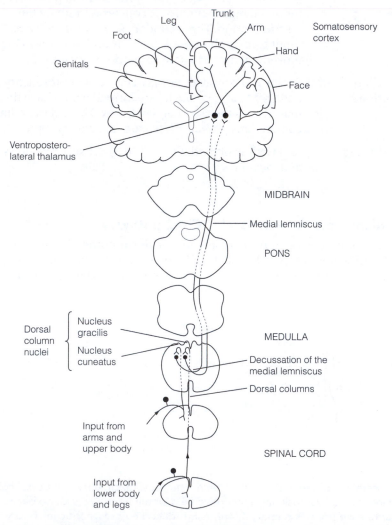

Figure 2. The dorsal column–medial lemniscal system. All neurons shown are excitatory.

(a)

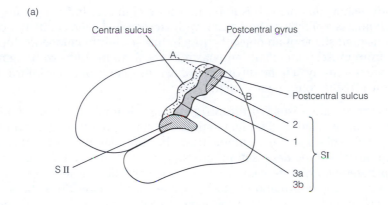

(b)

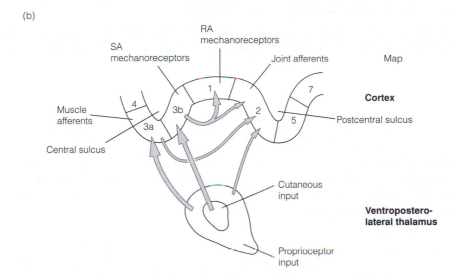

Figure 3. The somatosensory cortex: (a) location of the primary (SI) and secondary (SII) somatosensory cortex of the left cerebral hemisphere, lateral aspect. Numbers refer to Brodmann areas; (b) interconnections of the thalamus and somatosensory cortex viewed across line A–B in (a). RA, rapidly adapting; SA, slowly adapting.

The general properties of the DCML system are:

- The great strength of its synaptic connections.

- The properties of its neurons are matched to those of the sensory receptors supplying them, so the dynamic features of stimuli are transmitted with high fidelity.

- Somatotopic mapping preserves localization at every stage. Body maps are found in DCN, VPL, and each of the four regions in SI. Cutaneous and proprioceptor input are partly separately streamed (Figure 3b).

Neurons in SI are organized into **columns** aligned at right angles to the brain surface. Each column gets input from just a single type of receptor, and from a specific location. Adjacent locations are represented in adjacent columns in a somatotopic manner. Extensive neural connections exist within a column; connections between columns are sparse.

Lesion studies show that area 3b is important for tactile discrimination, area 1 is concerned with analysis of texture, and area 2 with **stereognosis**, the ability to perceive the three dimensional shape of an object by touch. In addition to cutaneous input, area 2 gets input from muscles and joints and has reciprocal connections with the motor cortex. These are not involved in modifying ongoing movements but may inform the motor system of the sensory consequences of moving.

The secondary somatosensory cortex (SII) gets input from the thalamus and SI. Many neurons in SII have bilateral RFs, that is, stimuli in corresponding regions on both sides of the body will evoke a response. Inputs from the contralateral body surface arise as a direct consequence of the **decussation** (crossing over) of the medial lemniscus. Inputs from the ipsilateral body surface enter SII from the contralateral side via the corpus callosum. By integrating information from both sides of the body, SII is the first stage in forming a whole body perceptual experience. It enables tactile discrimination learned using one hand to be easily performed with the other.

SII is important in controlling movement in the light of somatosensory input via its connections with the motor cortex. In addition, SII has inputs to the limbic cortex enabling tactile learning.

Descending connections

The somatosensory cortex has reciprocal connections with all of the subcortical structures which relay sensory input to it. The descending pathway is made by the corticospinal (pyramidal) tract either directly or via its connections with the brainstem reticular nuclei. These back projections have a somatotopic mapping precisely in register with the ascending DCML pathway. They are probably the vehicle by which somatosensory input can be selectively filtered as an attention mechanism. For example, much of the time we are not aware of the sensations caused by clothes.

F3 Pain

Key Notes

Definition of pain

Pain is the unpleasant sensory and emotional experience that accompanies stimuli that cause tissue damage. Nociceptive (acute) pain has a protective role. Pathological (clinical) pain is associated with disease and nervous system dysfunction.

Anterolateral pathways

Anterolateral pathways convey temperature, pain, and crude touch sensation.

Small-diameter, high-threshold, afferents enter the dorsal horn to terminate on nociceptor specific laminae I and II cells and on lamina V cells. Nociceptor afferents convey either fast or slow pain signals. Large-diameter, low-threshold afferents terminate on cells in laminae IV and V. Most axons of these neurons cross the midline to ascend in the anterolateral columns.

Pain pathways

The spinothalamic tract conveys nociceptor signals from laminae I, II, and V, via the thalamus to the somatosensory cortex, for high fidelity localization and discrimination of fast pain. The spinoreticular tract conveys nociceptor signals via brainstem and thalamic reticular nuclei to cortical regions involved in emotional and cognitive aspects of pain. The spinoparabrachial tract conveys slow pain signals from lamina I to hypothalamus and amygdala to affect autonomic-appetitive aspects of pain and fear learning. Visceral pain is mediated by the dorsal column–medial lemniscal system.

Crude touch sensation

Activation of lamina V cells by mechanoreceptor afferents alone generates a well localized but poorly discriminated sense of touch.

Discriminative pain sensation

Noxious stimuli excite both lamina I cells, which carry the pain signal, and lamina V cells which allow the pain to be localized.

Spinal anti-nociception

At the spinal cord the gate control theory describes how input into the spinothalamic tract from small diameter primary afferents is inhibited by concurrent activity in large diameter mechanoreceptor afferents, via enkephalinergic interneurons in lamina II.

Supraspinal anti-nociception

Anti-nociception systems descend from the periaqueductal gray (PAG) matter to activate the serotonin and noradrenergic neurons in the brainstem which inhibit nociceptor input. The PAG gets input from the limbic system. Activity in the pain pathways switches on the anti-

nociception systems via inputs to reticular nuclei and brain
aminergic pathways. Stress-induced analgesia (SIA) that
occurs as a result of pain engages these systems. Both opioid-
dependent and opioid-independent SIA occur. The non-
opioid variety requires endogenous cannabinoids.

Related topics   (F1) Sensory receptors   (F4) Pain pathology
                      (F2) Touch

Definition of pain

Pain is the unpleasant sensory and emotional experience associated with **noxious stimuli**, those which can cause tissue damage. The protective role of **nociceptive pain**, felt
only in the presence of acute injury or inflammation, is shown by the injury that individuals with congenital insensitivity to noxious stimuli unwittingly inflict on themselves.
Pathological (**clinical**) **pain** is associated with chronic inflammatory disease (e.g., arthri-
tis), or to damage or dysfunction of the nervous system (e.g., phantom pain) and has no
obvious purpose.

Anterolateral pathways

The anterolateral pathways convey temperature and pain sensations, and poorly discriminated touch sensation (Figure 1). The primary afferents are small diameter high
threshold dorsal root ganglion (DRG) cells driven by nociceptors and low threshold DRG
cells excited by thermoreceptor or mechanoreceptor input.

The small diameter fibers, situated laterally in the dorsal roots, enter **Lissauer's tract**,
where they divide into ascending and descending branches that enter the dorsal horn
within a couple of segments. Nociceptive Aδ afferents synapse with projection neurons
in laminae I and II, and with the distal dendrites of lamina V cells, although visceral and
muscle nociceptors do not project to lamina II. Nociceptive C fibers synapse with other
laminae I and II neurons. Large diameter fibers from mechanoreceptors enter the dorsal
horn medially and synapse with neurons in deeper Rexed laminae (IV–VIII). Axons of
dorsal horn cells either cross the midline within 1 or 2 segments, or remain ipsilateral, to
ascend in the **anterolateral columns**.

Pain pathways

The projection neurons in laminae I and II are nociceptor specific. Conduction in Aδ
mechanical nociceptor afferents is faster than in C fiber polymodal nociceptor afferents
so a painful blow produces a sharp **fast pain** initially, followed by an aching **slow pain**.
Fast and slow pain are processed separately. There are three pain pathways:

- The **spinothalamic tract** (**STT**) mediates fast, well-localized, well-characterized pain.
  It arises from neurons in laminae I, II, IV, and V, axons of which cross over and ascend
  on the contralateral side to the **posterior** (I and II) and **ventroposterolateral** (IV and
  V) **nuclei** of the thalamus. Thalamic neurons project to the primary somatosensory
  cortex (SI). The somatosensory cortex is assumed to mediate pain localization and discrimination, though it has no pain map.

The largest number of spinothalamic tract cells are in lamina V. Their dendrites extend
into lamina I and receive inputs from Aδ (fast pain) afferents. But lamina V neurons

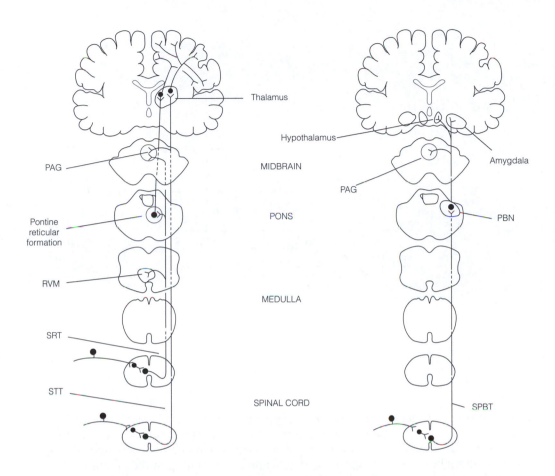

Figure 1. Anterolateral pathways. PAG, periaqueductal gray; PBN, parabrachial
nucleus; RVM, rostroventral medulla; SPBT, spinoparabrachial tract; SRT, spinoreticular tract;
STT, spinothalamic tract.

also receive connections from large diameter (Aβ) mechanoreceptor afferents so are
said to be **wide dynamic range** (**WDR**) cells, because they get inputs from low threshold *and* high threshold afferents (Figure 2). Hence most spinothalamic tract neurons
can be excited by *both* innocuous and noxious stimuli. In contrast, lamina IV cells get
input only from large diameter (Aβ) mechanoreceptor afferents, and not from nociceptor afferents.

- The **spinoreticular tract** (**SRT**) arises from lamina V–VII and makes bilateral connections with reticular nuclei in the brainstem, home to second order neurons that project
to reticular (intralaminar) nuclei of the thalamus. These nuclei project to SI, and also to
the insula, anterior cingulate cortex and prefrontal cortex—regions involved in emotional and cognitive responses to pain. Interestingly, brain imaging shows that parts
of the insula and anterior cingulate are activated by watching pain inflicted on a loved
one. Activation of brainstem reticular nuclei generates autonomic responses to pain
(e.g., increases in heart rate and blood pressure) and changes in ventilation. In particular, by activating noradrenergic neurons in the locus coeruleus the SRT increases
arousal. Finally, the thalamic nuclei make connections with the basal ganglia, influencing motor activity.

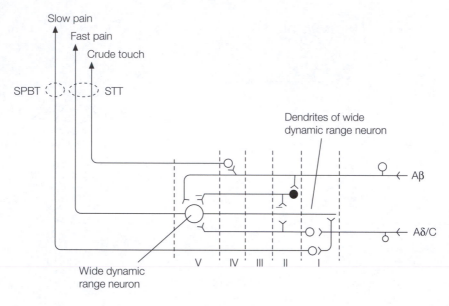

Figure 2. Pain circuitry in the spinal cord. SPBT, spinoparabrachial tract; STT, spinothalamic tracts; ●—<, inhibitory neuron

- The **spinoparabrachial tract** (**SPBT**) arises from lamina I cells that get their input from C fiber (slow pain) afferents and synapse with cells in the parabrachial nucleus that project to the hypothalamus and amygdala. Via these connections pain activates pituitary–adrenal axis stress responses, affects appetitive (eating and drinking) behavior and sleep, and mediates fear learning.

All three pain pathways receive input from visceral afferents as well as skin nociceptors, but visceral nociceptors also terminate on dorsal horn cells that project to dorsal column nuclei. Pain pathways mediate visceral reflexes but visceral pain perception is exclusively due to the dorsal column–medial lemniscal system.

Convergence of visceral and somatic nociceptor afferents onto the same neuron means that stimulation of nociceptors in internal organs is perceived as pain in muscle and skin some distance away. This is **referred pain** and the areas to which it is referred (**Head's zones**) are reliable enough to be useful in diagnosis (Figure 3).

Neurons equivalent to laminae I and V cells in the spinal cord are also found in the **spinal nucleus of the trigeminal nerve**, which get their inputs from the primary afferents of the face and head. These project to the ventroposteromedial and posterior thalamic nuclei and are the face and head equivalent of the STT.

Crude touch sensation

Lamina V (and lamina IV) cells when activated only by large-diameter low-threshold afferents convey crude touch sensation. These cells have small receptive fields and their projections are organized topographically so are able to localize stimuli precisely. However, convergence of inputs means they cannot discriminate well between different mechanoreceptor submodalities.

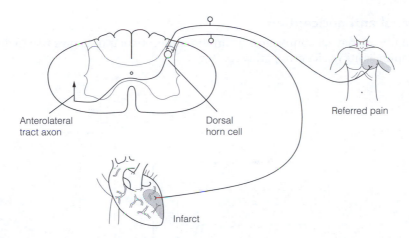

Figure 3. Referred pain. Typically the pain of ischemic heart disease is referred to the chest and left arm.

Discriminative pain perception

Any noxious stimulus will activate both lamina I and lamina V cells (stimuli exciting *only* lamina V cells are perceived as innocuous). Lamina I cells signal pain, but because they have large receptive fields, and relay to posterior and reticular thalamic nuclei which do not have precise topographic projections to the cortex, they cannot localize it well. The localization of painful stimuli depends on the simultaneous firing of lamina V cells, which project in somatotopic fashion to the ventroposterolateral thalamus and so to the primary somatosensory cortex.

Spinal anti-nociception

Mechanisms to reduce nociceptor input operate at both the spinal and supraspinal level. At the spinal cord level, whether a stimulus is perceived as painful or not depends on the relative activity of large and small diameter fibers. High large/small diameter fiber activity reduces nociceptor input. A modern version of this **gate control theory** circuitry is shown in Figure 2.

Wide dynamic range neurons in lamina V get convergent excitatory input from large (Aβ) mechanoreceptor afferents and small (Aδ and C) nociceptor afferents. The Aβ fibers inhibit lamina V neurons via inhibitory interneurons in lamina II. The nociceptor afferents also excite the lamina V cells by way of excitatory interneurons, and these can be inhibited by the Aβ-driven inhibitory interneurons. Hence, nociceptor afferents excite the lamina V cells both directly *and* via excitatory interneurons. However, large diameter afferent activity reduces nociceptor input to lamina V cells by increasing inhibition from the lamina II interneurons. Hence co-activation of low and high threshold afferents closes the pain gate.

The gate control theory accounts for **counter-stimulation analgesia** in which pain is reduced by stimulating low threshold afferents; we instinctively rub the site of a painful blow, stimulating mechanoreceptors. **Transcutaneous electrical nerve stimulation (TENS)** delivers high frequency, low intensity currents, sufficient to stimulate Aβ fibers, and also Aδ fibers that activate endogenous opioid-using supraspinal anti-nociception pathways. It is used mainly during childbirth and physiotherapy.

Supraspinal anti-nociception

Brainstem nuclei that get connections from the spinoreticular tract give rise to descending supraspinal anti-nociceptor pathways (Figure 4).

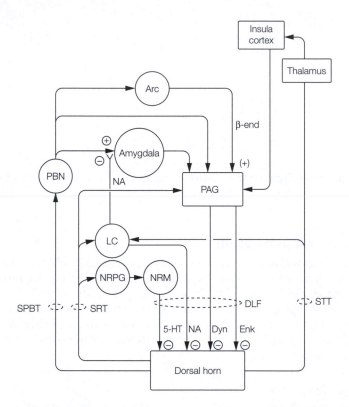

Figure 4. Supraspinal anti-nociception pathways. Arc, arcuate nucleus of hypothalamus; DLF, dorsal longitudinal fasciculus; LC, locus coeruleus; NRM, nucleus raphe magnus; NRPG, nucleus reticularis paragigantocellularis; PAG, periaqueductal gray; PBN, parabrachial nucleus; SPBT, spinoparabrachial tract; SRT, spinoreticular tract; STT, spinothalamic tract; 5-HT, serotonin; β-end, β-endorphin; Dyn, dynorphin; Enk, enkephalin; NA, noradrenaline (norepinephrine).

The **periaqueductal gray** (**PAG**) surrounds the cerebral aqueduct in the midbrain. Stimulation of the PAG causes a powerful suppression of pain responses, **stimulus-induced analgesia**. This has been exploited by neurosurgeons who implant chronic electrodes into the PAG of patients with intractable pain. The PAG exerts its anti-nociception effects by activating brainstem aminergic neurons via inhibitory enkephalinergic neurons acting on inhibitory GABAergic interneurons (inhibiting an inhibition is an excitation). The PAG gets inputs from the limbic system, hypothalamus, amygdala, and cortex (insula), allowing emotional states to modulate anti-nociception.

Neurons in the **locus coeruleus** (**LC**) and the **nucleus raphe magnus** (**NRM**) in the medulla, which use noradrenaline (norepinephrine) and serotonin respectively, send axons into the dorsal horn. Here the amines reduce transmission from nociceptors into the lamina I dorsal horn cells by:

- Presynaptic inhibition

- Stimulating postsynaptic receptors that open K^+ channels causing hyperpolarization of the lamina I cells

- Stimulating lamina II enkephalinergic and GABAergic inhibitory interneurons that hyperpolarize the lamina I cells by opening K^+ channels or closing Ca^{2+} channels

There is debate about whether the effect of serotonergic transmission is pain specific rather than a general effect on all somatosensory input, such as that seen in sleep.

The descending anti-nociceptive pathways appear to be brain analgesia systems. In the event of pain, stress, and/or exercise they produce the natural counterpart of stimulus-induced analgesia termed **stress-induced analgesia** (SIA). This occurs in individuals who have sustained injuries when they are in threatening or highly arousing situations (e.g., warfare or sport), is rapid in onset and wears off after a few hours. The analgesia can be localized to the injured area or generalized. The selective advantage of SIA is that it allows an individual to continue to function so they can remove themselves from danger. SIA can be endogenous opioid dependent or non-opioid dependent. It seems that opioid analgesia occurs at higher intensity stress than the non-opioid. Opioid SIA requires enkephalinergic neurons in the PAG, while the non-opioid SIA involves the release of endogenous cannabinoids that act on CB_1 cannabinoid receptors in the PAG.

Brain analgesia is switched on by nociceptor activity in the pain pathways. The spinoreticular tract signals to the **nucleus reticularis paragigantocellularis** (NRPG) which stimulates the serotonergic cells of the nucleus raphe magnus. The spinothalamic and spinoreticular tracts activate the locus coeruleus noradrenergic neurons. Both these aminergic pathways inhibit the lamina I cells that transmit nociceptor signals. Central noradrenergic pathways also shut down the spinoparabrachial tract by presynaptic inhibition of glutamate release from parabrachial terminals in the amygdala. The mechanism here is unusual. The noradrenaline (norepinephrine) acts at presynaptic α_2 receptors, liberating the $\beta\gamma$ subunits of their associated G protein. The $\beta\gamma$ subunits then block vesicle release sites at the active zone of the synapse, curtailing transmitter release.

Electro-acupuncture, which delivers electrical stimulation through needles inserted through the skin, relies on stimulation of small diameter afferents from skeletal muscle and joints for its analgesic effect. Low frequency stimulation activates β-endorphin-using neurons in the arcuate nucleus of the hypothalamus. These project to the PAG, activating an enkephalinergic pathway that inhibit dorsal horn cells (DHCs). High frequency stimulation activates the parabrachial nucleus which in turn activates a PAG pathway that uses dynorphin-secreting neurons to inhibit DHCs. Since enkephalin acts on μ and δ opioid receptors while dynorphin acts on κ opioid receptors, alternating between low and high frequency electro-acupuncture has synergistic effects on the DHCs and results in stronger analgesia than either alone.

Opiate drugs such as **morphine** and **pethidine** are full or partial agonists at one or more of the opioid receptors, and produce analgesia by mimicking the actions of the endogenous opioids in the pain-modulating pathways. The effects of opioid analgesics can be rapidly reversed by the competitive antagonists such as **naloxone**.

F4 Pain pathology

Key Notes

Altered pain states

Lowered threshold of sensory neurons can render nociceptor stimuli more painful (hyperalgesia) or can make innocuous mechanical and thermal stimuli painful (allodynia). They are produced by sensitization. Neuropathic pain results from damage to pain pathways.

Peripheral sensitization

A wide variety of substances released by injury increase the excitability of polymodal nociceptors. Neurogenic inflammation occurs due to substance P release from peripheral nociceptor terminals via axon reflexes.

Central sensitization

In central sensitization synapses on dorsal horn cells in the spinal cord are facilitated by protracted nociceptor input. Summation of slow epsps due to substance P causes large depolarization, activation of NMDA receptors and wind-up. NMDA receptor activation produces long-term changes in excitability. Dorsal horn cells also become hyperexcitable to low threshold primary afferent input resulting in tactile allodynia. Central sensitization occurs in chronic inflammation and involves microglial cells. Central sensitization also occurs in other regions in pain pathways.

Phantom pain

Re-wiring of the somatosensory cortex may be responsible for phantom pain that follows amputation of limbs.

Placebo effect

Agents or procedures that are without any physiological or pharmacological action reduce pain perception, the placebo effect. It may be that the expectation of pain relief activates the endogenous opioid-using anti-nociception pathways.

Related topics

(D8) Purines and peptides	(F2) Touch
(F1) Sensory receptors	(F3) Pain

Altered pain states

At the site of an injury sensory responses can be exaggerated:

- Noxious stimulation becomes more painful than usual (**hyperalgesia**)

- Mechanical or thermal stimuli that are normally innocuous become painful (**allodynia**)

- Pain can occur in the absence of any stimulus (**spontaneous pain**)

When transient these have a protective role in encouraging guarding of the site of injury, but when prolonged they are pathological. These states arise because neurons become more excitable—their thresholds are lowered—a process called **sensitization**. Inflammation is often a major factor (**inflammatory pain**).

Damage to pain pathways *per se* is termed **neuropathic pain**.

Peripheral sensitization

Peripheral sensitization is caused by hyperexcitability of primary nociceptor afferents. At an injury site a wide variety of substances are released including; H^+, adenosine, serotonin, histamine, prostaglandins, and numerous peptides such as bradykinin and cytokines. C polymodal nociceptors respond to these with second-messenger-mediated alterations to proteins, such as ion channels, enhancing excitability. For example, prostaglandins phosphorylate sensory terminal Na_v channels. This reduces the depolarization needed to activate them and makes the channel open longer, increasing the responsiveness of the cell.

A consequence of nociceptor activation is **neurogenic inflammation**. Primary nociceptor afferents release both glutamate (which acts as a fast transmitter) and peptides, particularly **substance P**. The peptides amplify and prolong the effects of glutamate. Glutamate and peptides are co-released from the peripheral terminals as well as the central endings of the nociceptor axon. Hence, stimulus-evoked action potentials are conducted both centrally and, in what is termed an **axon reflex**, also **antidromically** along neighboring branches to stimulate secretion from their peripheral terminals (Figure 1). This contributes to the classic signs of inflammation at an injury site since substance P produces vasodilation (heat and redness), increased permeability (swelling), and stimulates mast cells to release histamine that excites itch C fibers.

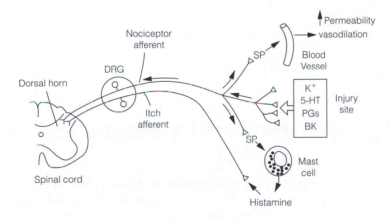

Figure 1. Neurogenic inflammation generated by axon reflex. Antidromic action potentials in nociceptor afferent branches stimulate the release of substance P (SP) from peripheral terminals. DRG, dorsal root ganglion; 5-HT, serotonin; PGs, prostaglandins; BK, bradykinin.

This sequence of events can be modified by drugs at a variety of points. For example, prostaglandins lower the threshold of nociceptors to other inflammatory mediators such as serotonin and bradykinin. **Aspirin** and other **non-steroidal anti-inflammatory analgesics** (**NSAIDs**) work by inhibiting cyclooxygenase 2 (COX2), an enzyme involved in prostaglandin synthesis.

The molecular receptors in thermal and polymodal nociceptors which sense noxious heat stimuli (T > 43°C) are **capsaicin** (**vanilloid**) **receptors** (**TRPV1**). These are ligand-gated Ca^{2+} channels and so called because they respond to **capsaicin**, the compound that gives chili peppers their pungency. Activation of TRPV1 triggers transmitter release (glutamate, ATP, substance P) from nociceptor terminals. In peripheral sensitization TRPV1 numbers increase, and individual receptors are made more sensitive, by actions of prostaglandin and bradykinin. Attempts to develop capsaicin receptor antagonists as analgesics are being thwarted by the involvement of TRPVs in core temperature regulation; TRPV antagonists cause hyperthermia.

Central sensitization

In the spinal cord **central sensitization** arises from changes in the behavior of dorsal horn cells (DHCs) in response to prolonged nociceptor traffic. This strengthens (facilitates) the synapses onto the DHCs. The release of substance P from C fiber nociceptor terminals generates slow epsps. If C fibers are stimulated at a constant frequency above 0.3 Hz these epsps summate to produce ever-larger depolarization of the DHCs, which activates NMDA receptors. This is a type of temporal summation called **wind-up** (Figure 2). This activity-dependent synaptic facilitation takes only seconds to induce but NMDA receptor activation switches on a cascade of kinase activation that produce persistent increases in excitability by, for example, recruiting more AMPA glutamate receptors, and the formation of new synapses. Central sensitization has many similarities to long term potentiation, the synaptic plasticity that underlies memory.

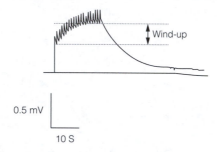

Figure 2. Wind-up in extracellular recording of spinal cord local field potential in response to C-fiber stimulation at 1 Hz.

The wide dynamic range DHCs also become hyperexcitable to low threshold (Aβ fiber) input and increase their receptive field size. This underlies tactile allodynia in which touch is painful. Under normal circumstances stimuli applied near the edge of the receptive field of a DHC usually elicit subthreshold responses (epsps). However, if a cell is very depolarized it is more sensitive to excitatory input. Previously subthreshold stimuli at the edge of its receptive field now become *suprathreshold*; in effect the receptive field is enlarged.

Analgesic cover in surgery is optimal if given preemptively and topped-up frequently enough that no pain occurs. This prevents wind-up and central sensitization.

Central sensitization is not confined to the spinal cord. It also occurs in the rostroventral medulla, thalamus, amygdala, and cingulate cortex.

Central sensitization in chronic inflammatory disorders involves glial cells (Figure 3). Peripheral nerve injury activates microglia in the local spinal cord. This up-regulates microglial $P2X_4$ receptors, one of a family of ligand-gated ion channels which respond to ATP. Stimulation of the $P2X_4$ receptors by the excessive ATP released by damaged primary afferents (ATP is a fast transmitter in many afferents) causes the microglia to release cytokines which increase the sensitivity of lamina I dorsal horn cells. This happens because

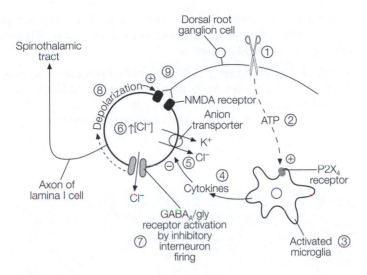

Figure 3. A model for central sensitization in response to peripheral nerve injury. Numbers indicate the sequence of events. Glutamatergic input from remaining intact fibers cause prolonged firing of the lamina I cell because the voltage-dependent blockade on its NMDA receptors has been removed by the abnormal depolarization (8).

the cytokines down-regulate an anion exporter in the lamina I cell bringing about a large rise in intracellular Cl⁻ concentration. These cells now respond to activation of their $GABA_A$ and glycine receptors by a large *depolarization*, caused by Cl⁻ *leaving* the cell down its concentration gradient. The depolarization allows NMDA receptors to be opened by any incoming excitation, triggering potentiation.

Microglia are also involved in central sensitization in the thalamus. Here a cytokine released by thalamic neurons activates COX2 in the glia so they synthesize and secrete prostaglandin E_2. This binds to prostaglandin receptors on neurons and increases their excitability by depolarization.

Phantom pain

After amputation of a limb patients experience the sensation that the limb is still present and can be painful (**phantom pain**). Phantom sensations can also follow loss of other body parts (e.g., mastectomy). Children born without limbs have powerful, non-painful, phantom limb sensations. This implies that complete somatotopic representations can exist in the absence of peripheral inputs, and this is the situation in amputees. The cause of phantom pain is unclear. However, it almost certainly involves re-wiring of the somatosensory cortex so that neurons that have lost their original inputs acquire functional connections from previously silent synapses established by neighboring cells. This accounts for how phantom limb sensations can be elicited by touching other parts of the body, commonly the face. With continuing cortical re-organization phantom sensations may fade over time.

Placebo effect

The **placebo effect** refers to the therapeutic efficacy of agents or procedures that are without any physiological or pharmacological action. In general, the more elaborate the placebo treatment the better its effectiveness. There is evidence that the placebo effect, at least when harnessed to reduce postoperative pain, can be blocked by naloxone and so depends upon endogenous opioid neurotransmission.

Functional magnetic resonance imaging (MRI) shows that placebo analgesia is associated with decreased activity in the brain regions associated with pain but *increased* activity in the dorsal lateral prefrontal cortex, an area involved in cognition. A working hypothesis is that the expectation of pain relief activates a top-down recruitment of supraspinal opioid pathways.

F5 Balance

Key Notes

Vestibular functions

Receptors in the inner ear detect the position and motion of the head in space and this information is used to maintain body posture and allow gaze to be controlled independently of head movement.

Vestibular labyrinth

The vestibular labyrinth of the inner ear houses the utricle and saccule (the otolith organs), and three mutually perpendicular semicircular ducts. The otolith organs detect linear acceleration; in the upright position the utricle responding to tilt away from the horizontal and the saccule to acceleration due to gravity. The semicircular ducts detect angular acceleration of the head.

Vestibular fluids

The vestibular labyrinth is filled with endolymph, while the space outside it contains perilymph. Active transport of potassium ions into the endolymph creates a high electrochemical gradient for potassium needed to drive sensory transduction by hair cells.

Transduction in otolith organs

The macula of the otolith organs is a sheet of epithelial cells containing sensory hair cells overlain by the otolith membrane. Stereocilia (mechanosensory organelles) of hair cells are embedded in the otolith membrane. Force which displaces the otolith membrane bends the stereocilia, either opening or closing K^+ channels in the hair cells, making them depolarize or hyperpolarize respectively. Changing the membrane potential of a hair cell alters its release of transmitter, modifying the firing rate of its postsynaptic primary afferent.

Transduction in semicircular ducts

At one end of a semicircular duct is an ampullary crest, the hair cells of which have their stereocilia embedded in the cupula. Rotation of the head in the plane of the duct distorts the cupula because inertia causes the rotation of the endolymph to lag behind. Distortion of the cupula stimulates the hair cells, the transduction mechanism of which is identical to that of hair cells in the otolith organs.

Central vestibular connections

Vestibular primary afferents live in the vestibular ganglion and send their centrally directed axons through the 8th cranial nerve to form connections with neurons in four vestibular nuclei. The inferior nucleus projects to the contralateral ventral posterior thalamus and so to the vestibular cortex. This pathway mediates the conscious balance perception.

Related topics	(G7) Oculomotor control and visual attention	(H2) Anatomy and physiology of the ear
		(J5) Brainstem postural reflexes

Vestibular functions

The sense of balance uses receptors which detect the position and motion of the head in space. These are located in organs in the vestibular part of the **inner ear** (**labyrinth**) that lies within the temporal bone. Vestibular input is used to adjust posture as forces shift the body's center of mass by modifying the output to antigravity muscles. Conscious perception of balance is normally overshadowed by visual and proprioceptive cues to head position and motion.

Vestibular input is also used to execute eye movements which are independent of head movement. These vestibulo-ocular reflexes are one of several reflexes for fixing gaze.

Vestibular labyrinth

Within the bony labyrinth, which contains all the inner ear structures, lies the membranous labyrinth, a sensory epithelium subserving hearing and balance (Figure 1). The **vestibular labyrinth**, concerned with balance, consists of two **otolith organs**, the **utricle** and **saccule**, and three **semicircular ducts**. The sensory structure of the otolith organs, the **macula**, which detects linear acceleration, is horizontal in the utricle and vertical in the saccule for a person standing upright. In this position the utricle is sensitive to tilting of the head (pitch and yaw) whilst the saccule is sensitive to vertically acting forces such as gravity. The three semicircular ducts are approximately mutually orthogonal. Each contains a sensory structure, the **ampullary crest**, which detects angular acceleration in

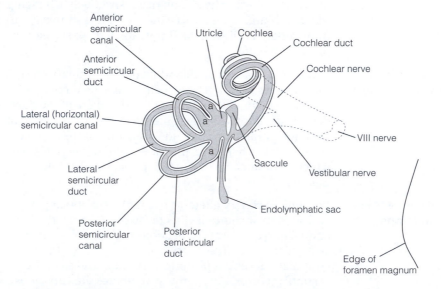

Figure 1. Left vestibular labyrinth viewed from above. The membranous labyrinth is shaded. a, ampullae of the semicircular ducts.

the plane in which the duct lies. Using the signals coming from all six semicircular ducts the brain computes the magnitude and direction of the angular acceleration of the head.

Vestibular fluids

The vestibular labyrinth is filled with **endolymph** which has a potassium concentration of about 160 mM and a sodium concentration of about 2 mM, and has a composition similar to intracellular fluid. It is secreted by a specialized epithelium, the **stria vascularis**, lining the outer wall of the **cochlear duct**, and drains into a venous sinus of the dura via the **endolymphatic sac**. The space between the bony and membranous labyrinths is filled with a CSF-like fluid, **perilymph**, secreted by arterioles of the periosteum (the connective tissue layer covering the bone), which drains into the subarachnoid space via the **perilymphatic duct**. Active transport by marginal cells maintains the high endolymph potassium concentration (Figure 2) and results in the endolymph having a potential difference of +80 mV with respect to perilymph. Since the resting potential of the hair cell is about −60 mV, the effective potential across its apical border is 140 mV. Hence there is a large electrochemical gradient favoring facilitated passive diffusion of K^+ across the hair cell. This is crucial for hair cell transduction.

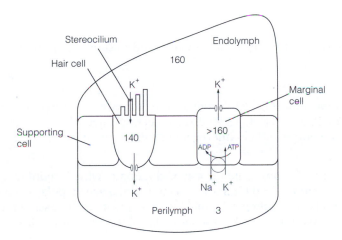

Figure 2. Simplified model of K^+ transport in the inner ear. Figures are approximate concentrations of K^+ (mM).

Transduction in otolith organs

The macula is an epithelial sheet of **supporting cells**, and sensory **hair cells**. Each hair cell is innervated at its base by two nerve fibers, a vestibular afferent and an efferent. The apical border of a hair cell has a single motile kinocilium, resembling a cilium, and 40–100 stereocilia, microvilli which are progressively shorter the further they are from the kinocilium (Figure 3). This defines an **axis of polarity** for a hair cell, with a direction going from the smallest stereocilium to the kinocilium. Stereocilia lying along this axis are connected at their tips by thin filaments called **tip links**. Stereocilia are the mechanosensory organelles of the inner ear since their tips contain stretch-activated potassium channels that are regulated by the tip links.

The kinocilium and stereocilia are embedded in a gelatinous matrix, the **otolith membrane**, containing tiny crystals of calcium carbonate, called **otoliths**. The hair cell is at

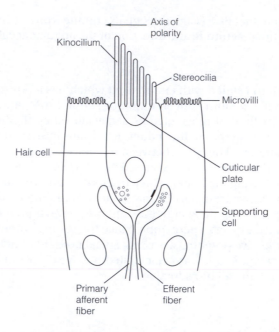

Figure 3. The hair cell of an otolith organ surrounded by supporting cells of the sensory epithelium.

rest if no force acts on the otolith membrane to cause the stereocilia to pivot. In this state the tension in the tip links (Figure 4) connecting adjacent stereocilia is slight so only about 10% of the potassium channels gated by them are open, causing a small depolarization.

This is sufficient to sustain tonic release of glutamate, which maintains baseline firing of the primary afferent. Head tilt in the direction of the axis of polarity causes the otolith membrane to pull on the stereocilia, making them pivot, so increasing the tension in the tip links. This opens stereocilia K^+ channels, allowing potassium influx to depolarize the

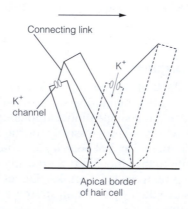

Figure 4. Transduction by otolith hair cells. Two stereocilia are shown with their connecting link. When pivoted to the right (dotted lines) the connecting link is stretched, opening K^+ channels to cause an influx of K^+.

hair cell, increasing glutamate release and raising the afferent firing rate. Tilt in the opposite direction reduces tip link tension so K$^+$ channels close, the hair cells hyperpolarize and primary afferent firing drops. Tilt which is perpendicular to the axis of polarity of a hair cell has no effect because stereocilia are not linked in this direction. Tilts in intermediate directions cause graded receptor potentials. The responses of individual otolith afferents are proportional to tilt angle and adapt only with prolonged stimulation. The axes of hair cells are orientated in an orderly pattern so a given stimulus will depolarize some hair cells and hyperpolarize others.

Transduction in semicircular ducts

Velocity is a vector quantity, it has a magnitude (speed) and direction. For circular motion, such as head rotation, even if the angular speed is kept constant, the direction in which the velocity vector is acting is continuously changing. Hence head rotation is an angular acceleration.

Both ends of each semicircular canal insert into the utricle. Within the canal is the endolymph-filled semicircular duct. At one end of each duct is a dilation, the **ampulla**; in which sits the ampullary crest. Vestibular hair cells in the crest have their stereocilia embedded in a gelatinous sheet, the **cupula**. Rotation of the head maximally stimulates hair cells in the canals lying in the same plane as the rotation. Rotation of the endolymph lags behind head rotation because of its inertia, so the endolymph exerts a pressure distorting the cupula, bending the stereocilia. The transduction mechanism is identical to that of hair cells in otolith organs. Because the cupula is not an ideal pressure transducer the signals transmitted by the duct afferents measure angular acceleration for slow and fast rotations, but encode velocity for mid-range rotation speeds. Semicircular ducts on each side lying in the same plane operate in pairs. Head rotation that causes depolarization of hair cells in the horizontal duct of the left ear will hyperpolarize hair cells in the horizontal duct of the right ear.

Central vestibular connections

The vestibular primary afferents (about 20 000 on each side) are pseudobipolar cells with their cell bodies in the **vestibular ganglion**. Their axons run in the **vestibulocochlear** (8th cranial) **nerve** to enter the vestibular nuclei which lie laterally in the medulla and pons. There are four vestibular nuclei.

The pathway for conscious balance perception are axons of the inferior vestibular nucleus which cross to the contralateral side, ascend close to the medial lemniscus to terminate in the ventral posterior thalamus. From here third-order neurons project to the vestibular cortex which lies at the temporo-parietal boundary and posterior insula though other projections go to the superior temporal cortex adjacent to the auditory area and to the frontal cortex.

G1 Attributes of vision

Key Notes

Sensitivity

The human eye responds to light between 400 and 700 nm over a 10^{11}-fold intensity range, though discrimination declines for higher light levels.

Visual processing

From two-dimensional retinal images the brain constructs a three-dimensional percept by which it can identify what is present in the world and where it is. Different aspects of the visual image (color, form, movement, and depth perception) are processed in parallel. Visual perception arises because the brain has internal representations which it compares with retinal images to make hypotheses about what they are. This allows objects to be recognized in a variety of contexts, even if the image is poor. Some internal representations are developmentally programmed but most are learned. Object recognition is facilitated by perceptual constancy in which parameters such as size and color are preserved over big differences in viewing conditions.

Depth perception

The perception of distance comes from monocular cues for distant objects and binocular cues for nearer objects. Monocular cues include motion parallax and perspective. Binocular vision, stereopsis, arises because each eye gets a slightly different view of the world so the image of a nearby object lies on different points in each retina. For modest degrees of retinal disparity the brain fuses the two images into a single percept and computes the object distance from the disparity.

Color vision

Color vision allows boundaries to be discerned simply on the basis of differences in the wavelength composition of reflected light. It requires a minimum of two types of receptor that respond over different wavelengths so that two values for brightness can be assigned for each pixel of an image. This dichromatic vision is the case for most mammals. Many primates, including humans, have trichromatic vision mediated by three types of receptor, which allows three brightness values to be ascribed to an object. The brain compares these values to give the perception of color.

Related topics

(G3) Photoreceptors
(G5) Early visual processing

(G6) Parallel processing in the visual system

Sensitivity

The human eye is sensitive to the electromagnetic spectrum between the wavelengths 400 nm (violet) and 700 nm (red), with maximal sensitivity at 550 nm (green). The range of light intensity to which we are exposed is huge, about 10^{11}-fold. Although the eye can respond to a single photon of light, 5–8 photons arriving within a short time are required to give the experience of a flash of light in the dark-adapted state. Because intensity is encoded by the visual system logarithmically, it is difficult to distinguish differences in intensity at high light levels.

Visual processing

Vision can be defined as the process of discovering from images what is present in the world and where it is. In humans it is estimated that almost half of the cerebral cortex is implicated in vision, more than is devoted to any other single function. This implies that vision is the most complex task that the brain performs.

The brain uses a two-dimensional shifting pattern of light intensity values on the two retinas (the light sensitive layers of the eyes) to form a representation of the form of an object, its color, movement, and position in three-dimensional space. Each of the visual channels (color, form, movement, and depth perception) are handled simultaneously by distinct (but interdependent) pathways. This is called **parallel processing**. (It contrasts with **serial processing** in which a task is segmented into several subroutines that are executed sequentially.) Parallel processing has the advantage of speed. The final visual representation is a unified percept in which all the channels are combined. How the brain does this is uncertain and is termed the **binding problem**, which applies to all sensory modalities.

The visual system abstracts key features from retinal images and this involves considerable data compression. Thus visual processing gives higher weight to regions of the visual world that are changing in time (movement) and space (contrast) than those that are constant. **Visual perception** requires the existence of internal representations of the visual world which allow the brain to make hypotheses about what the retinal image is. Internal representations account for the fact that vision allows **pattern completion**, it can generate a complete percept even when the raw sensory data is incomplete or corrupted by noise, and **generalization**, the ability to recognize objects from a wide variety of vantage points and contexts. Some internal representations are specified during development and are immutable, but most probably depend on early learning. Mental images of objects are thought to be manifestations of the internal representations of the objects and can be manipulated by most people in predictable ways. Whenever an unresolvable mismatch occurs between the sensory input and the internal representation the result is a **visual illusion**.

Perceptual constancy is a key property of vision. Visual perception can be invariant over wide differences in the properties of the retinal image. For example, with **size constancy**, familiar objects do not diminish in size in proportion to the reduction of the retinal image, but appear larger than they should. **Color constancy** preserves the colors of objects in the face of alterations in the wavelength composition of the light source. Perceptual constancy permits successful object recognition under a wide variety of ambient conditions.

Depth perception

The retinal image is two dimensional, but from it the visual system can infer the three-dimensional structure of the world. There are both monocular and binocular cues to depth perception. Monocular cues are most important for distant objects, where binocular cues cannot be used, and include the following examples:

- Movement of the head causes an apparent movement of near objects with respect to distant ones (motion parallax); the closer the object the bigger this apparent movement.

- Parallel lines appear to converge with distance (geometric perspective); artists from the early 1400s onwards used this as a major depth cue in painting.

- The relative sizes of objects of known dimensions.

- More distant objects can be partly hidden by nearer ones (occultation).

- Distant objects are faded and bluer because of intervening haze (extinction).

- The ability to resolve fine detail falls off with distance.

- Neural signals that correspond to how much the visual system has had to change the shape of the lens to keep the object in focus (accommodation).

The binocular cue to depth perception is called **stereopsis**. Because the eyes are about 6.3 cm apart each has a slightly different view of the world. (This can be visualized by viewing a scene first through one eye, then the other, when nearby objects appear to jump sideways; **binocular parallax**.) This has the effect that the image of a nearby object falls onto different horizontal positions on the left and right retinas, a phenomenon called **retinal (binocular) disparity**. When the eyes converge to fixate on a nearby point the two retinal images of the point are perceived as fused into a single point. All other points at the same distance are fused because their images lie at the same corresponding positions

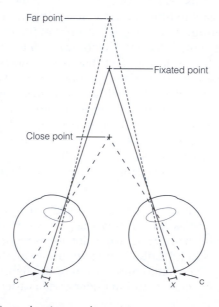

Figure 1. Stereopsis. The fixated point produces images on corresponding positions, c, on the retinas so that the images are fused. The images of the far point are displaced from the corresponding points by a distance x giving a binocular disparity of 2x. A similar argument applies to the close point images.

on left and right retinas. Points in space that lie either closer or further away than these will form images at different corresponding retinal positions and hence different binocular disparities (Figure 1). The closer the object, the bigger the disparity. Images of these points will also fuse provided that disparity is not too great (> 0.6 mm or 2°). Beyond this either two images are seen, **double vision** (**diplopia**), or information from one eye is completely rejected by the visual cortex. Clearly stereopsis is only possible for the field of view in which the two monocular visual fields overlap. The brain is able to compute depth from disparity simply by comparing where the same pattern lies on left and right retinas. Stereopsis does not require form, movement, or color.

Color vision

Color vision permits boundaries to be seen between regions that have equal brightness, provided the spectrum of wavelengths they reflect is different. The spectrum of light reflected from an object depends on the wavelength composition of the illuminating light and the reflectance of the surface, but color vision is not just a matter of measuring all the wavelengths in the reflected light.

Color vision requires a minimum of two types of receptor that respond over different wavelength ranges. This is dichromatic color vision and is the case for all mammals except old world monkeys, apes, and humans. With two receptors the visual system can assign two brightness values for each pixel of the visual field. By comparing these values, colors may be perceived. For example, if a pixel reflects more short-wavelength light it will appear brighter to a short-wavelength receptor than a long-wavelength receptor, and will be seen as blue. If a pixel reflects more long-wavelength light it will be seen as red. In the case that a pixel reflects equal amounts of short- and long-wavelength light it will appear monochrome, either white or shades of gray depending on the intensity of the light.

Human color vision is trichromatic because the eye has three populations of receptors (cones) that can function in daylight, each sensitive to a different (but wide and overlapping) range of wavelengths. The three types of cones have *maximum* absorptions corresponding approximately to violet, green, and yellow light. The wavelength of the light does not affect the character of the response of the cone. A given cone simply has a higher probability of absorbing a photon which is close to its peak wavelength. This means that the visual system has no way of detecting the absolute wavelength composition of any light. The trichromatic visual system abstracts three brightness values for an object and comparisons of these values determines its color.

Color vision has several remarkable properties. It shows **color constancy**. An object can be viewed under a variety of light sources with different spectral compositions, for example, neon lighting, sunlight, or tungsten light, and appear to be the same color even though the wavelengths of light it reflects in each case will be quite different. While some colors in the same pixel of visual space perceptually mix to produce other color categories (e.g., blue and green mix to give cyan) complementary colors (e.g., red and green) do not perceptually mix; reddish-green colors are never seen. This is **perceptual cancellation**. **Simultaneous color contrast** is the perceptual facilitation of complementary colors that occurs across boundaries. For example a gray disc within a red background looks slightly green while a gray disc in a green background appears slightly red. Each of these features can be accounted for in terms of visual system physiology.

G2 Eye and visual pathways

Key Notes	
Structure of the eye	The eye has three layers. The outer sclera maintains the shape of the eye and provides attachment for extraocular muscles. The choroid is pigmented to prevent light being reflected within the eye. Innermost is the light-sensitive retina. At its front the sclera becomes the transparent cornea responsible for most of the refraction of light rays entering the eye. The front of the choroid forms the ciliary body and iris. The biconvex lens is attached to the ciliary body by a suspensory ligament. The iris is a diaphragm surrounding the pupil and contains smooth muscles which act as pupillary sphincter or dilator.
Organization of the retina	Light-sensitive photoreceptors synapse with retinal interneurons including bipolar cells which synapse with ganglion cell, axons of which enter the optic nerve. There are 100-fold fewer ganglion cells than photoreceptors so the retina must do a great deal of visual processing and data compression. Only ganglion cells are excitable, all other retinal neurons signal via passively conducted synaptic potentials. The central fovea has the highest visual acuity.
Anatomy of visual pathways	The optic nerves meet at the optic chiasm where nerve fibers from the nasal half of each retina cross to the other side. Optic tract fibers go to the pretectum, which controls pupil and accommodation reflexes, and the superior colliculus, which mediates many visual reflexes, but most go to the lateral geniculate nucleus of the thalamus. From here the optic radiation goes to the visual cortex.
Visual reflexes	Light shone in one eye causes pupil constriction in both eyes by stimulating the pretectum, which activates the parasympathetic system that supply the pupillary sphincters. For close objects greater refraction is needed to focus the image. This is achieved by the accommodation reflex. In response to image blurring parasympathetic activity contracts the ciliary body allowing the lens to become more spherical.
Related topics	(G1) Attributes of vision (G5) Early visual processing (G7) Oculomotor control and visual attention (L5) Autonomic nervous system (ANS) function

Structure of the eye

Three layers enclose the contents of the eye, the sclera, the choroid and the retina (Figure 1). The **sclera** is a thick, stiff, outer layer of connective tissue. At the front of the eye it becomes the cornea. At the back it becomes the dura mater covering the optic nerve. The sclera maintains the shape of the eyeball and provides attachment for the extraocular muscles. The curvature of the transparent **cornea** refracts incoming light and provides most of the focusing power of the eye. The **choroid** is a thin, highly vascular layer, dark brown in color because of the presence of **choroidal pigment cells**. By absorbing light it limits total internal reflection within the eye.

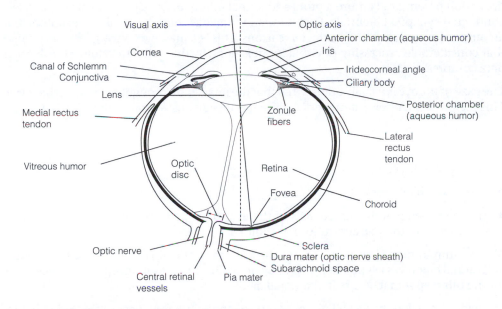

Figure 1. Horizontal section through the right human eye.

At the front the choroid becomes the **ciliary body** and the **iris**. The ciliary body gives rise to numerous, thin **zonular fibers** which attach to the capsule of the lens as the **suspensory ligament**. Inside the ciliary body lies the ciliary muscle composed of smooth muscle fibers arranged in both radial and circular directions.

The iris is a diaphragm surrounding a central hole, the **pupil**. The iris contains two intraocular muscles which act in concert to control the size of the pupil. Innermost is a flat ring of circularly arranged smooth muscle fibers, the **pupillary sphincter**. Surrounding the sphincter is a layer of radially organized myoepithelial cells which form the **pupillary dilator**.

Aqueous humor is actively secreted by the epithelium of the ciliary body into the **posterior chamber**. It percolates through the pupil to the **anterior chamber** from where it drains into the venous system via the **canals of Schlemm** located in the **irideocorneal angle**. Aqueous humor supplies metabolic substrates for the lens and cornea which have no blood supply, and maintains eyeball pressure. **Vitreous humor** is a gel of extracellular fluid which contributes to refractive power.

The biconvex **lens** of the human eye has a diameter of 9 mm. It is encapsulated within an elastic connective tissue membrane which is attached to the suspensory ligament.

Organization of the retina

The retina is the innermost layer of the eye. It contains five distinct neuron types, interconnected in circuits that are repeated millions of times.

The light-sensitive photoreceptors, which lie at the back of the retina adjacent to melanin-containing **pigmented epithelial cells**, form synapses with a variety of retinal interneurons; bipolar cells, horizontal cells, and amacrine cells. Bipolar cells synapse with the output neurons of the retina, **ganglion cells**, the axons of which form the optic nerve, and only become myelinated as they leave the eye via the optic disc. Although all the cells in the retina (except pigment cells) are neurons, only the ganglion cells are able to fire action potentials. Photoreceptors and retinal interneurons signal by way of passively conducted synaptic potentials. Each human retina has roughly 10^8 photoreceptors, but an output via only about 10^6 optic nerve axons. This is a massive convergence and shows that considerable processing of visual input is done by the retina to achieve this level of data compression.

The gaze of the eye is usually adjusted so that the images are brought to focus at the **fovea**. This region of the retina (diameter 1.5 mm) has the greatest visual acuity. This is due to:

- Its high density of photoreceptors

- The displacement of overlying layers of the retina to the side so that light hits the photoreceptors directly

- The lack of blood vessels

- The fovea being at the optical axis of the eye so image distortion by the optics (e.g., spherical or chromatic aberration) is minimal

About 4 mm from the fovea towards the nose lies the optic disc, where optic nerve fibers and retinal blood vessels pierce the retina. This region lacks photoreceptors and accounts for the **blind spot** that occurs in the visual fields.

For regions outside the fovea light passes through the full thickness of the retina before striking the light-sensitive photoreceptors. **Muller glial cells** offset some of the disadvantages this poses. These funnel-shaped cells have their broad end at the retinal surface and extend long slender processes to the photoreceptor layer. Muller cells act as light guides, rather like fiber-optic cables, in that they channel light through the neuronal layers of the retina to the photoreceptors at the rear. They do this more efficiently for light being brought to focus than for light that has suffered multiple reflections within the eye. Hence they enhance the signal-to-noise ratio. Muller cells are tuned to visible light so that radiation in the near IR and UV leak out to be absorbed by surrounding neurons rather than excite photoreceptors. In addition, Muller cells refract light so that blue light is brought to the same focus as red light, correcting a lens defect called chromatic aberration.

Anatomy of visual pathways

The optic nerves meet in the midline at the **optic chiasm** (Figure 2). Here, 53% of optic nerve fibers, those from the nasal halves of the retina, cross to the contralateral side in the **optic decussation**. Axons from the temporal halves of the retina remain on the ipsilateral side. Retinal axons leave the optic chiasm to enter the **optic tracts** from where they go to three targets. A small proportion go to the **pretectum** of the midbrain which controls pupil and accommodation reflexes. Others go to the **superior colliculus** in the tectum of the midbrain which organizes several visual reflexes, and a visual pathway runs to the

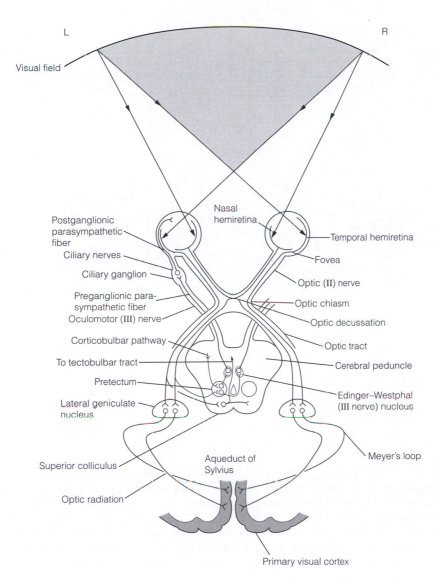

Figure 2. Visual pathways. Reflex pathways are shown complete only on the left side. The direction of light rays from left and right halves of the visual field into the eyes is depicted. Note that light from the left visual field falls on the right halves of each retina (nasal hemiretina of the left eye, temporal hemiretina of the right eye) and light from the right visual field goes to the left hemiretinas. Binocular vision is possible only in the shaded region.

hypothalamus to entrain circadian rhythms. The great majority of axons go to the **lateral geniculate nucleus** (**LGN**) of the thalamus. From here, the **optic radiation** sweeps to the medial aspect of the pole of the occipital cortex, most axons terminating in layer IV of Brodmann area 17, the **striate** or **primary visual cortex** (**V1**). The retina–LGN–visual cortex pathway is responsible for visual perception. Visual defects characterized clinically can provide clues to the site of the lesion within the visual system (Figure 3).

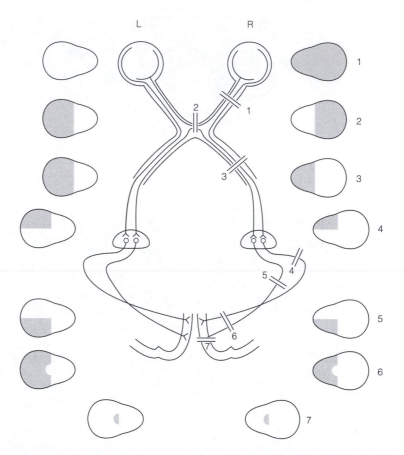

L   R

1
2
3
4
5
6
7

Figure 3. Visual defects arising from damage to visual pathways. Lesion 2 commonly arises from compression of the central part of the optic chiasm by a pituitary tumor. Optic tract lesions (3) are rare. Optic radiation damage is usually due to an infarct or a tumor in the temporal lobe (4) or parietal lobe (5). Lesion 6 is usually caused by occlusion of the posterior cerebral artery. The fovea area is spared because it is supplied by the middle cerebral artery. Damage to one occipital lobe pole is usually caused by trauma and, since the fovea has by far the largest representation, selective loss of foveal vision is typical (lesion 7).

Visual reflexes

The **pupil light reflex** controls the amount of light entering the eye by altering pupil size. This ranges between 1.5 and 8 mm in diameter, being maximal in complete darkness. Although this allows only a 30-fold change in light entry (which is small compared with the range of light intensity the visual system can experience) the reflex is useful because it operates over the light levels typically encountered during daylight. Light shone in one eye produces pupil constriction of the same eye (the **direct** reflex) *and* of the contralateral eye (the **consensual** reflex) because of reciprocal crossed connections in the midbrain. The reflex pathway is shown in Figure 4. Optic nerve axons synapse in the pretectum which sends output to preganglionic parasympathetic fibers in the **Edinger–Westphal (accessory) oculomotor nucleus**. These autonomic fibers travel in the oculomotor nerve to the ciliary ganglion which lies in the orbit. Postganglionic fibers from there go to the pupillary sphincter. Light stimulation of optic nerve fibers excites the parasympathetic

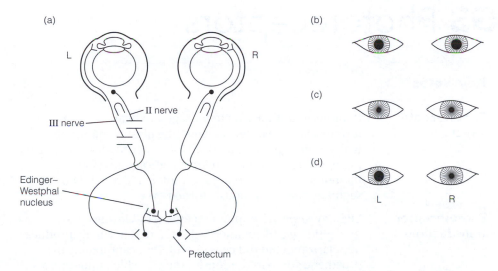

Figure 4. Altered pupil reflexes following damage to either optic (II) or oculomotor (III) nerves on the left side: (a) reflex pathway; (b) optic nerve damage, left eye stimulated; (c) optic nerve damage, right eye stimulated; (d) oculomotor nerve damage, left or right eye stimulated.

terminals to release acetylcholine, which contracts the sphincter. Lesions of the optic and oculomotor nerves, or of the midbrain can be diagnosed by examining defects in the pupil light reflex (Figure 4).

For close objects, light rays are diverging as they enter the eye and so greater refraction is needed to bring them to focus at the fovea. This is achieved by the **accommodation reflex**. Contraction of the ciliary muscles pulls the ciliary body forwards and inwards, easing the tension in the suspensory ligament and lens capsule, allowing the lens to become more spherical and reducing its focal length. The stimulus for the accommodation reflex is blurring of the retinal image (large retinal disparity). This is monitored by the visual cortex which projects to the pretectum via the corticobulbar pathway. Via connections between the pretectum and the Edinger–Westphal nucleus, parasympathetic fibers are activated which contract the ciliary muscles. Accommodation occurs in both eyes equally.

Observing a close object also causes convergence of the visual axes of both eyes, the **vergence reflex**. This enables both eyes to fix their gaze on an object. In addition, the degree of convergence provides a cue for stereopsis, since the closer an object is, the greater the convergence must be. Vergence can be triggered by a blurred retinal image or by consciously altering gaze to a point at a different distance. The circuitry is from the visual cortex to the **frontal eye fields** in the frontal cortex concerned with the planning and execution of eye movements.

G3 Photoreceptors

Key Notes	
Photoreceptor structure	Photoreceptors have an inner segment bearing a synaptic terminal, and an outer segment, the plasma membrane which is invaginated into deep folds to form discs. The visual pigments, for example rhodopsin, are situated in the disc membrane. Photoreceptors cannot divide but their outer segments are continuously turned over.
Photoreceptor transduction	Light hyperpolarizes photoreceptors by closing cyclic nucleotide-gated channels (CNGCs) that normally produce depolarization. In rod cells photons are captured by the prosthetic group in rhodopsin, retinal, which undergoes photo-isomerization. This allows rhodopsin to couple with a G protein, transducin, thereby stimulating a phosphodiesterase that hydrolyzes cyclic guanosine monophosphate. It is the fall in cGMP which closes the CNGCs. Subsequently the photo-isomerized retinal dissociates from the rhodopsin leaving the pigment bleached. Dark adaptation is partly due to the regeneration of rhodopsin.
Rod cells	Rod photoreceptors, situated throughout the retina except the fovea and optic disc are extremely sensitive and used in dim light vision. In daylight they are unresponsive. Rod cell vision has a low acuity because many rod cell signals converge, which, while maximizing light sensitivity, causes loss of information about location.
Cone cells	Cone cells are concentrated at the fovea. They are 1000-fold less sensitive to light than the rods, fail in dim light, and are responsible for daylight vision. Daylight vision has high acuity because there is little convergence of cone signals. There are three populations, distinguished by the range of wavelengths to which they are sensitive; short (blue), medium (green), and long (red) wavelength cones They are randomly but patchily distributed, so color vision is coarse.
Related topics	(D2) G-protein-coupled receptors (G1) Attributes of vision (G4) Retinal processing

Photoreceptor structure

Rod and cone photoreceptors have similar structures (Figure 1). Photoreceptors have diameters ranging from 1 to 4 μm, being smaller at the fovea, enhancing visual acuity there. The inner segment contains the nucleus and has an axon-like process connected to a synaptic terminal. The outer segment in the cone cell has its plasma membrane

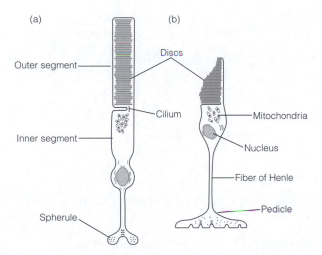

Figure 1. Photoreceptors: (a) a rod cell; (b) a cone cell.

invaginated into numerous closely packed parallel folds, forming discs. In rod cells the discs are pinched off the plasma membrane to become completely intracellular. The disc membrane is densely packed with **visual pigment**. In rod cells this is **rhodopsin**. Each type of cone cells has its characteristic **cone opsin**. The outer segment is continually regenerated from the base, whilst its apical tip is phagocytosed by pigment epithelial cells at the rate of 3–4 discs per hour. Photoreceptors are incapable of mitotic division.

Photoreceptor transduction

The resting potential of the photoreceptor plasma membrane in the dark is quite low, about -40 mV. Light produces a hyperpolarizing receptor potential, the amplitude of which is related to the light intensity (Figure 2).

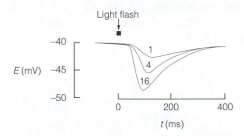

Figure 2. Cone receptor potentials in response to light flashes of increasing relative intensity (1, 4, and 16).

The hyperpolarization is produced by the light-evoked closure of cyclic nucleotide-gated cation channels that are open in the dark. The normal, relatively depolarized, state of the photoreceptor is caused by the flow of a dark current, as shown in Figure 3.

The cyclic nucleotide-gated cation channel allows Na^+ and Ca^{2+} ions to flow into the outer segment in darkness. Na^+ ions are actively extruded by the Na^+-K^+ ATPase in the inner

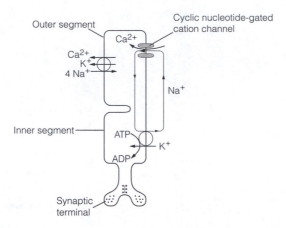

Figure 3. The ionic basis of the rod photoreceptor dark current.

segment. Ca^{2+} leaves the photoreceptor via a Na$^+$–K$^+$–Ca^{2+} transporter. When light pho-tons strike the outer segments a cascade of biochemical events is initiated which results in the closure of the cation channels, reducing the dark current, and hyperpolarizing the photoreceptor. The transduction process in rod cells is well understood. Rhodopsin con-sists of a G-protein-coupled receptor, **opsin**, and a prosthetic group, **retinal**, synthesized by retinol dehydrogenase from **retinol** (**vitamin A**). Retinol cannot be synthesized *de novo* in mammals and hence must be supplied in the diet.

In the dark, retinal is present as the 11-*cis*-isomer. Light causes photo-isomerization to the all-*trans* isomer. The isomerization occurs within a few picoseconds of the photon being absorbed and triggers a series of conformational changes in the rhodopsin to form photoexcited rhodopsin (R*). Photoexcited rhodopsin couples with a G protein, **trans-ducin** (**G$_t$**) and exchange of GDP for GTP occurs. The GTP-bound form of the transdu-cin alpha subunit activates a phosphodiesterase (PDE) which catalyzes the hydrolysis of 3′,5′-cyclic guanosine monophosphate (cGMP) to 5′-GMP. This reduces the concentra-tion of cGMP in the photoreceptor and so the cation channels, normally kept open by the cyclic nucleotide, close (Figure 4).

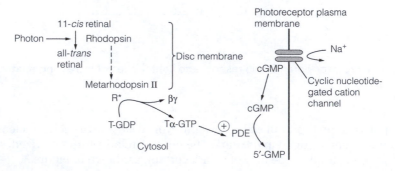

Figure 4. The role of transducin in photoreceptor transduction. T, transducin; PDE, phosphodiesterase.

This second messenger cascade has a large amplification. A single photon activates about 500 transducin molecules, closes hundreds of cation channels, blocking the influx of 10^6 Na^+ ions to cause a hyperpolarization of about 1 mV.

Several mechanisms act sequentially to terminate the cascade:

- Like other G proteins, transducin has an intrinsic GTPase which hydrolyzes the bound GTP to GDP, stopping the activation of PDE.

- Photoexcited rhodopsin is phosphorylated by **rhodopsin kinase**, and then binds **arrestin** which blocks the binding of transducin.

- Within a few seconds the bond between retinal and opsin in photoexcited rhodopsin spontaneously hydrolyzes and the all-*trans* retinal diffuses away from the opsin. In high light levels most of the rhodopsin exists in this dissociated state in which it is described as being **bleached** and the rod said to be **saturated**. Regeneration of rhodopsin occurs in the dark: retinal isomerase catalyzes the isomerization of the all-*trans* isomer to the 11-*cis* isomer which then reassociates with the opsin. This process underlies **dark adaptation**.

Restoration of the dark state, in addition, requires the synthesis of cGMP. This is catalyzed by guanylate cyclase.

Light adaptation, in which photoreceptors become less sensitive during light exposure, allows them to respond to levels of illumination that vary by as much as four orders of magnitude (Figure 5). Light-evoked closure of the cation channels reduces Ca^{2+} influx, so the Ca^{2+} concentration in the rod outer segment falls. Since Ca^{2+} normally inhibits the guanylyl cyclase needed for cGMP synthesis, this drop in Ca^{2+} concentration increases the production of cGMP, offsetting its destruction by the light.

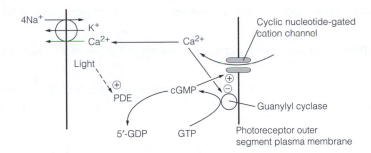

Figure 5. The role of Ca^{2+} in photoreceptor light adaptation.

Rod cells

Rod cells are 20-fold more numerous than cones and are distributed across the entire retina except for the fovea and the optic disc (Figure 6). They are 1000 times more sensitive to light than cone cells and used for **scotopic** (dim light) vision. Indeed, rod cells become unresponsive in daylight. The high sensitivity of rod cells is partly because they integrate responses to incoming photons over a long period (~100 ms).The disadvantage of this is that rods are unable to discern flickering light if the flicker rate is faster than about 12 Hz. Rod cell sensitivity is also due to their great amplification of the effect of incoming photons. Scotopic vision has a low acuity for two reasons. Firstly, the image formed on the peripheral retina is distorted. Secondly, many rods converge onto a single bipolar

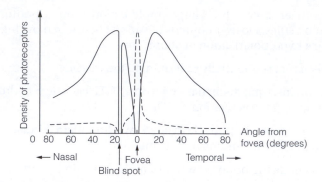

Figure 6. Distribution of cone (- - - -) and rod (——) photoreceptors in the human retina.

cell. Although this maximizes the chance of a rod bipolar cell responding to a dim light signal, because it can capture light signals from a large area of retina, by the same token information about localization is less precise.

Cone cells

Cone photoreceptors are most dense at the fovea and their numbers fall off sharply beyond 5° of it. They have a low sensitivity to light and do not saturate except in very intense light, so are used for **photopic** (daylight) vision. Photopic vision has high acuity because there is little or no convergence between cone cells and bipolar cells. Cone cells integrate photon responses over a short time and so are able to resolve a flicker frequency of less than about 55 Hz.

At low light levels (e.g., dusk) rod cells operate alongside cones, boosting the cone cell signal by means of electrical synapses to preserve color perception. This is termed **mesopic** vision.

There are three populations of cone cells which differ in their spectral sensitivity (Figure 7).

Although more properly called short (S), medium (M), and long (L) wavelength cones they are often referred to as blue, green, and red cones respectively, although their peak

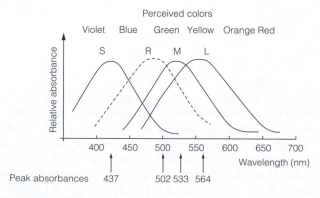

Figure 7. Spectral sensitivities of photoreceptors: S, short-wavelength cones; R, rods; M, medium-wavelength cones; L, long-wavelength cones.

sensitivities are not best described by these colors. The S cones are sensitive to wavelengths down to 315 nm. However, the normal eye does not see wavelengths shorter than 400 nm (i.e., ultraviolet) because they are absorbed by the lens.

Color vision requires comparisons of the relative strengths in the outputs of the S, M, and L cones. S cones constitute only about 5–10% of the total number of cones and are absent from the center of the fovea. This is because the lens suffers from **chromatic aberration**, in which short-wavelength light is not brought to focus at the same point as longer wavelengths, causing image blurring. This would compromise high acuity vision. Consequently, color vision at the central fovea is dichromatic and furthermore, M and L cones are distributed randomly leaving patches in which there is only one population of cone. These features mean that color vision is coarse grained and cannot resolve fine detail.

Scotopic vision is achromatic because all rod cells have the same spectral sensitivity curve. They are unable to distinguish between wavelengths on the rising and falling limbs of the spectrum that excite the cell to the same extent. Under scotopic vision the sensitivity of the eye is determined by the rod cells and peaks at around 500 nm. Under photopic conditions the wavelength sensitivity is governed by cones and is maximal at 550 nm. This shift in wavelength sensitivity during mesopic vision is called the **Purkinje shift**. It means that as dusk falls red fades first and the last color to be lost is green.

When moving from bright to very dim light the sensitivity of the retina to light increases a million-fold over a period of 30 min or longer. This is called dark adaptation and is a property of photoreceptors. **Dark adaptation** has two phases (Figure 8). The first is due to cone cells which increase sensitivity about 100 fold, the second longer phase is due to rod cells.

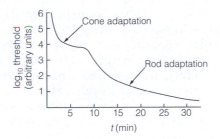

Figure 8. The time course of dark adaptation.

Light adaptation occurs when going from dim to brightly lit conditions. It is very much faster than dark adaptation.

G4 Retinal processing

Key Notes

Bipolar cells and on and off channels

Photoreceptors synapse with two types of bipolar cell, those which depolarize and those which hyperpolarize in response to light. Ganglion cells respond in the same way to light as the bipolar cells that drive them. This gives rise to two channels: on channels in which light increases the firing of on ganglion cells, and off channels in which light silences off ganglion cells. On channels signal the presence of local bright patches and off channels local dark regions of an image.

Horizontal cells and lateral inhibition

Bipolar and ganglion cell receptive fields are circular and divided into center and surround. Light stimulation has opposite effects on the cell depending on whether it falls on center or surround. This is lateral inhibition and its effect is to enhance contrast at boundaries. It is mediated by GABAergic horizontal cells that form reciprocal connections with adjacent photoreceptors. Because horizontal cells are heavily interconnected, the surround signal they generate represents light intensity averaged over an area of retina.

Ganglion cells

Ganglion cell axons form the optic nerve. They come in three types. P ganglion cells are small, get their input from single cone types, and mediate form and color vision. They exhibit color opponency, a sort of lateral inhibition in which they are excited by one type of cone but inhibited by one or both of the others. K ganglion cells are small, some are blue sensitive, others are motion sensitive. M ganglion cells are large, rapidly conducting, get input from middle and long wavelength cones together, and show transient responses. M ganglion cells detect brightness (but not color) contrast, and movement.

Rod signaling

In daylight only the cone channels are functional. At dusk rod cells become sensitive but transmit their signals via gap junctions to cone cells, boosting their function to maintain high acuity and color vision. When it is very dark cone cells fail and the rods cells signal exclusively through their own pathway via depolarizing (rod) bipolar cells and amacrine cells.

Amacrine cells

These interneurons, with neurites that have properties of both axons and dendrites, are extremely diverse. Different types are involved in rod cell pathways, surround inhibition, and in signaling the direction of movement of a stimulus.

Related topics

(E2) Coding of modality and location

(G1) Attributes of vision
(G3) Photoreceptors

Bipolar cells and on and off channels

Photoreceptors synapse with bipolar cells. Two types of bipolar cell can be distinguished on the basis of both morphology and physiological responses. Those with processes that form **triad ribbon synapses** (Figure 1) deep in the photoreceptor terminal are **invaginating bipolar cells** and they depolarize in response to light striking the photoreceptor. Triad ribbon synapses are so called because they have three postsynaptic components, the bipolar cell dendrite and dendrites of two horizontal cells. By contrast, **flat bipolar cells** form superficial **basal synapses** with photoreceptors and hyperpolarize in response to light.

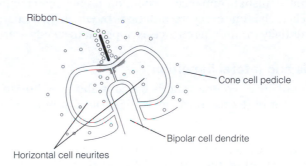

Figure 1. A triad ribbon (invaginating) synapse.

Cone cells form synapses with **midget bipolar cells** (so called because of their size) of either one or the other type. Midget bipolar cells synapse directly with **ganglion cells** which respond to light in the same sense as their bipolar cells. This arrangement gives rise to two labeled lines: **on channels** are formed by cone-depolarizing bipolar cells (on ganglion cells), whereas **off channels** are cone-hyperpolarizing bipolar cells (off ganglion cells). On ganglion cells are depolarized and increase their firing rate as a function of light intensity. Off ganglion cells are silenced by hyperpolarization (Figure 2).

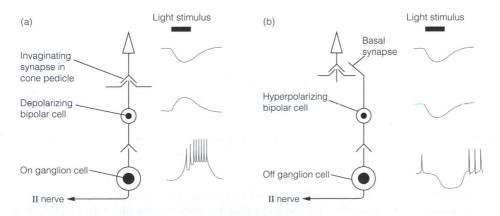

Figure 2. (a) On channel and (b) off channel in the retina. In each case electrophysiological responses of the cells to light stimulation recorded intracellularly is shown on the right. All cells depicted use glutamate as a neurotransmitter.

All photoreceptors use glutamate as a transmitter. The opposite responses of invaginating and flat bipolar cells come about because they have different glutamate receptors. For invaginating bipolar cells, tonic release of glutamate from photoreceptors in the dark is inhibitory. When light hyperpolarizes the photoreceptor glutamate release is suppressed, and inhibition lifted, so the bipolar cell depolarizes. For flat bipolar cells the response to tonic glutamate release is excitatory and light, by reducing that excitation, causes the bipolar cell to hyperpolarize.

On channels respond with increased firing to light levels that are greater than the local average. Off channels show increased firing in response to dark regions, that is, where light levels are lower than the local average. In this way the existence of separate on and off channels is a mechanism to enhance the boundaries between regions that reflect different amounts of light. It is one of several processes by which the visual system is adapted to respond preferentially to stimulus change as opposed to steady-state stimulation.

Horizontal cells and lateral inhibition

Lateral inhibition enhances contrast and is brought about by **horizontal cells**. Lateral inhibition can be seen in the receptive fields (RFs) of both bipolar and ganglion cells. These are circular and divided into an inner center and an outer surround. Stimulation of these two regions separately produces opposite effects on the cells. In the case of an on ganglion cell, for example (Figure 3), background firing rate is increased by center illumination but silenced by light on the surround. When light fills the whole of the receptive field there is little change in firing rate. Off ganglion cell RFs have the converse response, with center illumination producing inhibition and surround illumination excitation.

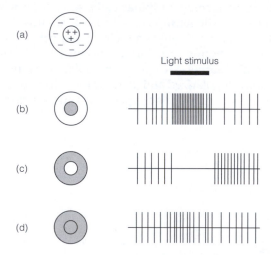

Figure 3. Extracellular recording from on ganglion cells: (a) receptive field; (b) central illumination; (c) surround illumination; (d) overall illumination.

Lateral inhibition arises because horizontal cells form reciprocal connections at triad synapses between neighboring photoreceptors. The details are a bit complicated. In the dark, horizontal cells are excited by glutamate release from photoreceptors, but themselves release GABA which tends to inhibit the photoreceptors. Light which hyperpolarizes surrounding photoreceptors causes them to secrete less glutamate so reducing horizontal cell excitation. This means that GABA release from the horizontal cells is lowered, which in turn allows the central cone to depolarize somewhat, so that it releases more glutamate. The final step in the sequence depends precisely on the type of bipolar cell the central cone synapses with. If it is a depolarizing (on) bipolar cell the increased glutamate will cause it to hyperpolarize, as glutamate is inhibitory at invaginating synapses. If it is a hyperpolarizing (off) bipolar cell the increased glutamate will make the bipolar cell depolarize. Note that in each case the bipolar cell response (and hence that of the ganglion cell with which it synapses) is the opposite for surround compared with central illumination (Figure 4).

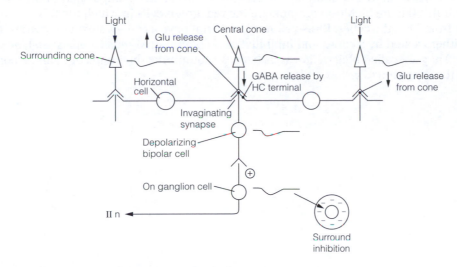

Figure 4. The mechanism of lateral inhibition by horizontal cells (HC).

Horizontal cells are extensively interconnected via gap junctions forming a network which spans an area of retina termed the **S space**. The S space horizontal cells provide the signal for surround inhibition and it is thought that this signal is a measure of the mean luminance over quite a wide area of retina.

Ganglion cells

Axons of ganglion cells form the optic (second cranial) nerve. Ganglion cells are the only retinal cells capable of firing action potentials. There are three ganglion cell types. **Parvocellular (P) ganglion cells** are small and by far the most numerous with about a million in each retina. **Magnocellular (M) ganglion cells** are large and number about 100 000 per retina. **Koniocellular (K) ganglion cells** are small and resemble P cells physiologically.

P and M types differ in several important respects:

● P cells have smaller RFs than M cells.

● Because of their smaller size, P cells have a slower conduction velocity than M cells.

● Whereas P cells often show sustained responses, M cells respond transiently to a prolonged visual stimulus.

● P cells are usually wavelength selective whereas M cells are not.

● M cells are much more sensitive than P cells to low-contrast stimuli.

From these differences it can be inferred that P cells must get their input from single cones, or from several cones with the same wavelength sensitivity (S, M, or L). By contrast, M cells get input from M and L cones together (but not S cones), and from rods. Hence P, but not M, cells must be involved in color vision. The rapid, transient responses of M cells makes them adapted for motion detection. The small RFs of P cells, and their sustained responses are suitable for fine form discrimination. The distinct functional properties of P, M, and K ganglion cells is the starting point for parallel processing in the visual system.

P cells are **color single opponent cells**. They have RFs that are excited by one type of cone cell but inhibited by another. Two types of P cell can be distinguished by the nature of their RFs (Figure 5). Most common are the **red–green cells** in which the RF compares input from M and L cones. **Blue–yellow cells** do not have a center–surround pattern but are either excited by S cones and inhibited by a combined M plus L cone signal, or *vice versa*. They are called blue–yellow cells because combining the inputs of M and L cones gives the sensation of yellow.

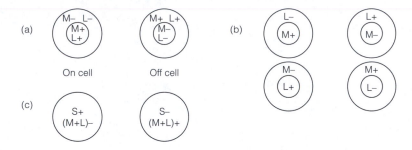

Figure 5. Receptive fields of retinal ganglion cells: (a) M ganglion cells; (b) concentric single opponent (red–green) P cells; (c) coextensive single opponent (blue–yellow) P cells.

The red–green cells respond differently to small or large spots of light. For example, a green-on/red-off cell will be *equally* stimulated by a small green or white spot that covers the center of the RF. This is because white light contains the wavelengths which excite the green cones. However, the same cell will be excited by a large green spot but not by a large white spot. This is because the white light contains the wavelengths which stimulate the red-off surround. A large red spot will silence the cell. So, in general, these cells are more wavelength selective for large stimuli than small ones. For small spots, they cannot distinguish red or green from white light, but they can signal brightness.

M cells get *combined* input from two types of cones so they are called **broad band cells** and their RFs measure only brightness contrast; M cells are color blind.

Some K cells are blue-on cells, others are motion sensitive and have low spatial resolution.

Rod signaling

Signaling by rod cells depends on the light intensity. At high light levels rods are saturated, and only cones operate. In the partially dark-adapted eye (e.g., at dusk) rod cells come on stream but signal through gap junctions to neighboring cones. This effectively augments cone cell function, so maintaining acuity and color vision (mesopic vision). However, when it is very dark (e.g., moonless night in remote countryside) cone cells fail even with signal boosting from rods. Rod cells in the dark will have a greater influx of Ca^{2+} via the dark current. One effect of this increase in Ca^{2+} is to close the gap junctions between the rod and cone cells. Rod signaling is now relayed via depolarizing (rod) bipolar cells which synapse with a population of amacrine cells. The effect of this is to increase contrast sensitivity.

Amacrine cells

Amacrine cells have no axons but their extensive neurites share properties of both axons and dendrites. They are a very diverse group morphologically, and most of the neurotransmitters identified in the nervous system are used by one or other of the 30 or so types of amacrine cell.

Amacrine cells are implicated in rod signaling, surround inhibition, and detecting the direction of motion of an object across the visual field. Dopaminergic amacrine cells are only about 1% of all amacrine cells but their long dendrites interconnect, possibly via gap junctions to form a network. These cells get input from cone bipolar cells so the network is able to signal average illumination which is used to produce surround inhibition. In the dark-adapted eye, ganglion cells become much more sensitive to light because dopamine surround inhibition is turned off. Some ganglion cells are sensitive to the direction of motion of a stimulus. Direction sensitivity is conferred by amacrine cell circuits.

G5 Early visual processing

Key Notes

Lateral geniculate nucleus

Fiber sorting in the optic chiasm means that the left LGN maps the right visual field. The primate LGN has six layers: two magnocellular layers get input from M ganglion cells and contain movement sensitive cells; four parvocellular layers are innervated by P ganglion cells, and contain wavelength selective cells. Each layer gets a precise retinotopic input from just one eye. The properties of LGN cells are similar to those of the ganglion cells which supply them and have circular receptive fields with surround inhibition. The LGN projects to the primary visual cortex. The LGN has a role in visual attention.

Primary visual cortex

The primary visual cortex (V1) gets a retinotopic projection from the LGN, with the fovea occupying a disproportionately big area. The M and P LGN cells project to different sublayers in layer 4C of the cortex, giving rise to distinct streams of information flow through the cortex. Most cells in V1 have elongated receptive fields and respond to linear features rather than spots. Simple cells are position sensitive. Complex cells are less sensitive to position than simple cells, many having a preference for linear stimuli moving at right angles to the long axis of their receptive fields.

Orientation columns

All cells lying within radial columns extending through V1 respond to linear features having approximately the same orientation. All orientations are represented for each point on the retina. These orientation columns are ordered so that a smooth gradient for orientation exists; columns with the same orientation are aligned in stripes across the cortex.

Binocular cells

Many cells in V1 get input from both eyes, but most are preferentially driven by one eye. These cells occur in discrete ocular dominance columns that are aligned in stripes across the cortex in which ipsilateral and contralateral eye dominance alternates. Binocular cells get input from corresponding positions on the two retinas and measure retinal disparity which allows binocular depth perception.

Hypercolumns

The volume of cortex in which every orientation is mapped for corresponding positions on both retinas is called a hypercolumn. It consists of a complete set of orientation columns and ocular dominance columns for a single pixel of the visual field.

Related topics

(A4) Organization of the central nervous system
(G1) Attributes of vision

(G2) Eye and visual pathways
(G6) Parallel processing in the visual system

Lateral geniculate nucleus

Pathways for visual perception start with the retinogeniculate fibers, axons of ganglion cells that end in the lateral geniculate nucleus (LGN). Because of the manner in which fibers are sorted in the optic chiasm, the left optic tract and left LGN carry axons from the left side of both retinas. Thus the left LGN represents the right side of the visual field (Figure 2 in Section G2).

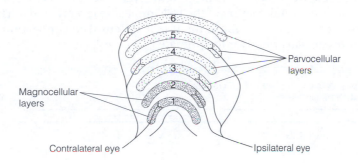

Figure 1. The structure and inputs of the lateral geniculate nucleus.

The primate LGN has six layers (Figure 1). The two most ventral are the **magnocellular** (large-cell) **layers** which receive input from M ganglion cells and are concerned with motion. Dorsal to these are the four **parvocellular** (small-cell) **layers** that are innervated by P ganglion cells and encode form and color. Interleaved between these major layers are **koniocellular layers** containing small **K cells**. These receive input from K ganglion cells and usually have small receptive fields. There are several classes of K cells. Some get input from blue-on K ganglion cells and relay to the primary visual cortex (V1). Others are motion sensitive and bypass V1, projecting to the medial temporal cortex which is also the major target for M cells conveying motion information.

LGN cells have circular RFs with surround antagonism. They show little or no response to diffuse light covering the whole receptive field. Each layer in the LGN gets input from only one eye and no cells show binocular responses (responses to both eyes). It is not until the visual cortex is reached that input from both eyes is combined. The responses of the LGN cells match those of the ganglion cells which supply them, so on and off channels remain independent and P cells display precisely the same color opponency properties as retinal ganglion cells. There is a very precise topographic (retinotopic) mapping in perfect register onto each of the layers of the LGN, with the fovea taking up about half of the space.

Responses of LGN cells are smaller than can be explained by classic center–surround processing. The responses are suppressed by the retina and possibly by inhibitory interneurons in the LGN itself. This suppression may act to allow the wide range of contrasts seen in the natural world to be accommodated by the limited dynamic range of neurons.

LGN output is via **geniculostriate neurons** that project to the primary visual cortex (V1). Only about 10% of the synapses on these are from retinal ganglion cells. Other synapses are made by back projections from V1, the superior colliculus (to koniocellular layers), the parabrachial nucleus, and the thalamic reticular nucleus. These are probably involved visual attention; that is, in selecting which retinal input is transmitted through to the visual cortex.

Primary visual cortex

The fibers of the optic tract terminate in the **striate cortex** (Brodmann's area 17) on the medial surface of the tip of the occipital lobe. This region is the **primary visual cortex (V1)**. Precise retinotopic mapping is maintained up to V1 with the fovea having a disproportionately large representation.

There are at least three parallel streams of information into the primary visual cortex. The movement sensitive M LGN cells input into layer 4Cα, the P LGN cells go to layer 4Cβ, whereas the koniocellular layers of the LGN project to layers 2 and 3. These streams remain quasi-independent throughout the visual system. The connections of the primary visual cortex are illustrated for the primate in Figure 2.

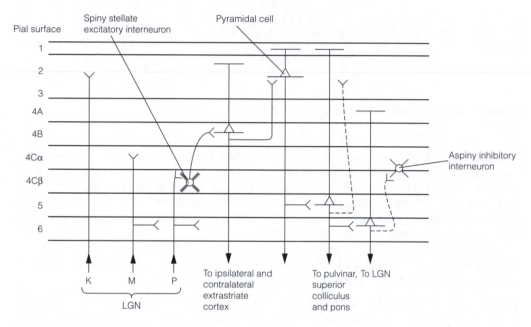

Figure 2. Canonical circuitry in the primary visual cortex illustrated for parvocellular (P) LGN input. Magnocellular (M) circuitry (not illustrated) has its input to 4Cα. Spiny stellate cells here send axons to pyramidal cells in 4B and these send collaterals to pyramidal cells in layers 5 and 6 directly rather than via more superficial layers. Koniocellular input is directly to pyramidal cells in blobs of 2 + 3. Feedback collaterals are dashed. Cortical layers are designated by Arabic numbers.

The great majority of cells in V1 have elongated receptive fields (RFs) with both inhibitory and excitatory regions, and respond to bars, slits, edges, and corners rather than spots of light. Most fall into two categories based on their RF properties: simple or complex cells. Both are orientation selective, in that they respond to linear features in only a narrow range of orientations:

- **Simple cells** are pyramidal cells found mostly in layers 4 and 6. They are highly sensitive to the position of a stimulus on the retina. They have small oval RFs with center–surround antagonism (Figure 3). A simple cell gets its input from a linear array of LGN cells having the same RF properties, so the RF of the simple cell emerges as a consequence of the RFs of the LGN inputs.

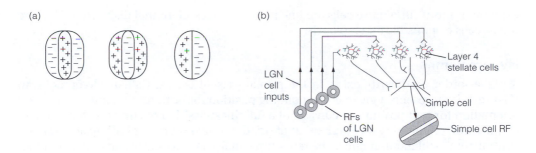

Figure 3. (a) Receptive fields of three simple cells; (b) a diagram depicting how lateral geniculate nucleus (LGN) cells contribute to the simple cell receptive field (RF), four on-center LGN cell RFs generate an on-center simple cell RF.

- **Complex cells** are most abundant in layers 2, 3, and 5. They have larger RFs than simple cells and, lacking distinct inhibitory or excitatory regions, a stimulus of the appropriate orientation anywhere in the RF evokes a response. Hence, complex cells are much less fussy about position than simple cells. Many complex cells show a preference for movement at right angles to the long axis of the stimulus. Some complex cells receive their inputs from simple cells but others get their input directly from the LGN.

Orientation columns

In common with other sensory cortex, the primary visual cortex is divided into radial columns 30–100 µm across. In each of these, all cells respond preferentially to linear features with a given orientation so they are called **orientation columns**. The cortex is organized so that adjacent columns have an orientation preference that differs by only about 15°; in other words, orientation is represented in a systematic way across the cortex. Columns which have the same orientation are arranged in stripes across the cortex. The obvious inference, that orientation selectivity is how the visual system represents straight line segments which can be built up to give the form of an image, need not be true. Computer modeling shows that orientation selectivity is a property of neural networks that learn the curvature of curved surfaces from their shading. Hence orientation selectivity might, counter intuitively, be concerned with representations of curves rather than linear features in the visual world.

Binocular cells

V1 is the first region in which input from both eyes is combined. Many cells, particularly in layers 4B, 2, and 3 show binocular responses in that they can be driven by either eye. This is a necessary condition for stereopsis. Most **binocular cells** show a preference for one eye, a phenomenon referred to as **ocular dominance**. Cells which have the same ocular dominance (e.g., those that are driven preferentially by the ipsilateral eye) occupy **ocular dominance columns** that are situated in long stripes about 500 µm across. Columns representing ipsilateral and contralateral input alternate regularly over the cortex and, when visualized at the level of layer 4C, look like the pattern of stripes on a zebra.

The receptive fields of binocularly driven cells resemble those of simple or complex cells, lie in corresponding positions in the two retinas, have identical orientation properties and have similar arrangements of excitatory and inhibitory regions. Similar input from both eyes into arrays of binocular cells is needed for perception of a fused image. To the extent that inputs into these cells are unequal they measure retinal disparity and so are the V1 cells responsible for stereopsis.

Hypercolumns

A higher-order modularity exists in the primary visual cortex. Called a **hypercolumn** (Figure 4) it represents a given corresponding position for both retinas, and maps every orientation for that position. It consists of a full thickness slab of cortex with an area of about 1 mm² containing a complete set of orientation columns for both ipsilateral and contralateral ocular dominance. The retinotopic map in V1 occurs because adjacent pixels of the retina map to adjacent columns in an orderly fashion.

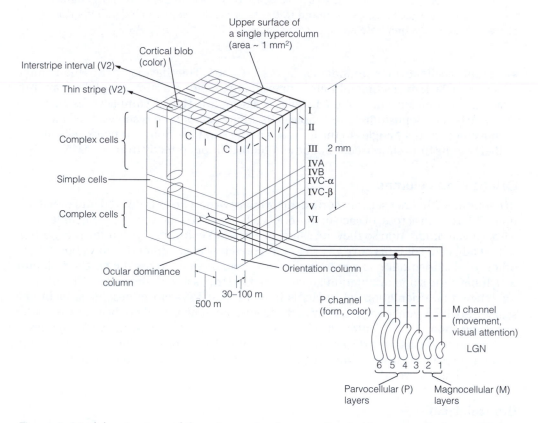

Figure 4. Modular structure of the primary visual cortex. Cortical layers are designated by Roman numerals. I, ipsilateral; C, contralateral (blobs are described in Section G6).

G6 Parallel processing in the visual system

Key Notes

Parallel processing in V1

There are three relatively independent pathways for processing information in the visual system. The magnocellular (M) pathway gets its input from M LGN cells, is color blind, and involved in analysis of moving stimuli, control of gaze, and stereopsis. The parvocellular (P) systems arise from P LGN cells and bifurcates into two systems. The parvocellular–interblob pathway is concerned with form, its cells are orientation selective and binocular. The parvocellular–blob pathway mediates color vision; wavelength-selective cells in this pathway have double opponent receptive fields in which the center is excited by some cones and inhibited by others, whilst the converse situation occurs in the surround.

Extrastriate visual cortex

All cortical areas involved in vision, other than V1, are together referred to as extrastriate visual cortex. It includes regions of the occipital, parietal, and temporal cortex. The secondary visual cortex (V2) receives input from V1 and projects to other extrastriate cortex. The three streams of visual information remain segregated in V2. The M pathway goes via thick stripes in V2 to V3 and then V5. Destruction of human V5 causes a loss of ability to see objects in motion. The parvocellular–interblob pathway goes via V2 interstripes to V3 and V4, whilst the parvocellular–blob pathway goes from V2 thin stripes to V4, cells of which show color constancy. Destruction of human V4 causes loss of color vision.

Where and what streams

Beyond V5 and V4 information is divided into two streams. From V5 a dorsal stream to the medial superior temporal cortex and the posterior parietal cortex is concerned with object location. The ventral stream from V4 to the inferotemporal cortex is concerned with object recognition. The two streams are called "where" and "what" streams respectively.

Related topics (G1) Attributes of vision (G5) Early visual processing

Parallel processing in V1

Three relatively independent pathways, each of which processes different aspects of vision in parallel can be delineated in the primary visual cortex.

The **magnocellular pathway** from M ganglion cells to M LGN cells has its input to spiny stellate cells in layer 4Cα (Figure 2 in Section G5). These excitatory interneurons synapse with pyramidal cells in layer 4B which show orientation and direction selectivity. These cells send axon collaterals to pyramidal cells in layers 5 and 6. Layer 5 cells project to subcortical regions, the pulvinar (a thalamic nucleus involved in visual attention), the superior colliculus, and pons. Layer 6 pyramidal cells go to the extrastriate cortex. The M pathway is specialized for analysis of motion. Its outputs via layer 5 are important in visual attention and gaze reflexes. Some cells in the M pathway are binocular so it contributes to stereopsis. Because it originates with ganglion cells which combine input from two classes of cone cell it is not wavelength selective; the M system is color blind.

There are two **parvocellular pathways**. They arise from P ganglion cells via P LGN cells which synapse with spiny stellate cells in 4Cβ. Like the M pathway the interneurons connect with pyramidal cells in 4B. However, in the parvocellular paths, 4B cells (which are orientation-selective simple cells) synapse with pyramidal cells in layers 2 and 3 which then relay with deep pyramidal cells in layer 5. Segregation of the two parvocellular pathways occurs in layers 2 and 3. When stained for the mitochondrial enzyme, **cytochrome oxidase**, layers 2 and 3 show pillars of high activity, **blobs**. Each blob is centered on an ocular dominance column. Between the blobs lies the **interblob** region. Cells in the interblob region are orientation selective, binocularly driven, complex cells. They are not wavelength selective or motion sensitive. They are part of the **parvocellular–interblob** (PI) pathway which processes high resolution analysis of form in the visual world. By contrast cells in the blobs are wavelength selective, show poor orientation selectivity and are monocular. The **parvocellular–blob** (PB) pathway mediates color vision. Blob pyramidal cells get direct input from blue-on K cells in the koniocellular LGN layers, but the function of this input is not yet understood.

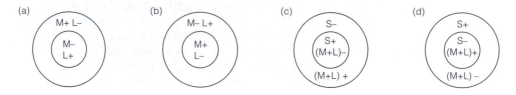

Figure 1. Double opponent cells in V1 blobs. Preferred stimuli: (a) red spot, green surround; (b) green spot, red surround; (c) blue spot, yellow surround; (d) yellow spot, blue surround.

Wavelength selective blob cells are **double opponent cells** with receptive field (RF) properties derived from their inputs, the single opponent parvocellular LGN cells. Double opponent cells have center–surround antagonist RF configuration, they signal color contrast and come in four classes categorized by their preferred stimuli. The left cell in Figure 1 is excited by L cones in the center and inhibited by L cones in the surround. In addition, it is inhibited by M cones centrally but excited by M cones in the surround. The preferred stimulus for this cell is a red spot in a green background. However, the cell gives **off-responses** if exposed to a green spot in a red background (Figure 2).

Unlike single opponent cells which are excited by small spots of white light, double opponent cells are unaffected by white light stimuli of any size, so they are more selective detectors of color contrast. The organization of double opponent cell RFs explains some of the properties of color vision described in Section G1:

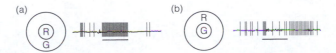

Figure 2. Responses of the double opponent cell in Figure 1(a): (a) preferred stimulus; (b) off response which might account for successive color contrast (see text for details).

- **Color constancy**. The way in which the brain computes color constancy is not understood in detail but is partly accounted for by the behavior of double opponent cells. A shift in the wavelength composition of light will produce equal but opposite effects on the responses of the center and surround of double opponent cells. There will be little effect overall on the RF of the cell which will continue to signal the same color. On the scale of the entire visual field, color constancy is thought to involve comparing red–green brightness, blue–yellow brightness (from color single opponent cells), and total brightness (added outputs of S, M, and L cones) over large areas of retina.

- **Perceptual cancellation** is explained by the way in which color opponency happens to be organized as red (R) versus green (G) and yellow (R^+G) versus blue channels. Since mutual antagonism occurs between red and green or between yellow and blue only one color in each pair can be seen at a single pixel of the retina at any time.

- **Simultaneous color contrast** can also be accounted for by the properties of double opponent cells. For example, the cell in Figure 2 cannot discriminate between a green stimulus to its surround or a red stimulus to the center; the response is the same for both. So a gray disc viewed in a green background is interpreted as red. A similar mechanism explains the complementary after-images that appear after staring at a uniform patch of color (Figure 2b).

Extrastriate visual cortex

The segregation of visual information for motion, form, and color in V1 is maintained in the **extrastriate visual cortex**, which is a term applied to all of the visual cortex except V1. The extrastriate cortex of primates contains about 30 regions that can be differentiated on the grounds of cytoarchitecture, connections, and physiological properties. Most have a retinotopic map of some aspect of the visual world. It includes not only occipital cortex areas 18 and 19 but also areas of parietal and temporal cortex. The location and connections of the major visual cortical areas are depicted in Figure 3a and 3b respectively.

Most outputs from V1 go to **V2**, the **secondary visual cortex**, which occupies part of area 18. V2 shows a characteristic cytochrome oxidase staining pattern, alternating thick and thin stripes running at right angles to the V1/V2 border. Pathway tracing techniques and electrophysiological studies reveal how the magnocellular and parvocellular pathways continue into V2 and beyond.

Cells in the V2 thick stripes are motion sensitive and binocular, being driven by a preferential retinal disparity. The thick stripe receives inputs from layer 4 of the interblob region of V1 and sends much of its output, via **V3**, to the **medial temporal (MT) visual cortex, V5**. Human V5 lesions result in loss of the ability to perceive motion. Hence, the V2 thick stripe–V3–V5 (MT) connection is the extension of the magnocellular pathway (Figure 3b), and concerned with motion and depth perception.

The interstripe region of V2 gets its inputs from the V1 interblob regions (layers 2 and 3) and sends outputs to V3 and then to **V4**. Many cells in V3 and V4 are orientation selective.

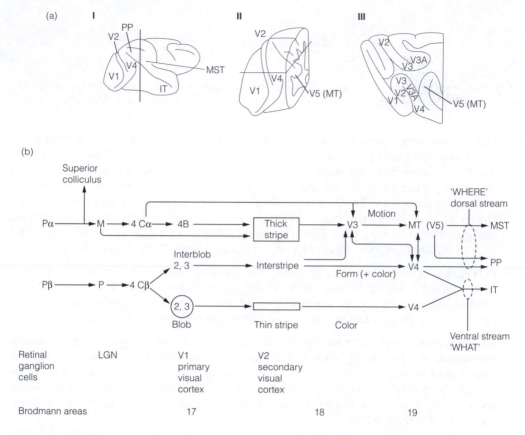

Figure 3. Parallel processing in the visual system. (a) Anatomy of the visual areas in the macaque: (i) left cerebral hemisphere; (ii) coronal section through the posterior third of the hemisphere; (iii) horizontal section. Adapted from Kandel ER, Schwartz J & Jessell T (eds) (1991) *Principles of Neural Science*, 3rd edn. With permission from McGraw-Hill. (b) Flow diagram of M and P channels in the primate visual system. IT, inferotemporal cortex; MST, medial superior temporal cortex; MT, medial temporal cortex; PP, posterior parietal cortex.

Hence this route is a continuation of the PI parvocellular pathway concerned mainly with form perception.

The blobs of V1 project to the thin stripe of V2 which in turn sends outputs to visual area V4. That this route is the extension of the PB parvocellular pathway for color vision is supported by the loss of color vision that occurs in patients with damage to human V4.

The M, PI, and PB pathways are not completely independent. Reciprocal pathways exist between V3 and V4, and V5 and V4 which presumably allow interactions between M and PI systems, both of which contribute to stereopsis. Interaction of motion and form analysis is probably required for the identification of moving objects. There seems to be no cross talk, however, between M and PB pathways. The M system is color blind and for equiluminant stimuli (those varying in color but not in brightness), which can only be perceived by the PB system, the perception of motion vanishes. Moreover, although the PI system uses color contrast to localize borders, form information is not available to the PB pathway: when looking at equiluminant blocks of color they appear to "jump around" because the PB system cannot localize boundaries.

During saccades the M, but not the P system, is shut down. Thus the M (motion) system is not confused by the rapid eye movements. The response times of the P system are sufficiently slow that they are unaffected by the shifting image.

Where and what streams

Parallel processing beyond V5 and V4 results from the segregation of information into two streams. The dorsal stream, largely from MT, goes to the **medial superior temporal (MST)** and **posterior parietal (PP) cortex**. PP cells show selectivity for size and orientation of objects, their firing depends on where an animal is looking and they fire as an animal grasps an object. Lesions to MST and PP in primates results in **optic ataxia**, in which visuospatial tasks are profoundly affected, but object recognition is unaffected.

By contrast, the ventral stream from V4 to the **inferotemporal cortex (IT)** is crucial for object recognition. IT cells are sensitive to form and color but are relatively unfussy about object size, retinal position or orientation. Many respond selectively to specific objects such as hands or faces. Unusually for visual cortex, the IT area has no retinotopic map. Lesions of the IT cortex cause **visual agnosia** in which animals fail at tasks requiring object recognition. Visuospatial tasks are unaffected. Because of their different functions the dorsal and ventral streams are often referred to as **where** and **what** streams respectively. Clinical data suggests that a similar dichotomy exists in humans.

Bilateral loss of V1 causes total loss of visual perception. However, there are primates and humans with this damage who are able to avoid obstacles whilst moving through space much better than by chance. This phenomenon is called **blindsight**. Humans with it report that they are completely unaware of the visual world and do not understand how they are able to navigate through space. Blindsight may be mediated by a pathway that goes directly from the magnocellular LGN to the thick stripe of V2. This provides input to the where system. The implication of this condition is that V1 is required for conscious visual perception.

G7 Oculomotor control and visual attention

Key Notes

Eye movements

Eye movements either keep the gaze fixed on an object when the head is turning, or shift the gaze to follow a moving object. Gaze fixation is brought about by the vestibulo-ocular reflex, which relies on the inner ear, and the optokinetic reflex which depends on visual input. Gaze shift can be produced either by saccades (fast), smooth pursuit (slow), or vergence.

Extraocular eye muscle control

The actions of three pairs of muscles allow the eyes to be rotated about three principle axes. During conjugate eye movements in which both eyes move in the same direction, eye muscle activity in one eye is complementary to eye muscle activity in the other. The extraocular muscles are controlled by motor neurons in the nuclei of the oculomotor, trochlear, and abducens nerves, which in turn are driven by brainstem reticular and medial vestibular nuclei.

Gaze stabilization

Rotation of the head, detected by the semicircular ducts, causes a well-matched opposite rotation of the eyes to keep the retinal image stationary, the vestibulo-ocular reflex. For large horizontal head rotations, once the eyes have rotated as far as possible they are rapidly reset to frontwards gaze—nystagmus. Vestibulo-ocular reflexes adapt in response to alterations in visual input. Slow head rotation causes images to move across the retina. This triggers eye movement in the opposite direction, an optokinetic reflex.

Saccades

Fast movements taking the fovea to a new point in visual space are saccades, produced reflexly by visual, auditory, or somatosensory stimuli. Burst cells in brainstem reticular nuclei are the direct drivers for saccades but they are produced by the superior colliculus and frontal eye fields (FEF). The superior colliculus has sensory and motor maps so each point represents a location in sensory space and specifies the saccades necessary to direct gaze towards it. The FEF can generate saccades via the superior colliculus or by other pathways.

Smooth pursuit and vergence

These are used to voluntarily track an object that is moving in the visual field. The velocity of the object is signaled by the "where" system to neurons in the pons. Signals for vergence include blurring of the retinal image or the degree of accommodation and require the visual cortex. Fast vergence movements are made during saccades.

Visual attention	Attention selectively filters out irrelevant information in favor of the relevant, which may be consciously perceived. Attention involves increasing the signal-to-noise ratio of neurons encoding particular features and is achieved, for example, by increased firing, greater synchronicity in firing, and temporary alterations to receptive fields. The circuitry of visual attention is coupled to that involved in oculomotor control and includes pathways from FEF to the LGN (via V4 and V1) and from the thalamic reticular nucleus to the LGN. Here attentional control is exerted at an early stage of visual processing. In addition the pulvinar of the thalamus exerts attentional control at a later stage by widespread modulation of cortical inputs. By synchronizing oscillations of widely distributed cortical neurons that encode different visual features the pulvinar may be responsible for binding these into a single percept.	
Related topics	(F5) Balance	(K5) Cerebellar function
	(G2) Eye and visual pathways	(K7) Basal ganglia function
	(K1) Cortical control of voluntary movement	

Eye movements

Eye movements either stabilize gaze by keeping the eyes fixated on an object during head rotation, or shift gaze so that the fovea is brought to bear on an object, or track a moving object. Five types of eye movements, each controlled by a distinct neural system, bring about these aims.

Gaze stabilization is controlled by the vestibulo-ocular and optokinetic systems. Rapid head rotation, detected by the semi-circular ducts provides input for vestibulo-ocular reflexes (VOR), whereas optokinetic reflexes depend on visual input to monitor slow head rotations. For both systems their output causes conjugate eye movements in the opposite direction to the head rotation, so that retinal images do not shift.

Three systems organize **gaze shift**. The saccadic system generates extremely rapid eye movements, saccades, which move the gaze from one point in the visual field to another, bringing new targets onto the fovea. The smooth pursuit system permits gaze to follow a moving target, so that its image remains on the fovea. Finally, for animals with binocular vision, the vergence system allows the eyes to move in opposite directions (disjunctive movements); either both converge or both diverge, so that both eyes can remain directed towards an object as it gets closer or recedes.

The output of all five eye motor systems is via oculomotor neurons in the brainstem, the axons of which run in three pairs of cranial nerves to the skeletal muscles that move the eyes.

Extraocular eye muscle control

Each eye is moved by three pairs of **extraocular eye muscles**. Two pairs of rectus muscles (superior and inferior, medial and lateral) originate from a common annular tendon attached at the back of the orbit. These muscles insert into the sclera in front of the equa-tor of the eyeball. The third pair are the oblique muscles (superior and inferior) which insert into the sclera behind the equator of the eyeball (Figure1).

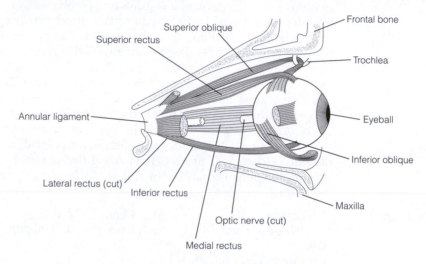

Figure 1. The right orbit showing the extraocular muscles.

Working in concert these muscles act to rotate the eye about three principal axes (Figure 2). The actions of the medial and lateral rectus muscles are simple. They cause the eye to rotate about the vertical axis so that the gaze moves horizontally. The medial rectus brings about rotation towards the midline (adduction) while the lateral rectus causes lat-eral rotation (abduction). The other two pairs of muscles produce rotations that have components along two of the principal axes, and the components change depending on the horizontal position of the eye. These actions are summarized in Table 1.

Eye muscles act in complementary fashion in the two eyes during conjugate movements in which the two visual axes move in parallel. Thus, contraction of the lateral rectus in one eye is coupled with contraction of the medial rectus in the other eye for a conjugate horizontal shift in gaze (Table 1).

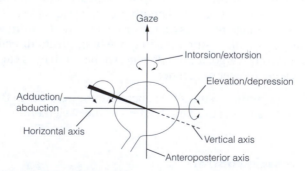

Figure 2. Principal axis for rotation of the eye, shown for the right eye. In health, torsional movements (rotation about the anteroposterior axis) are small.

Table 1. Actions of extraocular eye muscles*

Muscle	Innervation	Movement	Contralateral eye complementary muscle
Lateral rectus	Abducens (VI)	Abduction	Medial rectus
Medial rectus	Oculomotor (III)	Adduction	Lateral rectus
Superior rectus	Oculomotor (III)	Elevation and intorsion	Inferior oblique
Inferior rectus	Oculomotor (III)	Depression and extorsion	Superior oblique
Inferior oblique	Oculomotor (III)	Extorsion and elevation	Superior rectus
Superior oblique	Trochlear (IV)	Intorsion and depression	Inferior rectus

* For the muscles moving the eye vertically the action is different depending on whether the eye is also abducted or adducted. For example, the superior rectus elevates the eye if the lateral rectus is active at the same time, but causes intorsion if the eye is adducted by the medial rectus.

The extraocular muscles are innervated by motor neurons in the nuclei of the oculomotor (III), trochlear (IV), or abducens (VI) cranial nerves. These neurons are the final common path for the output of all five eye movement systems and are driven by brainstem reticular and medial vestibular nuclei axons that run in the **medial longitudinal fasciculus**. Eye motor neurons fire both statically, in a manner relating to eye position, and dynamically, reflecting eye velocity. To hold the eye steady in a given position requires tonic discharge by a particular set of motor neurons. The set will be different for different positions.

Eye movements are brought about by pulses of action potentials with a firing frequency directly proportional to the velocity of the movement. This brings the eye to a new position, which needs the generation of a new position signal. This is probably achieved by integration of the velocity signal by the vestibulocerebellum and prepositus nucleus of the brainstem reticular system.

Gaze stabilization

Head rotation detected by the semi-circular ducts triggers equal and opposite rotation of both eyes, a **vestibulo-ocular reflex** (**VOR**). For large head rotations the eyes cannot continue to rotate but must be reset to a central position by rapidly moving in the *same* direction as the head. This gives rise to **nystagmus**, eye movements characterized by a slow phase that stabilize the retinal image, and quick phase that resets the eyes. By convention the direction of the nystagmus is the direction of the quick phase (Figure 3).

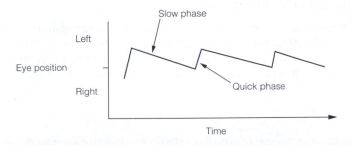

Figure 3. Leftward nystagmus during head rotation.

The horizontal semi-circular ducts are effectively wired to the medial and lateral rectus muscles to produce the eye movements that counter the head rotation (Figure 4).

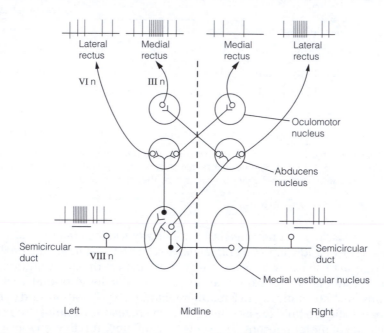

Figure 4. Circuitry for the vestibulo-ocular reflex. Stimulation of the horizontal semi-circular ducts by the leftward rotation of the head excites the ipsilateral medial rectus and the contralateral lateral rectus and inhibits their antagonists. Excitatory neurons, open circles; inhibitory neurons, filled circles. Firing patterns are shown for cranial nerve neurons.

The gain of the VOR (the magnitude of the eye rotation divided by the magnitude of the head rotation) is quite close to one for fast head rotations. This means there is a good match between eye and head movements which makes for a stable retinal image. The VOR can be modified by visual experience. When human subjects wear magnifying lenses, which means that they should make bigger eye movements to match head rotations, the gain of their vestibulo-ocular reflexes increases appropriately over the next few days. The cerebellum is required for this adaptation to occur, but not for it to be maintained once established. The instability of the retinal image acts as an error signal which is relayed to the cerebellum from the inferior olivary nucleus. The cerebellum learns to minimize the error and alters its drive to the extraocular muscles. This is an example of motor learning. Lesions of the vestibulocerebellum impair the ability to maintain steady gaze, causing inappropriate nystagmus.

Slow rotation of the head causes an apparent movement of the visual world in the opposite direction called **retinal slip**, This is detected by large, movement sensitive retinal ganglion cells and used to produce eye movements which are equal in speed but of opposite direction to the retinal slip. This is the **optokinetic reflex**. As in the VOR, nystagmus occurs for large head rotations.

Saccades

Saccades are fast conjugate eye movements that move the fovea to target a different point in visual space. The saccade system uses visual, auditory, and somatosensory input to determine the eye rotation required to realign the gaze. Horizontal saccades are controlled by the **paramedian pontine reticular formation** (**PPRF**) while vertical saccades are organized by the **rostral interstitial nucleus** which lies in the medial longitudinal fasciculus. Both of these contain burst cells which code for the size and direction of the eye movement and are excitatory to oculomotor neurons.

The signals that trigger saccades come from the superior colliculus and the frontal eye fields of the frontal cortex. Each can generate saccades independently of the other.

The **superior colliculus** is divided into superficial, intermediate, and deep layers. The superficial layer gets information from the retina and visual cortex and has a map of the contralateral visual field. The deep layer gets auditory and somatosensory input and so has an auditory map of the location of sounds in space and a somatotopic map in which the body parts closest to the eyes gets the greatest representation. The intermediate layers have a motor map. Neurons here are called **collicular saccade-related burst neurons** because they fire a high frequency burst of action potentials about 20 ms before a saccade. Each one has a **movement field** (the equivalent of a receptive field) that covers the sizes and directions of the saccades it participates in. The direction of any given saccade is encoded by a population of neurons in exactly the way in which the primary motor cortex uses population coding to determine the direction of movements.

The superior colliculus turns sensory coordinates into motor coordinates. All of its four maps are in register so each point on the superior colliculus represents a specific location in sensory space and the saccades necessary to direct gaze towards it. However, visual input to the superficial layers does not automatically lead to firing of cells in the intermediate layer. This is because superficial layer cells relay through the **pulvinar** of the thalamus and the visual cortex before influencing the intermediate layer cells. The relay may function to allow saccades only to stimuli with high salience and hence is a visual attention mechanism. Figure 5 is a diagram of the circuitry involved in saccades.

In addition to generating saccades the superior colliculus causes head rotation, by way of the **tectospinal tract**, to neck muscle motor neurons. This allows orientation towards a stimulus, so called **orienting responses**.

The **frontal eye field** (**FEF**) triggers saccades directly by stimulating the intermediate layer burst cells in the superior colliculus, and operates the oculomotor basal ganglia-thalamocortical circuit to select appropriate eye movements.

Damage to the superior colliculus causes temporary impairment in producing saccades. Recovery occurs because the FEF can trigger saccades by its direct connections with the pons and midbrain. Damage to the FEF causes transient paralysis of gaze towards the opposite side, but reflex saccades soon return, produced by the superior colliculus. However, loss of the FEF prevents intentional saccades.

Smooth pursuit and vergence

Intentional tracking of a moving object so that its image remains on the fovea is done by the smooth pursuit system. Smooth pursuit movements differ from optokinetic reflexes in being voluntary, and in attending to movement over a small part of visual space. By contrast, optokinetic reflexes are involuntary and are responses to movement of the entire visual world.

Signals relating to the velocity (i.e., speed *and* direction) of the target are generated by the medial temporal cortex of the visual "where" system which analyzes motion. Lesions of this cortex impair pursuit movements. These signals are transmitted to the **dorsolateral**

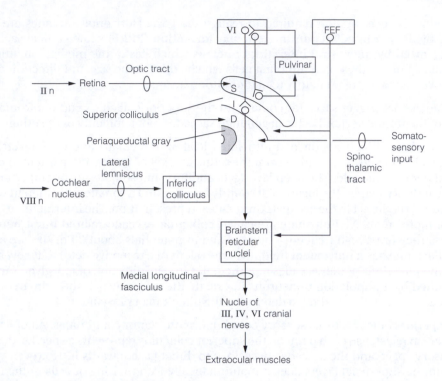

Figure 5. Circuitry for saccades. The superior colliculus has three layers: S, superficial; I, intermediate; D, deep. FEF, frontal eye fields; VI, primary visual cortex.

pontine nucleus (**DLPN**), which translates the target velocity into the motor commands for the pursuit movement. The DLPN projects to the vestibulocerebellum and its output to the medial vestibular nuclei drives the smooth pursuit movements.

Vergence is the only disjunctive eye movement. For example, shifting gaze to a closer target requires adduction of both eyes, which is achieved by contracting both medial rectus muscles. Signals to produce vergence include blurring of the retinal image by large degree of retinal disparity, the extent of accommodation, or monocular cues to distance. These all require the visual cortex. Fast vergence movements occur during saccades.

Visual attention

Attention selectively filters out irrelevant information in favor of the relevant, which may be consciously perceived. When this happens to sensory information it is called **sensory attention** and this has been extensively studied in vision.

Visual stimuli can be attended to by moving the head or eyes, but attention to any stimuli can also be made without moving, by alterations to the internal state (**covert attention**). Covert visual attention shifts *independently* of eye movements. It can be studied in humans and primates by spatial cueing tasks.

Attentional mechanisms are thought to involve increasing the signal-to-noise ratio and increased gain (signal amplification) of those neurons that encode the selected representation. The neural correlates of this include increased firing frequency, greater synchronicity in firing or synaptic activity within neuron populations, reduced response variability, and transient modulation of receptive fields. These sorts of variable can be studied in conscious humans (by EEG and brain imaging) and by electrophysiology in awake, behaving animals.

For example, visual attention is accompanied by alterations in the receptive field properties of some visual neurons. During attentional tasks these cells become transiently biased to ignore distractor stimuli that would, in the non-attending state, elicit a response. It is as if in attention the RFs shrink to encompass only the target stimulus, so that the cell responds to it alone. This is a raising of the signal-to-noise ratio. Not all visual system neurons are modulated in visual attention. Only cells with large enough RFs to respond to several objects (target plus distractor) are involved. These are generally found in regions responsible for late rather than early visual processing.

However, receptive field properties place constraints on visual attention. For example, target objects that differ in apparent depth and movement from the surrounding distractors "pop out" because there are visual system neurons that encode both binocular disparity *and* movement. In contrast, cells that respond to both movement and color have never been identified and the visual system is presumably forced to analyze each object in turn for each of these stimulus features in order to identify the one with the target combination; this takes time.

The circuitry implicated in visual attention is closely related to that involved in oculomotor control (Figure 6). In monkeys, stimulation of the FEF at levels too low to produce saccades enhanced performance in an attention task. Similar effects have been found for the superior colliculus and areas of parietal cortex. Other studies show that activation of the FEF not only allows covert attention to the appropriate location in the visual field but also selects the appropriate saccades to fixate the target, so visual attention and oculomotor control are tightly linked. Several cortical areas generate attentional signals (V4, V2, V1, lateral prefrontal cortex) but the earliest activity in attentional tasks is seen in the FEF which implies that this is the source of attention-control signals.

Considerable evidence implicates three thalamic nuclei, the pulvinar, LGN, and thalamic reticular nucleus in visual attention. Electrophysiology on behaving monkeys, brain imaging, and lesions all support a role for the **pulvinar** in controlling visual attention at a cortical level. It gets input from the superior colliculus, LGN, and cortex, has topographic maps of the visual world and responds selectively to various aspects of visual stimuli including orientation, color, and motion, presumably because of its cortical connections. Motion-sensitive pulvinar neurons respond to movement in the real world but are silent for the apparent motion of a stimulus caused by eye movements because input from the superior colliculus allow them to subtract retinal signals produced by eye movements. The pulvinar connections with the cortex are widespread, include visual cortex, and are reciprocal. As a general rule if two areas of cortex are directly connected they are also *indirectly* connected via the pulvinar. In this way the pulvinar is thought to modulate the efficacy of direct inputs from one cortical area to another. This could be an attention mechanism.

However, the pulvinar may do more than this. It also controls intrinsic oscillations of cortical neurons and hence could synchronize firing of widely distributed cortical neurons representing different aspects (form, color, depth, motion) of the same object. This could be how the brain integrates information in parallel visual channels to produce a unified percept: a solution to the binding problem. Pulvinar lesions in humans can disturb binding of visual features.

Human brain imaging shows that visual attention enhances LGN responses to attended locations and attenuates those to ignored stimuli. There seem to be two routes by which this happens, with different functions:

- The **thalamic reticular nucleus** (**TRN**) has access to information from the prefrontal cortex, visual cortex, and superior colliculus, and makes retinotopic connections with the LGN. The TRN is inhibitory on the LGN and attention correlates with a decrease in this inhibition. This route seems to be responsible for an early LGN response, which is thought to be attention to a specific location in the visual field.

- Back projections from the FEF via V4 and V1 are retinotopic and enhance LGN responses. In this route, which is responsible for a later phase of attention (200 ms after a stimulus appears), P, M, and K streams are segregated so could allow specific features (e.g., form, motion) to be selected for attention even at the early LGN processing stage.

Hence the LGN acts as a "gatekeeper" selecting what retinal input is transmitted through to the visual cortex.

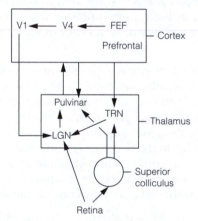

Figure 6. Circuitry implicated in visual attention. The pulvinar makes reciprocal connections with many cortical areas. FEF, frontal eye fields; LGN, lateral geniculate nucleus; TRN, thalamic reticular nucleus.

H1 Acoustics and audition

Key Notes

Sound waves	Sound is longitudinal pressure waves in a medium. The frequency of the wave is the pitch of the sound and given in Hertz (cycles per second).
Sound pressure amplitude	The change in pressure produced by a sound wave is the sound pressure amplitude (P). By comparing P with a reference amplitude at the threshold of human hearing a sound pressure level (SPL) can be calculated. The unit of SPL is the decibel. Differences in SPL are perceived as differences in the loudness of a sound. Loudness varies with frequency in a manner that is determined by the sensitivity of the ear.
Frequency response of human hearing	In the young the detectable frequency range is between 20 Hz and 20 kHz . Peak sensitivity (the ability to hear quiet sounds) and auditory acuity (the ability to discriminate tones) both lie between 1000 and 4000 Hz.
Related topics	(H2) Anatomy and physiology of the ear (H3) Peripheral auditory processing (H4) Central auditory processing

---

Sound waves

Sound is the oscillation of molecules or atoms in a compressible medium. The energy of the oscillations is transmitted as a longitudinal wave in which the medium is alternately compressed and rarefied, causing periodic variations in the pressure of the medium (Figure 1).

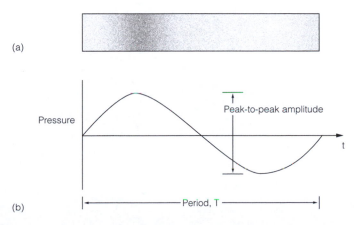

Figure 1. Sound waves: (a) density of air molecules during propagation of a longitudinal pressure wave; (b) sine wave representation of a pressure wave.

For a sine wave the **period**, T, is the time taken for one complete cycle. The **frequency** of the wave, the perceived **pitch** of the sound, is the reciprocal of the period (i.e., $f = 1/T$). The unit of frequency is the **Hertz** (**Hz**); one cycle per second.

Sound pressure amplitude

The amplitude of a sound wave is the total change in pressure that occurs during a single cycle. Because of the huge range in sound wave amplitudes, P, it is expressed in a logarithmic scale as a ratio of a reference pressure, P_{ref}.

$$\text{Sound pressure level (SPL)} = 20 \log_{10} P/P_{ref}$$

P_{ref} is 2×10^{-5} Pa, a sound pressure which is at the threshold of human hearing. The unit of SPL is the **decibel** (dB). Each 10-fold increase in SPL is equivalent to 20 dB. Sound pressure levels in excess of 100 dB can result in damage to hearing, and at 120 dB auditory pain results.

Differences in sound pressure level are perceived as differences in **loudness** which varies with frequency in a manner that is determined by the sensitivity of the ear.

Frequency response of human hearing

The **frequency response** of the human ear is from 20 Hz to 20 kHz optimally, but rapidly narrows with age, with most of the loss occurring at the higher frequencies. By 50 years the upper limit averages about 12 kHz. The highest **sensitivity** (the ability to detect quiet sounds) occurs at 1000–4000 Hz. This closely matches the frequency range of human speech, 250–4000 Hz. Greatest **auditory acuity** (the ability to discriminate between tones) is also found between 1000 and 4000 Hz.

H2 Anatomy and physiology of the ear

Key Notes

Middle ear

The middle ear converts pressure waves in the air to vibrations of perilymph in the inner ear. Sound waves striking the ear drum cause it to vibrate. This vibration is transmitted by three articulated middle ear bones, the malleus, incus, and stapes, to the oval window and so via the perilymph to the round window. A four-fold amplification of sound occurs across the middle ear. Two middle ear muscles, upon contraction, act on the middle ear bones to reduce sound transmission, affording protection against sustained loud sounds.

Inner ear anatomy

The auditory inner ear is the cochlea, a bony canal arranged in a coil. Within it lies the cochlea duct, a part of the membranous labyrinth, which divides the cochlea cross section into three compartments. The cochlea duct itself is the scala media and contains endolymph. On either side lie the perilymph-containing scala vestibuli and scala tympani. These are continuous with each other at the apex of the cochlea so that vibrations of the oval window are propagated through the scala vestibuli to the scala tympani and then to the round window. Pressure waves moving through the perilymph cause the basilar membrane on the floor of the scala media to oscillate. Hair cells in the organ of Corti, which lies on the basilar membrane, have their stereocilia embedded in the tectorial membrane.

Cochlea function

Vibrations of the basilar membrane cause it to shear with respect to the tectorial membrane and bend the hair cell stereocilia to and fro. This results in periodic hair cell depolarization and hyperpolarization by the same transduction mechanism that operates in vestibular hair cells. The periodic changes in the release of transmitter from the hair cells alter the firing of the auditory primary afferents with which they synapse. Because the basilar membrane differs in width, mass, and stiffness systematically along its length, different frequencies of sound make the membrane vibrate maximally at different distances along it. This is the basis of pitch discrimination.

Related topics

(F5) Balance   (H1) Acoustics and audition

The middle ear

The function of the middle ear is to convert pressure waves in the air to vibrations of the perilymph in the inner ear. Sound waves pass along the **external auditory meatus** striking the **tympanic membrane** (ear drum) which resonates faithfully in response. The ear drum is critically damped in that it stops vibrating the instant the sound ceases. A sound at hearing threshold causes the ear drum to vibrate with an amplitude of about 0.01 nm—one tenth the diameter of a hydrogen atom! The movement of the ear drum is transferred with an overall efficiency of about 30% to the fluid in the inner ear by a lever system, composed of three **ear ossicles**, lying in the **tympanic cavity** (middle ear) (Figure 1).

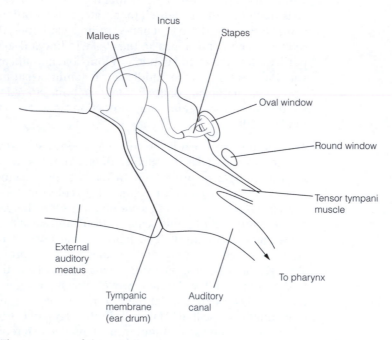

Figure 1. The anatomy of the middle ear.

The **malleus** (hammer) is fixed at its thin end (the handle) to the tympanic membrane. Its thick end (the head) articulates with the head of the **incus** (anvil) via a saddle-shaped joint. The long process of the incus makes a ball and socket joint with the head of the **stapes** (stirrup). The base of the stapes is attached by an annular ligament to the **oval window** (fenestra vestibuli). The malleus vibrates with the tympanic membrane. Inward movement locks the joint between the malleus and the incus, driving the long process of the incus inward, pushing the stapes in the same direction to exert a pressure on the perilymph beyond the oval window. This pressure wave is transmitted through the perilymph to cause a compensatory bulge of the **round window** (fenestra cochleae). Outward movement reverses these motions. Since the area of the oval window is 20 times smaller than the tympanic membrane, the pressure (force per unit area) at the oval window is proportionally greater. This is important because perilymph is incompressible and so must be driven to vibrate *en masse*. This needs more force than it takes to transmit sound waves through air. In addition it results in an amplification of the sound by about 20 dB, corresponding to a four-fold increase in loudness, by the middle ear.

There are two middle ear muscles, the **tensor tympani** and **stapedius**. When they contract together the handle of the malleus and the tympanic membrane are pulled inwards and the base of the stapes is pulled away from the oval window. This reduces sound transmission by 20 dB, especially for low frequencies. Reflex contraction of these muscles in response to loud noise may prevent damage to the inner ear but since the reaction time is 40–60 ms this **tympanic reflex** affords no protection against *brief* loud sounds. The **auditory canal** connects the middle ear to the pharynx which allows the air pressure to be equalized when changing altitude.

Inner ear anatomy

The auditory part of the inner ear is the **cochlea**, a bony canal 3.5 cm long, which spirals two and three quarter turns around a central pillar, the **modiolus**. Within the cochlea lies a tubular extension of the membranous labyrinth, the **cochlear duct**, attached to the modiolus and the outer wall of the cochlea. This divides the cochlea into three compartments, the **scala media** which contains endolymph, and the **scala vestibuli** and the **scala tympani** which contain perilymph and are continuous with each other via a small gap known as the **helicotrema** situated at the apex of the cochlea where the cochlear duct ends blindly (Figure 2).

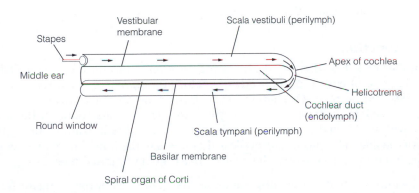

Figure 2. The cochlea, depicted unfurled. Arrows show the direction of propagation of sound waves through the perilymph.

Pressure waves generated at the oval window are propagated through the scala vestibuli into the scala tympani and to the round window where the energy dissipates. During their passage the pressure waves cause oscillations of the **basilar membrane**, the floor of the scala media on which rests the sensory apparatus, the **spiral organ of Corti** (Figure 3).

The spiral organ is a narrow sheet of columnar epithelium running the length of the cochlear duct. The epithelium contains sensory hair cells resembling those in the vestibular apparatus. A single row of 3500 **inner hair cells** form **ribbon synapses** with myelinated axons of large bipolar cells (type I) in the **spiral ganglion** of the cochlear nerve. Each inner hair cell is innervated by about 10 such axons, a large degree of divergence. There are about 12 000 **outer hair cells** arranged in three rows. These are innervated by an unmyelinated axon from small bipolar cells (type II) in the spiral ganglion, each of which synapses with 10 hair cells, representing considerable convergence.

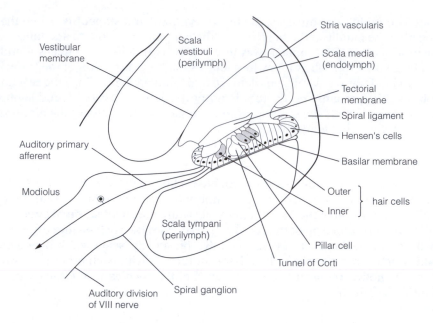

Figure 3. Transverse section through the cochlea showing the organ of Corti.

Cochlea function

Cochlea hair cells lose their kinocilia during development and the tips of their tallest stereocilia are embedded in the overlying **tectorial membrane**, a matrix of mucopoly-saccharides and proteins. Oscillations of the basilar membrane in response to a sound stimulus cause it to shear with respect to the tectorial membrane, bending the stereocilia first one way and then the other. This results in periodic depolarization and hyperpolarization of the hair cells, producing cyclical alterations in the tonic secretion of glutamate. The transduction mechanism for hair cells is like that of vestibular hair cells.

A sound stimulus causes a **traveling wave** (like that generated by twitching the free end of a rope fixed at its other end) to spread along the basilar membrane from base to the apex. High frequencies cause vibration at the basal end whereas low frequencies cause vibration towards the apex. This frequency sorting is a result of the continuous variation in the width, mass, and stiffness of the basilar membrane along its length. The basilar membrane is narrow (50 µm) and stiff at the base, wider (500 µm) and less stiff at the apex. The relationship between frequency and length is logarithmic. At a given frequency, increasing the SPL increases the amplitude of the vibration and the length of basilar membrane responding.

Outer hair cells (OHCs) contract in a voltage-dependent manner. Depolarization causes them to shorten. The speed with which they change length is so fast that they are able to follow the high frequency voltage changes produced by sound stimuli. By this means OHCs augment the vibrations of the basilar membrane, a process called **cochlear ampli-fication**. It probably contributes to the high sensitivity and fine tuning to frequency exhibited by the basilar membrane, since these features are lost when OHCs are selectively damaged by **aminoglycoside antibiotics** such as streptomycin. Cochlear amplification causes vibrations of perilymph that are transmitted to the oval window across the middle ear in the "wrong" direction to the tympanic membrane which now acts as a loudspeaker producing inaudible **otoacoustic emissions**. These are not necessary for normal audition but provide insights into ear function.

H3 Peripheral auditory processing

Key Notes	
Primary auditory afferents	Primary auditory afferents have their cell bodies in the spiral ganglion and send their central axons to the pons via the auditory (VIII cranial) nerve. Auditory afferents fire tonically and increase firing in response to a tone. Most have their highest sensitivity for a narrow range of frequencies.
Coding of sound frequency	Frequency is coded in two ways. For frequencies above 3000 Hz the frequency response of an afferent depends on where along the basilar membrane it is from. This is tonotopic mapping. For lower frequencies, afferents fire during a particular phase of a sound wave. This is called phase locking. A population of afferents encodes the entire waveform.
Coding of sound level	An auditory afferent responds only to a limited range of sound pressure levels. The full range is encoded by afferents with different dynamic ranges. Afferents that display the highest rate of spontaneous firing are the most sensitive. Efferents from the superior olivary complex make inhibitory synapses with outer hair cells, reducing the sensitivity of auditory afferents with increasing sound pressure levels.
Related topics	(H1) Acoustics and audition (H4) Central auditory processing

Primary auditory afferents

The primary afferents have their cell bodies in the **spiral ganglion** located in the modiolus. Their centrally directed axons project through the vestibulocochlear (VIII) nerve to synapse in the cochlear nuclei of the lower pons. In humans, about 30 000 type I afferents from inner hair cells provide the bulk of the output from the cochlea. Three quarters of the hair cells (the OHCs) send their output to only about 3000 type II afferents. The nature of type II cell signaling is unknown. Auditory afferents fire tonically.

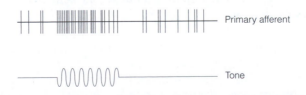

Figure 1. Firing of auditory primary afferent (upper trace) in response to a tone (lower trace).

In response to a tone, type I afferents show an increase in firing which adapts. When the sound stops, firing ceases for a brief period. Hence they exhibit both dynamic and static responses (Figure 1). Responses of type I afferents plotted as **tuning curves** (Figure 2) show that they are sharply tuned at low sound pressure levels. The frequency to which the unit is most sensitive is the **characteristic frequency** (**CF**). At high SPLs the primary afferents respond to a much wider frequency range.

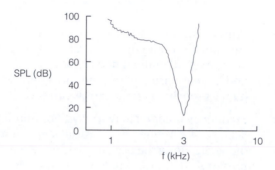

Figure 2. Tuning curve for a type I cochlear afferent. The plot shows the minimum SPL required to evoke a response over a range of frequencies.

Coding of sound frequency

The nervous system has two ways of encoding the frequencies of a sound. **Tonotopic mapping**, in which frequency maps to position along the cochlea, occurs because afferents with successively lower CFs are found closer to its apex. This is most important for frequencies above 1–3 kHz. For lower frequencies coding uses the property that afferents fire with greatest probability during a particular phase of a sound wave, **phase-locking**. It is only necessary that an individual afferent fires during some cycles if a group of cells is involved. Moreover, if different groups phase-lock onto different parts of the cycle then a whole population of cells acting in concert can encode frequency.

Coding of sound level

Auditory afferents have a **dynamic range** of about 30 dB, beyond which further increase in SPL has no additional effect. The full range of SPL (0–100 dB) is signaled by afferents with different sensitivities. Cells with the same CF may differ in threshold SPL by 70 dB. Afferent sensitivity correlates with its spontaneous rate (SR) of firing. High SR cells have the greatest sensitivity. Low SR cells are least sensitive and encode the frequencies of loud sounds.

The sensitivities of afferents can be modified by efferents which have their cell bodies in the **superior olivary complex** (**SOC**) and form inhibitory synapses on OHCs. These neurons alter the gain of the cochlear amplifier, reducing the sensitivity of type I afferents as sound pressure levels increase.

H4 Central auditory processing

Key Notes

Central auditory pathways

Primary auditory afferents terminate in the cochlear nuclei in the pons. Ventral cochlear axons go to the superior olivary nucleus on both sides. This structure projects to the nuclei of the lateral lemniscus, and is primarily concerned with localizing the direction of a sound source. The dorsal cochlear nucleus projects directly to the contralateral nucleus of the lateral lemniscus. The nuclei of the lateral lemniscus sends axons to the inferior colliculus, which in turn projects to the medial geniculate nucleus (MGN). The auditory radiation which originates in the MGN goes to the primary auditory cortex, and is responsible for conscious sound perception. Although the largest auditory pathway is contralateral, extensive connections across the midline ensures interactions between sides.

Cochlear nuclei

Distinct cell types within the cochlear nuclei are able to process different features of a sound stimulus. Bushy cells signal exact timing information to the superior olivary nucleus which, by comparing input from both ears, is able to localize sound. Stellate cells signal sound level. Many cells are finely tuned to particular frequencies and show lateral inhibition which sharpens this tuning.

Tonotopic mapping

Maps in which frequency is represented in a systematic way occur in all auditory structures. In the primary auditory cortex isofrequency columns are found perpendicular to the cortical surface. These are arranged in bands which form an ordered tonotopic map.

Sound level

Cells responding to differences in sound level are found throughout the auditory system. Some are finely tuned to a characteristic sound level. In humans there are no maps of sound level.

Localization of sound

Elevation of a sound source is signaled by the delay caused by sound waves being reflected from the pinna. The azimuth of a sound source is measured in two ways. For higher frequencies, the difference in sound level between the ear nearest and that furthest from the sound source is computed by neurons in the lateral superior olivary nucleus. This projects to the tectum which controls eye and head reflexes in response to sound. For lower frequencies, cells in the medial superior olivary nucleus compute the phase difference that occurs because sound entering the ear furthest from the source is slightly delayed. In the auditory cortex most cells respond preferentially to input in the contralateral ear and are either excited or inhibited by ipsilateral input. Sound localization is broadly encoded. Cortical regions beyond the auditory cortex encode movement of sound sources.

Related topics	(E2) Coding of modality and location	(H1) Acoustics and audition
	(G) Vision	(H3) Peripheral auditory processing

Central auditory pathways

Primary auditory afferents bifurcate to terminate in both the ventral and dorsal **cochlear nuclei**. From the ventral cochlear nucleus axons run to the **superior olivary nucleus (SON)** on both sides and to the contralateral inferior colliculus via the **trapezoid body (TB)**. The SON compares input from the two ears to compute the whereabouts of a sound source. It projects to the nuclei of the lateral lemniscus. The dorsal cochlear nucleus sends axons directly to the contralateral nucleus of the lateral lemniscus (Figure 1).

The nucleus of the **lateral lemniscus** projects to the **inferior colliculus (IC)**. The IC relays with the **medial geniculate nucleus (MGN)** of the thalamus which sends its output via the **auditory radiation** to the **primary auditory cortex, AI** (Brodmann's areas 41 and 42) located in the superior temporal gyrus. This is the pathway for conscious auditory

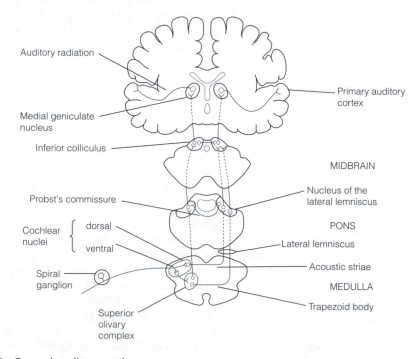

Figure 1. Central auditory pathways.

perception. The biggest auditory pathway is contralateral. However, reciprocal connections between the nuclei of the lateral lemniscus (via Probst's commissure) and between the ICs, ensure extensive interactions between the input from both ears.

The ascending sensory pathways are matched by descending projections which modify the sensitivity of cochlear afferents.

Cochlear nuclei

Central auditory pathways process three features of sound input in parallel; tone, loudness, and timing. From the last two the brain calculates the location of the sound in space. Parallel processing begins in the cochlear nuclei.

Several neuron types are present in the cochlear nuclei that can be distinguished by their shape and responses. Numerous **bushy cells** reproduce the firing pattern of the primary afferents faithfully, including phase-locking. Their output, which precisely signals the timing of sound, goes to the superior olivary nucleus which compares the input from both ears to compute the location of a sound source. By contrast **stellate cells** have a much greater dynamic range than bushy cells and signal sound level. Hence, timing and sound level are processed in parallel.

Receptive fields of auditory neurons are called **response maps** and are plotted in the same way as the primary afferent tuning curve. Five classes of cell can be distinguished in the cochlear nuclei on the basis of their receptive fields (RFs). Type I cells have a purely excitable RF that precisely matches primary afferent tuning curves, but all the other types have inhibitory responses which arise by lateral inhibition and which fine tune their frequency response (Figure 2). Type IV cell axons are the main output of the dorsal cochlear nucleus.

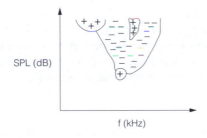

SPL (dB)

f (kHz)

Figure 2. Response map of a type IV cell in the dorsal cochlear nucleus. Excitatory region, +; inhibitory region, −.

Tonotopic mapping

Tonotopic maps are found in the cochlear nuclei, superior olivary nucleus, inferior colliculus, and auditory cortex. Some structures have multiple maps. The cochlear nucleus is divided into isofrequency strips, each containing cells with similar characteristic frequencies (CFs). Strips representing increasingly higher frequencies are found progressively more posteriorly. In the primary auditory cortex, **isofrequency columns** running through the entire thickness of the cortex are arranged in isofrequency strips running mediolaterally, with low frequencies represented rostrally and high frequencies caudally (Figure 3). There are at least three other tonotopic maps in the auditory cortex. Adjacent maps are always mirror images of each other. In humans there is no great over-representation of particular frequencies.

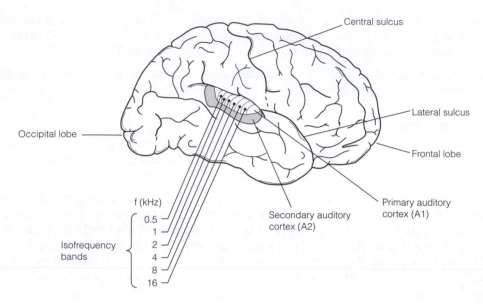

Figure 3. Tonotopic map of the human primary auditory cortex.

Some regions of auditory cortex, for example, **secondary auditory cortex (AII)**, are less well tonotopically organized and contain cells that respond to a wider range of frequencies.

Sound level

Cells throughout the auditory system respond to differences in sound level and fall into two broad classes. Monotonic cells have sigmoid plots of sound level against firing rates. Non-monotonic cells are more finely tuned with a maximum firing rate at a characteristic sound level. There are no maps of sound level in humans.

Localization of sound

The ability to localize the source of a sound in space is very important in avoiding danger. The coordinates of a sound source in vertical and horizontal planes are **elevation** and **azimuth** respectively. Different mechanisms are involved in determining these two coordinates.

The **pinna** (outer ear) and **ear canal** act as direction selective filters for locating sound elevation. Sound waves enter the ear either directly or reflected by pinna and ear canal and hence will arrive at slightly different times at the ear drum. The delay times will depend on the elevation of the sound source and the peculiarities of the external auditory system.

The superior olivary complex uses two methods to localize sound in the horizontal plane. Both compare input into the two ears (**binaural sound localization**) and allow azimuth to be pinpointed with a precision of about one degree of arc:

• **Interaural level differences (ILDs)**. If the head is orientated so that one ear is closer to the sound source, then the head forms a shadow which reduces the sound level entering the other ear. ILDs are unambiguous for frequencies > 1600 Hz for which the dimensions of the head are greater than the length of the sound waves. ILDs as low as one dB can be detected.

Neurons of the **lateral superior olivary nucleus** (**LSO**) have a tonotopic map restricted largely to high frequency input. These cells receive inputs from both ipsilateral and contralateral cochlear nuclei. However, the contralateral route is by way of a glycinergic inhibitory neuron. Equal sound level in both ears causes overall inhibition of the LSO neuron and increasing the sound level in the contralateral only serves to augment the inhibition. However, increased sound level to the ipsilateral ear causes LSO firing. Maximum firing rate is seen when ILD is 2 dB or more. Corresponding cells in the opposite LSO will show reverse responses to the same sound. The LSO projects to the ventromedial part of the IC central nucleus. The IC connects extensively with the deep layers of the **superior colliculus** to form an **auditory space map** in register with the retinotopic map. Hence the superior colliculus is implicated in the auditory reflexes organizing gaze and head rotation towards the sound source.

- **Interaural time differences** (**ITDs**). A sound wave enters the closer ear slightly earlier than the further one. For low frequencies (< 800 Hz) this results in a time delay less than one period which is analyzed by neurons capable of phase-locking. At higher frequencies input into the furthest ear is delayed by more than a single period, and this makes phase-locking unreliable, so ITDs cannot provide an unambiguous cue to location except by detecting delays in sound onset or offset (**group delays**). ITDs as short as 20 µs can be detected.

The neural system for measuring ITDs depends on cells in the **medial superior olivary nucleus** (**MSO**) acting as coincidence detectors. The MSO has inputs from bushy cells in both cochlear nuclei that phase-lock in response to low frequency stimuli. If a phase difference exists between the two ears then the bushy cells corresponding to the furthest ear fire slightly later. The MSO circuitry for transforming this timing difference to azimuth has been worked out for the owl, but the mammalian circuit is different and not yet understood.

Most cells in A1 are binaural and fall into two groups of cortical columns. Those in **summation columns** show bigger responses to input from both ears than from one. By contrast, cells in **suppression columns** have a preference for input from one ear (Figure 4). Summation and suppression columns are arrayed alternately and at right angles to isofrequency strips. The location of a sound source seems to be encoded by a broad cortical channel in each hemisphere, rather than numerous channels tuned to discrete spatial positions.

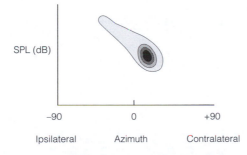

Figure 4. Location tuned neuron in a suppression column of the auditory cortex. Higher density of shading corresponds to a greater firing rate.

The motion of a sound source can be analyzed from rises and falls in sound pressure levels. Rising intensity activates a neural network responsible for space recognition, sound motion, and attention, which lies in the superior temporal lobe (planum temporale), right temporoparietal junction, right motor and premotor cortices, the left cerebellar cortex, and part of the midbrain. The special significance of rising intensity is that it results from approaching (i.e., potentially threatening) sources in natural environments.

I1 Olfactory receptor neurons

Key Notes

Olfactory epithelium	Bipolar olfactory receptor neurons lie in the olfactory epithelium. Their dendrites extend to the surface of the epithelium to form a cluster of olfactory cilia. Their axons form the olfactory (cranial I) nerve and synapse with neurons in the olfactory bulb.
Olfactory transduction	Odor molecules are detected by odorant receptors in the olfactory cilia. The receptors are a family of about 1000 G-protein-coupled receptors most of which are positively coupled to the cAMP second messenger system. Cyclic AMP opens a cyclic nucleotide-gated cation channel, resulting in a depolarizing generator potential that is graded according to the concentration of odorant. Each olfactory sensory neuron expresses just a single species of odorant receptor and each odorant receptor binds a range of related molecules.
Related topics	(D2) G-protein-coupled receptors (I3) Taste

Olfactory epithelium

A sense of smell allows us to detect spoilt food and noxious gases, motivates eating, and may play a part in sexual attractiveness. The olfactory epithelium lies in the dorsal nasal cavity. It consists of bipolar **olfactory receptor neurons** (ORNs) and supporting cells. A dendrite emerges from the ORN and extends to the surface of the epithelium to form a cluster of 6–12 immobile **olfactory cilia**. These extend into a layer of mucus secreted by the supporting cells. The centrally directed unmyelinated axon of the ORN runs through the olfactory (cranial I) nerve to synapse with cells in the **olfactory bulb**. Human olfactory epithelium contains about 10^8 ORNs.

Olfactory transduction

Odor molecules are usually small ($M_r < 200$ Da), lipid soluble, and volatile. Initially they bind to **odor-binding proteins** in the mucus which concentrate the odor molecules in the vicinity of the cilia. Odor molecules are recognized by **odorant receptors** in the cilia plasma membrane. These are G-protein-coupled receptors, and around 1000 have been identified in mammals. A given odor molecule can bind two to six different odorant receptors. Odorant receptors are relatively nonspecific, binding a range of related odor molecules, so although an individual ORN expresses just a single subtype of odorant receptor it responds to several odors. The mammalian nervous system is able to discriminate some 10 000 distinct odors on the basis of precisely which array of odorant receptors (and hence sensory neurons) are stimulated, and with what relative intensities.

Odorant receptors are coupled to $\mathbf{G_{olf}}$ proteins (related to G_s proteins) which usually stimulate adenylyl cyclase, and binding of an odor molecule causes a rise in cAMP in about

50 ms. This activates a cyclic-nucleotide-gated (CNG) channel, a nonspecific cation conductance permeable to Na$^+$, K$^+$, and Ca^{2+} ions (Figure 1). The resulting depolarizing generator potential is graded with an amplitude that signals the concentration of the odor molecule. However, maximal response is produced by the opening of only a small fraction (3–4%) of the CNG channels available. This means that the concentration range that can be signaled by firing of an ORN is narrow, about a 10-fold difference.

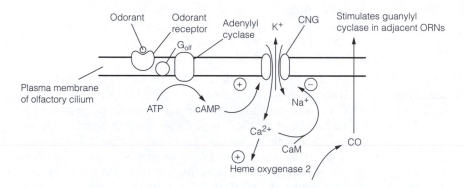

Figure 1. Olfactory transduction in olfactory receptor neurons mediated by receptors coupled to the cAMP second messenger system. CaM, calmodulin; CNG, cyclic-nucleotide-gated channel.

High odor concentration or prolonged exposure allows a high Ca^{2+} influx through the CNG channels. This ion has a number of modulatory effects in olfactory receptor neurons. Ca^{2+} activates heme oxygenase 2, an enzyme that synthesizes carbon monoxide (CO) which can activate **guanylyl cyclase** (GC) as shown in Figure 1. Because Ca^{2+} also inhibits GC, there is no overall activation of the cyclase in the target ORN. However, CO is freely diffusible so it can activate GC in adjacent *unstimulated* ORNs, producing cyclic guanosine monophosphate (cGMP) which binds to and opens the CNG channels. In this way odorant excitation spreads to a cluster of ORNs. Since neighboring ORNs respond to the same odors this increases sensitivity without loss of specificity.

ORNs show **adaptation** to protracted stimulation. Ca^{2+} binds to calmodulin (CaM) which can then bind to CNG channels, reducing the efficacy with which the cyclic nucleotides can open them. Hence Ca^{2+} attenuates the size of the generator potential.

I2 Olfactory pathways

Key Notes

Olfactory bulb

Olfactory receptor neurons synapse with mitral cells or tufted cells (M/T) and periglomerular cells within glomeruli in the olfactory bulb. Glomeruli are odor specific; each one receives terminals from receptor neurons that respond to the same set of odors. Odor discrimination is maximized by lateral inhibition which suppresses glomeruli responding to slightly different odorants.

Central olfactory connections

M/T neurons project to the olfactory cortex via the olfactory tract. Olfactory cortex is three-layered palaeocortex; the only cortex to receive sensory input directly rather than by way of the thalamus. Cortical projections to the hypothalamus and amygdala are important for emotional and motivational aspects of odors, a pathway to the hippocampus is concerned with smell memory, and output via the thalamus to orbitofrontal cortex mediates the conscious perception of smell. Brainstem aminergic pathways alter olfaction in the context of the behavioral state.

Related topics

(I1) Olfactory receptor neurons

Olfactory bulb

The axons of the ORNs run in the olfactory nerve to make excitatory synapses on the dendrites of **mitral cells** or **tufted cells** (M/T) and short axon inhibitory **periglomerular cells** in the olfactory bulb. M/T cells send their axons into the olfactory tract. Synapses between ORNs and M/T and periglomerular cells are found in **olfactory glomeruli**, spherical zones some 150 µm across. The olfactory bulb contains about 2000 glomeruli, each receiving the terminals of 25000 ORNs that respond to the same odors. Hence, glomeruli are odor-specific functional units (Figure 1). Low concentrations of a given odor molecule activate cells in the single glomerulus which gets input from the ORNs bearing odorant receptors with the highest affinity for the molecule. At higher concentrations, cells in other glomeruli are activated as their ORN odorant receptors' low-affinity binding sites for the molecule are occupied. Each glomerulus has dendrites from about 75 M/T cells. The M/T cells integrate weak inputs from a large number of ORNs within a glomerulus to generate a strong signal.

Lateral inhibition dampens responses from glomeruli with slightly different odor specificities so as to heighten odor discrimination. This is brought about by reciprocal dendrodendritic synapses between M/T cells and inhibitory interneurons termed **granule cells**. Via these synapses, M/T cells excite granule cells, which then inhibit the same, and adjacent, M/T cells.

There is a topographical organization to the fibers of the olfactory nerve and their projections to the olfactory bulb. Thin strips of olfactory epithelium running in an

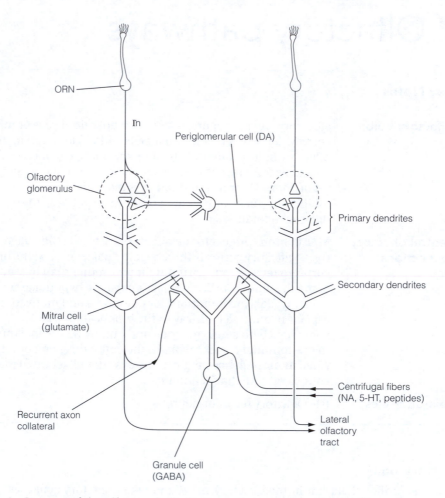

ORN

In

Periglomerular cell (DA)

Olfactory
glomerulus

Primary dendrites

Secondary dendrites

Mitral cell
(glutamate)

Centrifugal fibers
(NA, 5-HT, peptides)

Recurrent axon
collateral

Lateral
olfactory
tract

Granule cell
(GABA)

Figure 1. Circuitry of the olfactory bulb. Reciprocal synapses are denoted by ↔.
Neurotransmitters used by specific cell types are shown in parentheses: DA, dopamine;
GABA, γ-aminobutyric acid; 5-HT, serotonin; NA, noradrenaline (norepinephrine).

anteroposterior direction go to neighboring glomeruli. A given odor excites a particular
array of glomeruli across the olfactory bulbs, an **odor image**. The higher the concentration of the odor molecule the bigger the area activated.

Central olfactory connections

M/T cell axons project via the **olfactory tract** (OT) to the **olfactory cortex**. This cortex is unusual in two respects. Firstly, it is **palaeocortex** (old cortex) having only three
layers. Secondly, it is the only cortex to receive sensory input directly rather than via the
thalamus.

There are five regions of the olfactory cortex with distinct connections and functions but
all receive input from the olfactory tract. The **anterior olfactory nucleus** project axons
across the midline in the **anterior commissure** to the contralateral olfactory bulb (Figure
2). The **anterior perforated substance** (**olfactory tubercle**) sends output to the posterior
hypothalamus. This pathway, together with that to the **corticomedial amygdala**, which

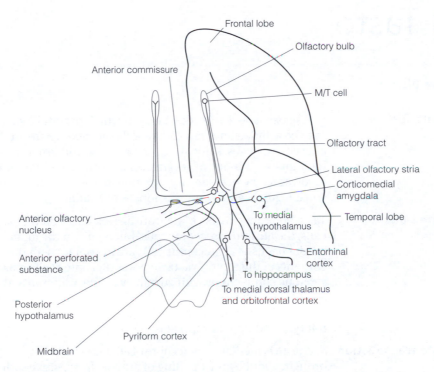

Figure 2. Connections of the left olfactory cortex viewed from below. M/T, mitral and tufted cells.

then projects to the medial hypothalamus, is concerned with the affective and motivational aspects of odors. A pathway from the olfactory tract to the entorhinal cortex, which sends its entire output to the hippocampus, presumably encodes olfactory components of episodic memories.

The **pyriform cortex** is concerned with olfactory discrimination. It projects, via the **medial dorsal thalamus**, to the **orbitofrontal cortex** which mediates the conscious perception of smell.

Olfactory processing is subject to considerable modulation. The olfactory bulb receives inputs from noradrenergic and serotonergic neurons in the brainstem and cholinergic neurons in the forebrain. In addition the anterior perforated substance receives a projection from the brainstem dopaminergic system. These various inputs are implicated in olfactory learning and modifying olfaction on the basis of appetite and arousal. Thus, mitral cell responses to food odors are modulated depending on whether an animal is hungry or sated.

I3 Taste

Key Notes

Gustation

The sense of taste provides a way in which harmful foods may be avoided and nutritious foods selected. There are at least five tastes; salty, sweet, sour, bitter, and umami (glutamate). It is one of several senses (others include smell) involved in the oral sensory experience of eating. Taste and smell regulate autonomic responses to feeding.

Taste buds

Taste buds are clusters of neuron-like excitable epithelial cells, the gustatory receptor cells. Microvilli on the apical border of gustatory receptor cells are in contact with the contents of the mouth via taste pores. Receptor cells make synaptic connections with gustatory primary afferents, the axons of which travel through the VII, IX, or X cranial nerves. Taste buds are found not only in the tongue but also in the pharynx and upper esophagus.

Taste transduction

Salt taste is mediated by receptor cell depolarization brought about by the opening of amiloride-sensitive sodium channels. Hydrogen ions (sour taste) produce depolarization by blocking voltage-dependent K^+ channels. The sensation of sweetness involves the taste molecule binding to a G-protein-coupled receptor that usually activates the cAMP second messenger system. This causes depolarization by closing a potassium channel. There are multiple transduction pathways for bitter molecules but all result in receptor cell depolarization. Umami is mediated by a metabotropic glutamate receptor.

Related topics

(D2) G-protein-coupled receptors

Gustation

The sense of taste, **gustation**, provides a means of avoiding potentially noxious food-stuffs or selecting for foods which have a high energy content. Five tastes are well defined—salty, sour, sweet, bitter, and umami (due to monosodium glutamate)—on the basis that no cross adaptation occurs between them. Plant alkaloids, some of which are toxic in high concentrations, are extremely bitter. A sour taste may signify a food degraded by microbiological action. By contrast a sweet food has a high content of sugars and so a readily available supply of metabolic energy. The sensory experience produced by having food in the mouth is called **flavor perception**, and relies on several sensory modalities. Apart from smell and taste, information about food texture is provided by mechanoreceptors and proprioceptors in the mouth and jaw innervated by trigeminal afferents. Flavor perception is important in triggering or modifying autonomic responses to feeding, for example, salivation, gastric secretion, and changes to gastrointestinal motility.

Taste buds

Gustatory receptor cells are epithelial cells but are excitable. They are organized into small clusters of 50–100 which, with supporting cells, form **taste buds**. Taste buds are located in the epithelium of the tongue, palate, pharynx, epiglottis, and the upper part of the esophagus. In the tongue they are present in small projections, **papillae**.

Microvilli on the apical border of each receptor cell project through **taste pores** in the gustatory epithelium, bringing them into contact with the contents of the mouth. The microvilli are the taste organelles (Figure 1).

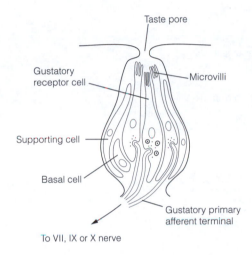

Figure 1. A taste bud.

Receptor cells form synaptic connections with primary gustatory afferents, each of which branches to synapse with receptor cells in more than one taste bud. The primary afferent axons from tongue, palate, and pharynx enter the facial (VII) and glossopharyngeal (IX) nerves. The few afferent in the epiglottis and esophagus are innervated by the vagus (X) nerve.

Taste transduction

Many of the ions or molecules responsible for taste sensation are hydrophilic and freely diffusible. Those which are hydrophobic include plant alkaloids which may bind to proteins in the saliva, equivalent to odorant-binding proteins, for presentation to gustatory receptor cells. Transduction involves changes in membrane conductance which causes a depolarizing generator potential, triggering action potentials and release of neurotransmitter which excites the gustatory primary afferents.

Salt taste is caused by Na^+ ions. Salt transduction (Figure 2a) occurs by the influx of Na^+ through an **amiloride-sensitive Na^+ channel**.

H^+ ions responsible for sour (acid) sensation causes a generator potential by blocking voltage-dependent K^+ channels in the apical membrane which at rest carries an outward, hyperpolarizing current.

Sugars, some amino acids and some proteins produce sweet sensations by interacting with G-protein-linked receptors coupled to second messengers. Sugars activate adenylyl

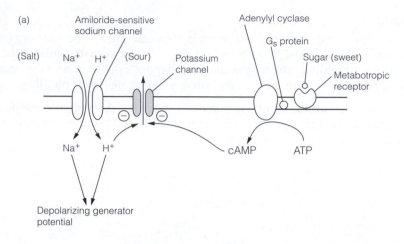

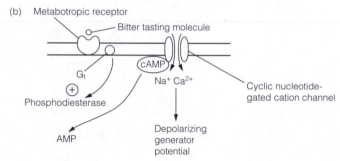

Figure 2. Taste transduction: (a) salt, sour, and sweet transduction mechanisms, note that an amiloride-sensitive Na$^+$ channel is implicated in both salt and sour taste; (b) one of several mechanisms involved in bitter transduction.

cyclase and the consequent rise in cAMP produces depolarization by closing a K$^+$ channel. Some compounds responsible for sweetness (e.g., artificial sweeteners) increase inositol trisphosphate (IP$_3$) levels and mobilize Ca^{2+} within receptor cells.

Multiple pathways mediate bitter taste transduction (Figure 2b). This reflects the wide diversity of molecules that are bitter flavored. Divalent salts and quinine block K$^+$ channels and so depolarize by reducing an outward potassium current. In a mechanism with striking parallels to phototransduction, some bitter tasting agents bind metabotropic receptors coupled to transducin (G$_t$) activating a phosphodiesterase that hydrolyzes cAMP. The fall in cAMP concentration causes cAMP to dissociate from a cyclic-nucleotide-gated (CNG) cation conductance, allowing influx of Na$^+$ and Ca^{2+} and so depolarization.

The umami taste sensation produced by L-glutamate seems to involve metabotropic glutamate (mGluR$_4$) receptors coupled via G$_i$ proteins to the inhibition of adenylyl cyclase.

14 Taste pathways

Key Notes

Anatomy of gustatory pathways	Gustatory primary afferent cell bodies lie in the ganglia of either the VII, IX, or X cranial nerves. Their axons terminate in the nucleus of the solitary tract (NST) in the medulla. Some NST cells project to the lateral hypothalamus for autonomic responses to feeding, others project via the thalamus to the ipsilateral cortex taste area I in the lateral sulcus for conscious taste perception. Taste area II in the insula is involved in emotional responses to taste.
Gustatory coding	Gustatory afferents are broadly tuned. Those in the facial nerve respond best to salt or sweet stimuli, those in the glossopharyngeal respond preferentially to sour and bitter stimuli, while vagal afferents measure the extent to which the ionic concentration differs from extracellular fluid. The classic taste sensations do not correspond to separate labeled lines, and are generated by opponent processing.
Related topics	(I3) Taste (G3) Photoreceptors

Anatomy of taste pathways

The gustatory primary afferents of cranial nerves, VII, IX, and X have their cell bodies in the **geniculate**, **petrosal**, and **nodose** ganglia respectively. Their centrally directed axons end in the rostral portion of the **nucleus of the solitary tract** (**NST**) which lies in the dorsal medulla (Figure 1). Taste primary afferents secrete glutamate and substance P.

Some NST cells project to the lateral hypothalamus which organizes autonomic responses to feeding. Other NST gustatory neurons project via the central tegmental tract to the ipsilateral **ventral posterior medial nucleus** (**VPMpc**) of the thalamus, terminating on a population of small cells distinct from those receiving somatosensory input from the tongue or pharynx. These cells send their axons to the ipsilateral cortex; unlike most sensory pathways, that for taste is uncrossed. Taste area I for conscious perception of taste is located in the lateral sulcus adjacent to the somatotopic mapping of the tongue. Taste area II is in the insula and is concerned with the affective aspects of taste.

Gustatory coding

Afferents in the VII nerve commonly exhibit preferences for either salty or sweet stimuli, whereas most of those in the IX nerve, supplied by the posterior tongue, are tuned to acids (sour) or bitter stimuli. Many vagal (X) afferents respond to distilled water. These neurons have their lowest firing rate in 154 mM NaCl, and increase firing as salt concentration increases or decreases from this value. Vagal afferents thus appear to measure to what extent the pharyngeal contents differ in ionic concentration from extracellular fluid.

The fact that gustatory neurons are generally quite nonspecific argues against the existence of labeled lines corresponding to the classical taste sensations. Hence distinctive

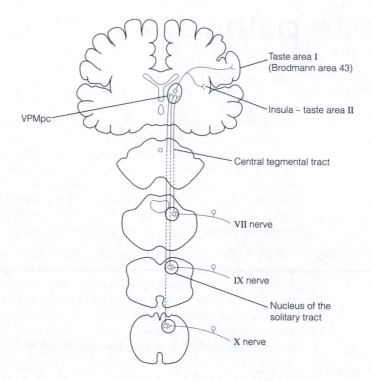

Figure 1. Central gustatory pathways. VPMpc, parvocellular part of the ventral posterior medial nucleus of the thalamus.

taste sensations arise from neurons with opponent receptive fields that compare the outputs of differently tuned populations of afferents. This is analogous to how color vision arises from opponent processing that compares output from just three populations of cone photoreceptors.

J1 Nerve–muscle synapse

Key Notes

Neuromuscular junction

The neuromuscular junction is the synapse between a motor neuron and a muscle fiber. Acetylcholine (ACh) released from the nerve terminal activates nicotinic cholinergic receptors (nAChR) in the postjunctional membrane, the endplate, causing it to depolarize. Spontaneous secretion of ACh from a single vesicle causes a miniature endplate potential (mepp) of 0.4 mV. Firing of the motor nerve releases ACh from hundreds of vesicles causing an endplate potential, a large depolarization that triggers muscle fiber action potentials and contraction.

Neuromuscular blocking agents

Muscle relaxants, which block neuromuscular transmission, are used to paralyze skeletal muscles during surgery. They fall into two categories. Nondepolarizing drugs are competitive antagonists of nAChR, and their effects can be reversed by acetylcholinesterase inhibitors which increase ACh concentrations in the cleft. Depolarizing drugs are nAChR agonists that produce blockade by inactivating muscle calcium channels and nAChR desensitization.

Related topics

(D1) Ligand-gated ion channel receptors

(J3) Elementary motor reflexes

Neuromuscular junction

The axon of a motor neuron divides into a number of branches at the surface of the muscle fiber. Each branch ends in a bouton which forms a synapse with the muscle fiber, called a **neuromuscular junction** (**nmj**). The cleft of the nmj (Figure 1) is about 50 nm across. The postjunctional membrane, the **endplate**, which is thrown into folds, has a high density of nicotinic acetylcholine receptors under the active zones where acetylcholine (ACh) is released. The cleft contains a collagenous basement membrane (**basal lamina**) to which is bound acetylcholinesterase (AChE). Soluble forms of the same enzyme are also secreted into the cleft.

Nicotinic acetylcholine receptors (nAChR) are ligand-gated ion channels that mediate fast ACh transmission. Binding of ACh opens the channel, allowing Na^+ influx and K^+ efflux. The reversal potential for this current is close to 0 mV, so activating nAChR causes depolarization.

Spontaneous release of a single quantum of ACh at the nmj causes a 0.4 mV depolarization at the endplate called a **miniature endplate potential** (**mepp**). The arrival of an action potential at the motor nerve terminal triggers the release of 200–300 quanta producing numerous mepps which sum to produce a depolarization to about −20 mV, an **endplate potential** (**epp**). This greatly exceeds the threshold for activating voltage-dependent

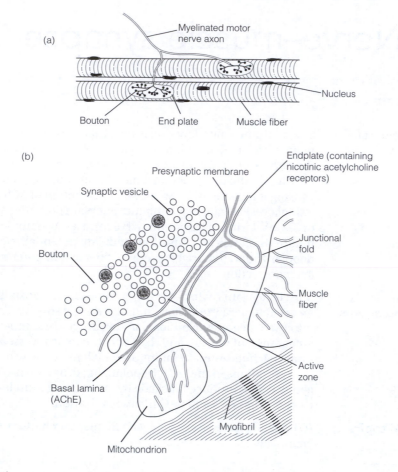

Figure 1. The neuromuscular junction: (a) a motor neuron forming synapses on two muscle fibers (×150); (b) a drawing of an electron micrograph of a neuromuscular junction.

sodium channels in the muscle membrane, so an action potential which is propagated over the muscle fiber membrane. The neuromuscular junction is unique among vertebrate synapses in that firing of the motor neuron almost invariably results in the triggering of muscle action potentials.

The concentration of ACh reaches 1 mM in the nmj within about 200 μs of the arrival of an action potential at the motor nerve terminal but within a millisecond or so the ACh concentration has fallen back to baseline levels because of the high activity of AChE in the cleft. The enzyme hydrolyzes ACh to choline and acetate. Choline is taken back into the nerve terminal via a Na$^+$-dependent transporter.

Neuromuscular blocking agents

Blockade of transmission at the nmj is used during surgery to produce relaxation of skeletal muscle. They are effective within one minute of injection. Muscle relaxant drugs fall into two categories depending on their mode of action and all have a structural resemblance to ACh.

Nondepolarizing drugs, such as the tetrahydroisoquinoline derivatives (e.g., the prototypical D-**tubocurarine**, atracurium) and the aminosteroids (e.g., pancuronium), are competitive antagonists of nAChR. The duration of action of these drugs ranges from 15 to 180 minutes. Their action can be rapidly reversed by AChE inhibitors, such as **neostigmine**, that cause a rise in the concentration of ACh which then out-competes the drug for the nicotinic receptor.

Depolarizing drugs, of which **succinylcholine** is the only agent of clinical importance, are nAChR agonists. Initially, agonist binding opens the nicotinic receptor channel causing persistent depolarization of the endplate. This first causes generalized disorganized contractions of muscles called **fasciculations**, and is followed by **flaccid paralysis** as muscle Ca^{2+} channels inactivate and the contraction mechanism fails as a result. This early stage in the action of depolarizing drugs (called **phase I** block) arises as a result of an ACh-like depolarization and so is augmented rather than reversed by AChE inhibitors. With continuing exposure **phase II** block occurs in which the nAChR either desensitizes, or suffers open channel blockade by the drug. Phase II block can be reversed by AChE inhibitors. Succinylcholine is rapidly hydrolyzed by circulating esterases, so its duration of action is only about 5 minutes.

J2 Motor units and motor pools

Key Notes

Motor unit

A motor unit consists of a motor neuron plus all the muscle fibers it innervates, which ranges from six to a few thousand. In mammals, each muscle fiber gets input from just one motor neuron. An action potential in a motor neuron causes a twitch, a single contraction, in all the fibers it supplies. At high firing rates individual twitches summate to produce tetanus, a prolonged maximal contraction. There are three types of motor unit. Slow twitch (S) units drive type 1 muscle fibers that are adapted for aerobic metabolism and capable of sustaining low forces for very long periods. These dominate in postural muscles. Fast twitch fibers (divided into fatigue resistant (FR) and fast fatigue (FF)) innervate type 2 muscle fibers and can produce large forces rapidly, but only for short periods.

Motor pools

A motor pool is the set of motor neurons that innervates a single muscle. The force of contraction of a muscle is determined by the firing frequencies of individual motor neurons, and by the number of motor neurons in the pool that are firing. Larger forces are generated by recruiting an increasing number of motor units. Recruitment generally (but not always) follows the size principle in which smaller motor neurons come on-line before larger ones. This gives the order S–FR–FF. Motor neurons are made hyperexcitable to glutamatergic input by monoaminergic neurons in the reticular system. This allows the generation of higher forces from the same input when arousal is high and during locomotion.

Related topics

(J1) Nerve–muscle synapse   (J3) Elementary motor reflexes

Motor units

A **motor unit** consists of a motor neuron and the muscle fibers it innervates. In mammals each muscle fiber is supplied by only one motor neuron. However, each motor neuron synapses with anything from six to a few thousand muscle fibers within a single muscle. The size of a motor unit is related to the precision of motor control required of a given muscle. Finely regulated muscles (e.g., extraocular eye muscles) consist of small motor units, less finely regulated muscles have larger ones. The fibers of a single unit are scattered widely throughout a muscle so no part of a muscle is controlled by just one motor unit.

A single action potential in the motor neuron causes a **twitch**, a single contraction, in all of the muscle fibers to which it is attached (Figure 1a). The contraction and relaxation of muscle fibers is very much longer than the muscle action potential of about 3 ms. If a volley of action potentials is fired and there is insufficient time for the muscle to relax between successive impulses the twitches summate to increase the force which oscillates about a plateau value. This is called **unfused tetanus** (Figure 1b). As the firing frequency increases the oscillations smooth out and the plateau reaches maximum force. This is **fused tetanus** (Figure 1c).

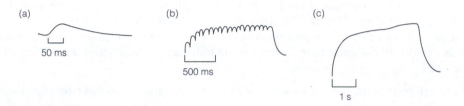

Figure 1. Muscle fiber contraction: (a) single twitch; (b) unfused tetanus (firing frequency 12 Hz); (c) fused tetanus (30 Hz). Note the increase in force of contraction in going from (a) to (c).

Three types of motor unit can be distinguished by the firing behavior of their motor neurons and the properties of their muscle fibers.

The most numerous are the **slow twitch** (**S**) motor units which take about 50 ms to develop peak force and show little decline in force even after an hour of repetitive stimulation. The motor neurons of S units are small, have a low conduction velocity and quite long refractory periods because they contain a high density of Ca^{2+}-activated K^+ channels which cause a long afterhyperpolarization. This limits maximum firing frequencies to quite low rates, but fused tetanus is achieved at low frequencies (15–20 Hz). The **type 1** muscle fibers of S motor units are rich in mitochondria, have high activities of Krebs cycle enzymes which fits them for high rates of aerobic metabolism and form **red muscle** because of their high myoglobin content. Slow twitch motor units are capable of exerting low forces for long periods. They form the bulk of the antigravity or postural muscles of the trunk and legs.

By contrast, fast twitch units contract maximally in 5–10 ms but cannot sustain the contraction for very long. With repetitive stimuli, **fatigue resistant** (**FR**) units can sustain moderate force for 5 minutes or so before a steady decline sets in that takes many minutes. **Fast fatigue** (**FF**) motor units can achieve the greatest force of the three types, but with repetitive stimuli the force falls precipitously after about 30 seconds. The motor neurons of both FR and FF units are large, with high conduction velocities. For brief periods they fire at high rates but action potential volleys are of short duration particularly for FF units. Fast twitch units contain **type 2** muscle fibers that form **white muscle** because of their low myoglobin content and require firing frequencies of 40–60 Hz to produce fused tetanus. **Type 2b** found in FF motor units are anaerobic, which explains why these fatigue so quickly. **Type 2a**, found in FR units, are intermediate between types 1 and 2b in terms of metabolism. Both FR and FF are adapted for producing rapid, large forces and so are found particularly in muscles involved in executing fast movements.

In motor units the properties of the muscle fibers and motor neurons are matched for optimal performance. This happens because muscle fiber properties are determined by the motor neurons which innervate them. If type 1 muscle fibers are denervated and the axon of an FF unit sprouts to establish new connections with the denervated fibers, they acquire the characteristics of type 2b muscle fibers.

Motor pools

Motor neurons that innervate the same muscle form a common **motor pool**. Motor pools are topographically localized in motor nuclei of the brainstem and spinal cord. Spinal motor nuclei extend over several spinal segments. Axons of motor neurons leave the ventral horn of the spinal cord to run in the spinal nerve of the same spinal segment. Sorting of fibers destined for the same muscle but originating from different spinal segments occurs in the **nerve plexuses**. Axon collaterals of motor neurons ascend and descend a few segments to influence the behavior of other motor neurons in the same pool.

The force of contraction of a muscle is determined by the motor pool in two ways; the rate at which individual motor neurons fire and the number of motor neurons in the pool that are firing. Small increases in force are met mostly by increased firing rate, but larger contractions involve increasing the number of active motor units, a process called **recruitment**. This is done in an orderly manner. In general, the earliest units to be recruited are S, followed by FR and finally FF, an order determined by the **size principle**. Two effects are at work here. Firstly, the *size* of the motor neurons, secondly how synapses onto them are organized.

How does size of a motor neuron determine the size principle? Small cells offer a bigger resistance to the flow of current than large ones. Ohm's law says that the relationship between the membrane voltage, V, and the current, I, flowing into a cell is given by:

$$V = IR$$

This means that a given current will produce a greater change in membrane voltage in a small cell (with large R) than it will in a big cell (with small R). Neurons in a motor pool are excited by common inputs. For a given sized synaptic current input into cells in the pool, the small cell body of an S motor neuron will have a bigger excitatory postsynaptic potential than the larger cell body of a fast twitch unit, because the S cell has the greater resistance (Figure 2). This means that the weakest inputs recruit the S units, because they have the lowest threshold for synaptic activation. As the inputs to the pool get progressively stronger the other motor neurons are excited in turn.

The second effect determining the size principle is that the synaptic inputs to the three classes of motor unit are weighted in such a way that as input strength increases so motor units are recruited in the sequence S–FR–FF. However, recruitment does not always obey the size principle. In some instances synapses are arranged so that large motor neurons get more excitation than small ones. For example, in humans, cutaneous afferents preferentially excite fast twitch motor units.

Motor neurons are modulated by monoaminergic neurons (NA, 5-HT) that project in the reticulospinal tracts. This enhances the response of the motor neurons to excitatory (glutamatergic) input. For example, normally S motor units fire tonically at 20–30 Hz, but with elevated monoamine input Ia afferents will produce firing rates in excess of 50 Hz. The firing rate of monoaminergic neurons increases with arousal so the modulation by NA allows the generation of higher forces from the same input when arousal is high, such as "fight or flight" situations. Serotonin neurons fire during locomotion and their firing rate increases with the speed of locomotion, driving motor neurons to fire at a higher

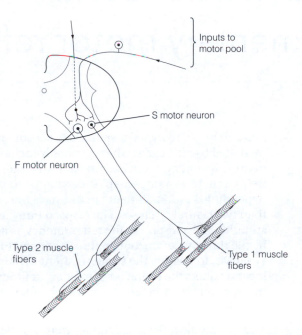

Figure 2. The size principle in recruitment. The smaller slow twitch (S) motor neurons are recruited before fast twitch (F) motor neurons because they have a bigger excitatory postsynaptic potential in response to a given input.

frequency for a given excitatory input than they otherwise would. Monoamines bring about motor neuron hyperexcitability largely by the enhancement of a depolarizing current through L-type calcium channels in their dendrites.

J3 Elementary motor reflexes

Key Notes

Properties of reflexes	Reflexes are stereotyped responses to sensory input mediated by reflex arcs, which generally include a sensory neuron, a motor neuron and one or more interneurons. Reflexes may be monosynaptic, disynaptic or polysynaptic depending on whether their circuits have one, two, or more than two central synapses. The elapsed time between a stimulus and a response, the reflex latency, is determined by the time for conduction in the circuit and the number of synapses. Increasing the stimulus strength or changing its location alters the character of a reflex. Reflexes are modified by experience in a variety of ways such as habituation, sensitization, and conditioning.
Muscle spindle reflexes	Muscle spindle reflexes are monosynaptic reflexes which cause a muscle to contract when it is stretched. They control muscle length by negative feedback. The sensory component of the reflex is the muscle spindle, which contains small intrafusal fibers that lie in parallel with the ordinary extrafusal fibers. Afferents from the intrafusal fibers synapse with motor neurons supplying the same and synergistic muscles. Intrafusal fibers receive input from γ fusimotor neurons. Contraction of the intrafusal fiber by γ fusimotor activity keeps it taut over all muscle lengths, so that it is kept sensitive to stretch whatever the muscle length. During a movement muscles shorten. To allow this, muscle spindle reflexes must be overridden. This is achieved by the simultaneous firing of both the α motor neurons and γ fusimotor neurons so that both extrafusal and intrafusal fibers shorten together.
Inverse myotatic reflexes	Inverse myotatic reflexes control muscle tension by negative feedback. Golgi tendon organs located in tendons measure muscle tension. Their Ib sensory neurons synapse, via dedicated Ib inhibitory neurons, with α motor neurons which go to the same and synergistic muscles. An increase in muscle tension causes inhibition of the α motor neurons, reducing the force of contraction of the muscle, and hence the tension.
Control of muscle stiffness	In many normal situations it is not possible to maintain a constant muscle length and constant muscle tension at the same time. Hence, rather than controlling muscle length and tension independently, CNS motor systems probably control muscle stiffness.

Related topics	(E1) Rate coding	(J2) Motor units and motor
	(J1) Nerve–muscle	pools
	synapse	(J4) Spinal motor function

Properties of reflexes

The simplest operation the nervous system can execute is the **reflex**, which couples sensory input to motor output. A reflex is a stereotyped response to a particular stimulus. When it involves the autonomic nervous system it is an **autonomic reflex** and the effector is typically cardiac or smooth muscle or a gland. When it occurs in the somatic nervous system it is a **motor reflex** and the effector is skeletal muscle. They are mediated by specified neural circuits sometimes called **reflex arcs** (Figure 1), which consist of a sensory neuron, a motor neuron and usually interneurons interposed between the two, which may be excitatory or inhibitory.

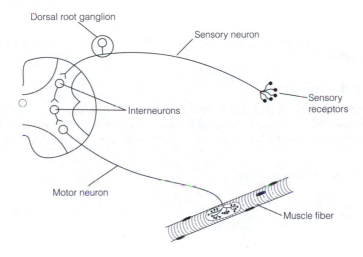

Figure 1. A polysynaptic spinal motor reflex arc.

Reflex arcs with two or more interneurons must have three or more central synapses (Figure 1) and are said to be **polysynaptic**. If there is only one interneuron the reflex is **disynaptic**. The only example of a **monosynaptic** reflex, where there is no interneuron, is the stretch reflex.

A sensory neuron will form synapses with several interneurons (or motor neurons in the case of a monosynaptic reflex). Usually the effect of afferent firing is to produce quite large epsps on a few neurons and more modest epsps in a bigger group, depending on the number of synapses. The connections made by several sensory neurons on interneurons overlap and this provides for the possibility of integration. This integration is often nonlinear; that is, the excitation of the motor neuron can be bigger than the sum of the individual inputs, **facilitation**, or dominated by one input and hence little affected by additional ones, **occlusion**.

The time between the stimulus and response is called the **reflex latency** or **reflex time**. It results chiefly from the conduction time along afferent and efferent fibers, but also includes the time taken for sensory transduction and for activation of the effector (muscle or gland). A small interval is taken up by the **synaptic delay**, usually between 0.5 and 1 ms. When conduction time is taken into account reflex latencies reflect the number of central synapses.

Increasing the stimulus intensity will change the amplitude of the reflex (e.g., the amount by which a limb moves) but may also alter the form of the response by recruiting additional muscles, this is called **irradiation**. The exact form of the reflex response depends on precisely where the stimulus is applied and so which afferents are excited. This is called the **local sign**.

Probably all reflexes can be modified by experience. The attenuation of a reflex by the repeated application of a constant innocuous stimulus is **habituation**. It is caused by synaptic depression. Any change to the stimulus (e.g., in its intensity) causes **dishabituation** in which the reflex returns to its baseline state. By contrast, repeated application of a noxious stimulus can enhance a reflex, by a decrease in latency, increased amplitude, or irradiation. This is known as **sensitization** and results from increased transmitter release. Both habituation and sensitization are examples of **nonassociative learning** because only one stimulus is involved. Some reflexes are capable of the more complicated associative learning, in which a response occurs if two stimuli are paired in time. These are **conditioned reflexes**.

Muscle spindle reflexes

The most elementary modulation of motor unit output is made by sensory input from the **muscle spindles** which measure the length and rate of change of length (velocity) of the muscle. Any attempt to stretch the muscle rapidly, for example by suddenly loading it, is

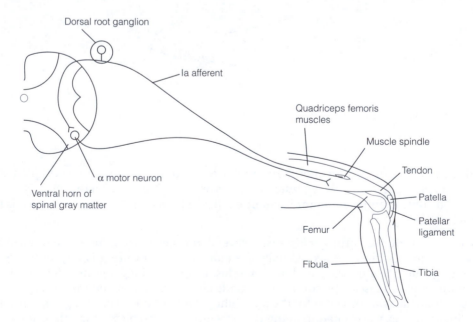

Figure 2. Basic circuit of a stretch reflex. Striking the patellar ligament excites a few hundred Ia afferents.

met by contraction. This is the **muscle spindle reflex** (**stretch reflex**, **myotatic reflex**) and is a negative feedback mechanism which defends a constant muscle length in the face of external forces which act to perturb it. A stretch reflex can be elicited from any skeletal muscle by sharply tapping its tendon. The resultant stretch causes the muscle to contract. The stretch reflex is most easily demonstrated by tapping the patellar ligament between its insertion into the tibia and the kneecap, causing the contraction of the quadriceps femoris, the powerful group of extensor muscles on the front of the thigh (Figure 2).

The sensory side of the stretch reflex consists of the **muscle spindle** and its afferents. Muscle spindles lie in parallel with the standard **extrafusal fibers** so any force acting on

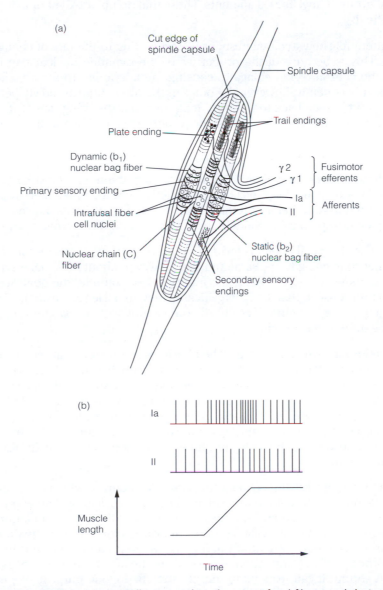

Figure 3. Muscle spindles: (a) a spindle opened to show intrafusal fibers and their innervations. A spindle normally contains one b_1, one b_2, and several c fibers; (b) responses of Ia and II afferents to muscle stretch.

the whole muscle acts in the same way on the spindle. Each muscle spindle is a fluid-filled capsule of connective tissue, 4–10 mm long and 100 μm in diameter, containing about seven modified muscle fibers called **intrafusal fibers** (Figure 3). Intrafusal fibers have contractile ends but their central regions are noncontractile. There are two types of intrafusal fiber, nuclear bag and nuclear chain.

Nuclear bag fibers (**b**) are swollen at their center, where their nuclei are clustered, and are innervated by large diameter myelinated (Ia) primary afferents, the ends of which spiral around the central region of the fiber. There are two sorts of nuclear bag fibers which can be recognized by whether in addition to primary afferent innervation they also receive secondary, group II, myelinated afferents. Those that do not are **dynamic** (b_1), those that do are **static** (b_2).

Primary afferents show dynamic responses, responding to the rate of change of length (velocity). This is because of the properties of the dynamic nuclear bag fiber. When stretched the central region elongates causing the Ia afferent to fire a volley of action potentials. Subsequently, however, the poles of the fiber elongate slowly permitting the central region to creep back to a shorter length so that the firing rate of the Ia afferent drops off. Primary afferents also show static responses, signaling muscle length, by virtue of their innervation of static (b_2) nuclear bag fibers which are stiffer than the dynamic fibers and hence elongate in proportion to muscle stretch.

Nuclear chain fibers (**c**) are of uniform diameter, are about half the size of b fibers and their central region contains a line (chain) of nuclei. They are innervated by primary and secondary afferents and being stiff (like the b_2 fibers) these afferents respond to muscle length. Typically a spindle will contain one b_1, one b_2, and three to five c intrafusal fibers.

The majority of Ia spinal afferents form synapses on homonymous motor neurons, that is, motor neurons going to the same muscle. However, about 40% make synapses with motor neurons which go to synergistic muscles. For example, the quadriceps femoris consists of four muscles that act synergistically (they are all leg extensors). Afferents from spindles in any one of them will establish connections with its own motor pool and the pools of the other three muscles.

A stretch reflex has two components. The phasic component is that seen by tapping the tendon of a muscle. It occurs rapidly, is brief, and occurs because of the dynamic activity of the Ia afferents. The tonic component is the much more sustained contraction brought about by the static activity of the Ia afferents and the secondary, group II afferents. This component is particularly important in maintaining posture. Standing in a moving vehicle, for example, the muscles in the legs and trunk that are stretched by the swaying will be contracted, so keeping the body upright. A sudden jolt will, of course, also trigger the phasic component.

Cell size is bimodally distributed in motor pools. The neurons which drive the extrafusal fibers to produce muscle contraction are Aα class with a cell body diameter averaging 80 μm, usually referred to as **α motor neurons**. In addition there is a population of smaller cells belonging to the Aγ class, called **γ motor neurons**, axons of which (**fusimotor fibers**) go to the muscle spindles. All intrafusal fibers have their contractile ends innervated by these γ motor neurons. Contraction of the ends of the intrafusal fibers keeps the central region taut so that it can respond to muscle stretch. So, one purpose of γ efferent discharge is to maintain the sensitivity of the muscle spindle to changes in length over a wide range of lengths. Without it, muscle contraction would cause the intrafusal fibers to slacken and fail to respond to stretch.

There are two categories of γ motor neuron that can be activated independently by the CNS; $γ_1$ (dynamic) innervate b_1 while $γ_2$ (static) innervate b_2 and c fibers. Stimulation of $γ_1$ fibers increases the sensitivity of the b_1 fibers, so that the primary afferent firing rate is higher in rapid than slow stretch. Stimulation of $γ_2$ fibers enhances firing of secondary afferents in response to constant stretch. In both cases the γ efferents are increasing the gain (sensitivity) of the spindle. Firing rates of γ efferents are raised when performing movements that are particularly complex.

Stretch reflexes must be overridden to allow the execution of a movement since the muscle must contract isotonically and shorten. This is achieved by descending motor pathways exciting both α and γ motor neurons at the same time. This is called **coactivation**. It makes the extrafusal and intrafusal fibers shorten together in such a way that the intrafusal fibers are always sufficiently taut to respond to stretch.

Inverse myotatic reflex

Located in the tendons, in series with muscle fibers are **Golgi tendon organs** (**GTOs**) which measure muscle tension. Increases in muscle tension activate a negative feedback reflex, the **inverse myotatic** (**Golgi tendon**) **reflex**, which opposes the increases in tension. It is brought about by GTO input activating inhibitory interneurons that synapse with α motor neurons supplying the muscle (Figure 4).

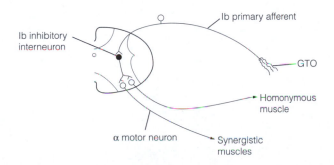

Figure 4. Circuitry of the inverse myotatic reflex. An increase in muscle tension causes the Golgi tendon organ (GTO) Ib afferent to fire at a greater rate.

The GTO is composed of the collagen fibers which join muscle fibers to tendons, interwoven through which are axon branches of a group Ib afferent neuron. The increased tension that occurs with muscle contraction stretches the collagen fibers, distorting the terminals of the Ib afferent which fires. Individual Ib afferents respond statically, reflecting the level of tension in response to the activation of a single motor unit. GTOs do not measure average tension of the muscle because less than one percent of fibers in motor units are coupled to GTOs.

The Ib afferents enter the spinal cord to synapse in the intermediate zone (Rexed laminae VI–VIII) on inhibitory neurons, which then synapses with motor neurons of the homonymous and synergistic muscles. These inhibitory neurons are all specific to this disynaptic reflex pathway (Figure 4) and so are designated **Ib inhibitory neurons** (**IbINs**). Inhibition of homonymous and synergistic muscle by GTOs is augmented by inputs from Ia spindle afferents, joint afferents, and cutaneous mechanoreceptor afferents onto IbINs. The functional importance of these connections is not clear. Descending motor systems may either excite or inhibit IbINs.

Control of muscle stiffness

The muscle spindle reflex for maintaining muscle length and the GTO reflex for keeping constant tension often work in opposition. When the loading on a muscle is altered either the muscle is stretched, in which case its *passive* tension increases (in the same way that tension increases in a rubber band as it is stretched), or it must contract isometrically to maintain a constant length, in which case the *active* tension rises due to contraction. It is impossible for both length and tension to be held constant at the same time. This suggests that overall, motor systems do not control length or tension independently, but **muscle stiffness**, k, a constant that describes how much the length of a muscle is altered, ΔL, by a change in load, ΔF.

$$k = \Delta F / \Delta L$$

It appears that the muscle spindle and GTO reflexes between them compensate for the complicated way in which muscle stiffness changes over the normal working range of muscle. This allows the organization of supraspinal motor systems to be simpler.

J4 Spinal motor function

Key Notes

Elements of spinal cord motor function

Considerable processing of sensory input takes place in the spinal cord before it influences motor output. Much of the circuitry of the spinal cord is either organized for the execution of motor reflexes, or into central pattern generators, which generate the cycles of muscle activity involved in locomotion. Three types of inhibition are important for spinal cord motor function.

Reciprocal inhibition

Reciprocal inhibition occurs between mutually antagonistic muscle motor neurons. Muscle spindle Ia sensory neurons synapse with Ia inhibitory interneurons which supply the motor neurons of antagonist muscles. This circuit allows for relaxation of antagonist muscle while the agonist is contracting. Reciprocal inhibition is important in allowing movement that needs alternating activity in agonist and antagonist muscles.

Presynaptic inhibition

Primary afferents inhibit the terminals of other primary afferents via inhibitory GABAergic interneurons that make axoaxonal synapses. The effect of GABA is to depolarize the terminal, inhibiting its release of transmitter in response to an invading action potential. Presynaptic inhibition is organized so that flexor afferents inhibit extensor afferents and *vice versa*. It is modified by descending motor systems.

Recurrent inhibition

Glycinergic Renshaw cells in the ventral horn activated by motor neurons produce inhibition of neighboring synergistic α motor neurons. This recurrent inhibition is the motor equivalent of lateral inhibition in sensory systems and sculpts economic movements from a broader range of motor neuron activity.

Flexor reflexes

Flexor reflex afferents (FRAs) elicit flexor reflexes in the ipsilateral limb, and other reflexes in the contralateral limb. The neural circuitry for these reflexes is recruited during normal limb movements, which are continuously modified by peripheral sensory input via FRAs. Flexor reflexes generated by nociceptor input enable a limb to be withdrawn from a noxious stimulus.

Central pattern generators

Locomotion involves alternate flexion and extension of limbs which are phased in specific ways depending on the manner of locomotion (e.g., walking or running). The basic rhythms of locomotion are produced by central pattern generators (CPGs), networks of spinal interneurons.

	Each CPG acts as an oscillator, driving a limb to flex then extend alternately. CPG activity is regulated via midbrain reticulospinal neurons.	
Related topics	(E1) Rate coding	(J5) Brainstem postural reflexes
	(J3) Elementary motor reflexes	

Elements of spinal cord motor function

About twice as many sensory fibers (half of them unmyelinated C fibers) enter the dorsal roots of the spinal cord as there are motor neurons in the ventral horn. Moreover, motor neurons make up less than 2% of the total number of neurons in the spinal cord, most of which are interneurons. This shows that massive processing of sensory input occurs before it influences motor neurons.

Organization of movement by the spinal cord depends on reflex circuits and central pattern generators (CPGs). With the exception of a few protective reflexes, such as the flexion withdrawal reflex which serves to remove a limb from a noxious stimulus, motor reflexes are not seen in isolation under physiological conditions. Motor reflexes are elements of circuits which operate coherently to allow descending motor control of muscles to be continually modified on the basis of proprioceptor input from muscles and joints, and input from the skin, enabling the smooth execution of the appropriate movement.

Locomotion involves cycles of activity in which muscle groups are made to contract in a precisely timed sequence. This requires neural networks that can generate the required rhythmic output. These networks, thought to be autonomous, though modifiable by reflexes and activated by supraspinal influences, are called central pattern generators (CPGs). The presence of CPGs can be inferred by experiment in a wide variety of vertebrates, including primates and humans, and putative circuits have been modeled using computer simulations.

Three types of inhibition contribute to spinal cord function, reciprocal, presynaptic, and recurrent.

Reciprocal inhibition

Axon collaterals of Ia afferents from muscle spindles synapse with glycinergic **Ia inhibitory interneurons** in lamina VII (IaINs) that project to motor neurons of antagonist muscles. This disynaptic circuit allows antagonist muscles to be relaxed during agonist contraction (Figure 1). The inhibition of mutually antagonistic muscle motor neurons is called **reciprocal inhibition**.

IaIN activity mediating reciprocal inhibition is modified by descending motor pathways (corticospinal, rubrospinal, and vestibulospinal) and locomotor networks in the cord. This happens:

- To facilitate rapid movements. Because muscle contractions are so long lasting, muscles follow faithfully only slowly changing neural input. Rapid changes in motor neuron firing cannot be translated into corresponding alterations in muscle tension. Hence to produce fast fluctuations in tension the motor system alternates contraction in agonist and antagonist muscles. This is helped by reciprocal inhibition (Figure 2).

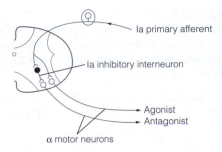

Figure 1. A disynaptic reflex for reciprocal inhibition of antagonist muscle.

- To adjust muscle stiffness so that it is appropriate for the load. The stiffness of a joint is increased by co-contraction of muscles with opposing actions at the joint. For example, co-contraction of the biceps and triceps increases stiffness of the elbow. Provided one muscle contracts more powerfully than the other the joint moves. Co-contraction stabilizes a joint, it provides better control when loads change unexpectedly because a given difference between expected and actual load will have a smaller effect on limb trajectory if a joint is stiffer. Co-contraction requires suppression of reciprocal inhibition.

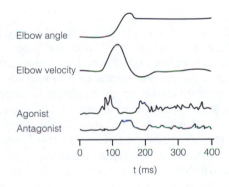

Figure 2. Pattern of activation of agonist (biceps) and antagonist (triceps) muscles to facilitate rapid movement (elbow flexion).

Presynaptic inhibition

Reflexes dependent on sensory input via Ia, Ib, and II afferents, can be modified by presynaptic inhibition from inhibitory GABAergic interneurons in the spinal cord. These interneurons make axoaxonal synapses on the afferent terminals. GABA secreted at these synapses acts on $GABA_A$ receptors to bring about *depolarization* because in these sensory neurons the membrane potential is more negative than the reversal potential of the chloride current through the $GABA_A$ receptor channels. The effect of this **primary afferent depolarization** (**PAD**) is inhibitory in that when an action potential sweeps into the terminal its amplitude gets smaller, less Ca^{2+} influx occurs, so transmitter release from the terminal is reduced.

Presynaptic inhibition of Ia terminals is organized on a reciprocal basis in which flexor afferents inhibit extensor afferents and *vice versa*. At the start of a movement presynaptic inhibition to Ia terminals going to agonist motor neurons is reduced whereas inhibition of antagonist muscle Ia terminals is increased. This means that spindle activity in the agonist is reinforced whereas that in the antagonist muscle is dampened. This makes sense; agonist action lengthens the antagonist muscle and this would provoke it to contract were its stretch reflex not suppressed.

Similar circuitry is involved in presynaptic inhibition of Ib and II afferents, allowing control of tension and tonic length signals independently of phasic length signals.

Recurrent inhibition

A population of interneurons called **Renshaw cells** in the ventral horn are activated by axon collaterals of α motor neurons and project to neighboring, homonymous, and synergistic α motor neurons. These cells fire high frequency bursts of action potentials that produce fast large ipsps. The effect of the inhibition is to silence motor neurons excited weakly and firing at low rates, and to dampen the firing frequency of strongly excited motor neurons. This is very like the lateral inhibition seen in sensory systems. It enhances contrast, and so enables economical movements. Renshaw cells are glycinergic and blockade of glycine receptors with **strychnine** results in convulsions due to the failure of recurrent inhibition. **Tetanus**, caused by infection with the organism *Clostridium tetani*, produces the same effect because its toxin blocks glycine release from Renshaw cell terminals.

Flexor reflexes

A variety of afferents, including group II and III muscle afferents, joint afferents, and skin mechanoreceptor and nociceptor afferents, elicit flexor reflexes in the ipsilateral limb and so are known as **flexor reflex afferents** (**FRAs**). These same afferents trigger an extension of the contralateral limb, the **crossed extensor reflex** or alternative reflex pathways. Different types of FRAs are connected to specific subsets of interneurons so the reflexes they excite differ in form and timing. Many normal limb movements consist of either flexion, or alternate flexion and extension, and are brought about by the actions of supraspinal motor systems on the interneurons targeted by the flexor reflex afferents. In one model to account for this these interneurons are organized into sets called **half centers**, between which reciprocal inhibitory connections exist. A flexor half center gets inputs from ipsilateral FRAs and excites flexor motor neurons, while an extensor half center is activated by contralateral FRAs driving extensor motor neurons (Figure 3). These half centers are also components of central pattern generators (see below).

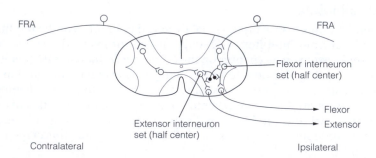

Figure 3. Organization of flexor reflex afferent (FRA) interneurons. Excitatory neuron cell bodies are open circles, inhibitory interneuron cell bodies are filled circles.

For the execution of a particular movement, descending motor axons are activated which project to the specific set of FRA interneurons that bring about the movement. The reciprocal connections ensure that the alternative set is inhibited. As the movement proceeds, FRA input from muscles, joints, and skin reinforces the movement and allows it to be fine-tuned.

Although flexor reflexes are usually recruited as elements of normal movement, the withdrawal flexor reflex, triggered by Aδ or C fiber (group IV) nociceptors is quite distinct in character. It overrides ongoing movement, involves flexor muscles throughout the limb so the response is dramatic, and it is long lasting. It is protective.

Central pattern generators

Locomotion is movement from place to place. There are numerous modes of locomotion (e.g., stepping, swimming, flying) but all depend on cycles of muscle activity, alternate flexion and extension of each limb, though the **gait** or manner of locomotion (e.g., walking or running) depends on the speed. The gait adopted is the one which minimizes the energy expenditure for that speed.

In human stepping each leg has a **stance phase**, when extensors are most active, and a **swing phase**, when the flexors are the most active. The same sequence in the opposite leg is out of phase, though during walking there is a brief overlap of the stance phase in both legs. As the speed increases the stance phase shortens until there is no overlap and the switch to running occurs.

The basic rhythms of locomotor activity are produced by **central pattern generators** (**CPGs**), networks of interneurons in the spinal cord that generate the precisely timed sequences of α motor neuron activation without the need for sensory input. Each limb has an array of CPGs. Each CPG is an oscillator with two half centers, one driving flexors, the other driving extensors, with reciprocal connections between them. Each half center produces rhythmic bursts of action potentials that are terminated in a time and manner determined by the intrinsic excitable properties of its constituent neurons. The cessation

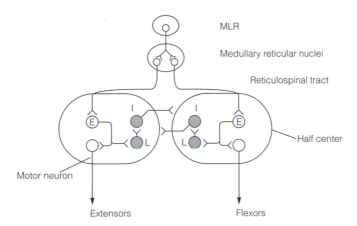

Figure 4. A simplified hypothetical model of a central pattern generator (CPG) based on studies of the lamprey. Each of the neuron symbols represents several cells. The excitatory (E) cells use glutamate and show burst firing in response to supraspinal input. The inhibitory (filled symbol) neurons are glycinergic; contralaterally projecting interneuron (I) inhibits the opposite half center. L, lateral interneuron; MLR, mesencephalic locomotor region.

of firing of one half center releases its opposite number from reciprocal inhibition, allowing it to fire a burst. In this way burst firing alternates between the two half centers.

CPGs have been studied in the lamprey, a primitive vertebrate, and it is likely that mammalian CPGs use similar principles (Figure 4). Depolarization of excitatory (E) cells in one half center activates their NMDA receptors. The resulting calcium influx prolongs the depolarization so that the E cell fires a burst of action potentials. The burst firing of the E cells stimulates motor neurons. Two features of the circuit allow it to flip-flop between bursting of first one half cell then the other. Firstly, the E cells stimulate L cells which inhibit the I neurons responsible for the reciprocal inhibition. This disinhibits the opposite half center. Secondly, the E cell burst ends because the calcium activates K_{Ca} channels, allowing K^+ efflux and hyperpolarization.

Locomotor activity is initiated by activity in the **mesencephalic locomotor region (MLR)** which projects to reticular nuclei in the medulla (Figure 4). Axons from here run in the **reticulospinal tracts** to the spinal cord. These reticular nuclei are excitatory, releasing glutamate to produce a large depolarization of the CPG neurons which then produce oscillating output for as long as the MLR input continues. CPGs are interconnected so that timing of events in all limbs is coordinated. The basic locomotor rhythms of the CPGs are extensively modified by the supraspinal motor systems.

J5 Brainstem postural reflexes

<table>
<tr><td colspan="2">Key Notes</td></tr>
<tr>
<td>Postural mechanisms</td>
<td>Postural mechanisms stabilize the body against forces which shift the center of mass. Posture is maintained largely by the action of antigravity muscles which include extensors of the back and legs (and in humans, arm flexors). Feedforward motor commands from motor cortex and cerebellum allow postural adjustments needed for an anticipated movement. Postural reflexes, organized by the brainstem in response to vestibular, proprioceptor, and visual input, use negative feedback to correct for unanticipated perturbations.</td>
</tr>
<tr>
<td>Vestibular (labyrinthine) reflexes</td>
<td>Vestibular input is used to maintain the orientation of the head in space in the face of tilting or rotation of the head and body as a unit. These vestibular reflexes achieve this by acting on neck muscles (vestibulocollic reflexes) or limb muscles (vestibulospinal reflexes) giving ipsilateral extension and contralateral flexion.</td>
</tr>
<tr>
<td>Neck reflexes</td>
<td>When the head moves with respect to the trunk, neck muscle stretch produces reflex contraction of neck muscles (cervicollic reflexes) to return the head to its previous position, and of limb muscles (cervicospinal reflexes) to produce ipsilateral flexion and contralateral extension. Cervicocollic and vestibulocollic reflexes act synergistically, cervicospinal and vestibulospinal reflexes are antagonistic.</td>
</tr>
<tr>
<td>Postural reflex pathways</td>
<td>The vestibular and reticular nuclei integrate vestibular and proprioceptor (e.g., muscle spindle) input and execute postural reflexes by way of the vestibulospinal and reticulospinal tracts of the medial motor system. Inputs from the cerebral cortex to reticular nuclei allow postural adjustments to be made during voluntary movements.</td>
</tr>
<tr>
<td>Related topics</td>
<td>(F5) Balance   (J4) Spinal motor function</td>
</tr>
</table>

Postural mechanisms

Postural mechanisms prevent the body from being destabilized by forces (including gravity and those produced by limb movements) acting to shift the center of mass. Muscles either oppose or assist gravity when contracting; those that oppose are described as **antigravity muscles**. Many antigravity muscles, such as the leg extensors and the short deep extensor muscles of the back (axial muscles), are involved in maintaining posture. In humans, the flexor muscles of the arms are also antigravity muscles. Since antigravity muscles are generally more powerful than muscles assisted by gravity, in human limbs the strongest muscles are the leg extensors and arm flexors.

Motor commands from the forebrain and cerebellum include those for the postural adjustments necessary during the execution of a movement. For example, if a standing dancer abducts her leg she must shift her center of mass above the other leg to avoid falling over. This needs *feedforward* postural adjustments to trunk and arms which anticipate the unbalancing forces that act during the movement. The adjustments (**postural set**) depend precisely on the initial position and nature of the intended movement and they must be learnt.

However, postural adjustments often have to be made to compensate for *unpredicted* disturbances in body position and movement. This cannot be done by feedforward since, by definition, the disturbance is not known ahead of time. So, these adjustments are made by **postural reflexes**, organized by the brainstem, which are *negative feedback* mechanisms (Figure 1). The sensory input to the reflex circuitry is from three sources:

- Vestibular, from the otolith organs

- Proprioceptor, from muscle spindles, Golgi tendon organs, and joint receptors

- Visual, from the superior colliculus

These inputs are highly integrated to recruit the required sequence of corrective muscle activity.

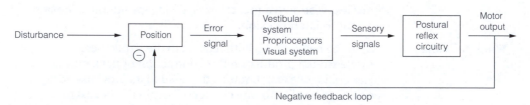

Figure 1. Postural reflex negative feedback. The circuitry produces a motor output which reduces the mismatch between the desired position of the body and its actual position. The error is detected by the sensory systems which feed information into the reflex circuitry.

The exact nature of postural reflex adjustments made by humans depends on context, that is, the initial position of the body and the size and direction of the destabilizing force. Swaying, produced by sudden displacement of the surface on which a person is standing, will activate quite different sets of muscles depending on the direction of sway, but in general distal muscles are excited before proximal ones, with most movement occurring at the ankle joint. Rotation or tilt of the surface, however, results in bending at the hips. *In extremis* postural reflexes try to maintain the center of gravity to prevent falling, or to put the limbs in a position to brace against falling.

Several distinct postural reflexes can be seen in animals by surgically transecting the brainstem (**decerebration**) and in humans who have suffered severe brain damage. These reflexes cannot *easily* be elicited in isolation in healthy behaving humans because motor functions are normally so highly integrated. They may be seen in newborn infants, in whom motor systems are immature.

Vestibular (labyrinthine) reflexes

Sensory input from otolith organs and semicircular ducts is used to stabilize the orientation of the head in space. Any tilting or rotation of the head and body as a unit activates motor neurons to muscles that maintain the head vertical with respect to gravity. These (mainly tonic) reflexes have a latency of about 40–200 ms. Those which activate motor neurons to neck muscles are called vestibulocollic reflexes, those to limb muscle motor neurons are vestibulospinal reflexes. **Vestibulocollic reflexes** act on the neck muscles to keep the head upright. If the body sways forward the neck extensors contract bringing the head up. If the body sways backwards, neck flexors are activated. **Vestibulospinal reflexes** act on limb muscles. They trigger contraction of arm extensor muscles and leg flexor muscles when falling, to reduce the impact of landing. Swaying sideways triggers extension of the ipsilateral limbs to brace against further tilt in that direction, and contralateral flexion.

Neck reflexes

Turning the head *relative to* the body excites spindles in neck muscles and afferents from the cervical vertebral joints. This evokes reflex contractions of neck muscles (cervicocollic reflexes) and limb muscles (cervicospinal reflexes) with both phasic and tonic components. **Cervicocollic reflexes** contract neck muscles that are stretched and so act to reorientate the head on the body. Cervicocollic and vestibulocollic reflexes are synergistic. **Cervicospinal reflexes** cause contraction of limb muscles in response to *rapid* head movement. In standing humans a force which throws the head backwards on the trunk activates all the limb extensors, whereas a force throwing the head forwards activates all limb flexors. Cervicospinal and vestibulospinal reflexes are antagonistic. If the head and trunk are tilted *as one* to the left, the vestibulospinal reflex causes left arm extension. But if the trunk *alone* is passively tilted to the left (with the head remaining fixed in relation to space) the cervicospinal reflex will flex the left arm. In the more usual situation that the head is tilted while the trunk remains stationary, these opposing reflexes cancel out.

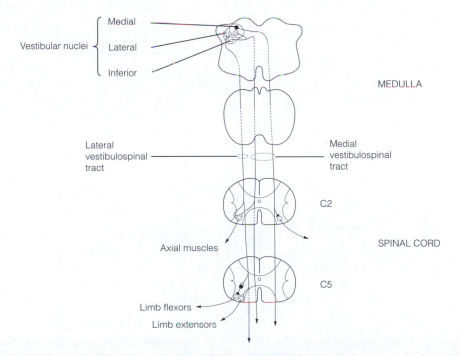

Figure 2. The vestibulospinal tracts. Filled circles and triangles represent the cell bodies and axon terminals, respectively, of inhibitory neurons.

Postural reflex pathways

Supraspinal descending control of movement uses two sets of pathways. Those mediating postural reflexes, which go from brainstem to spinal cord, are collectively called the **medial motor pathways** to distinguish them from the **lateral motor pathways** for making voluntary movements. The medial motor system comprises the vestibulospinal (Figure 2) and reticulospinal tracts (Figure 3) which terminate in the intermediate and

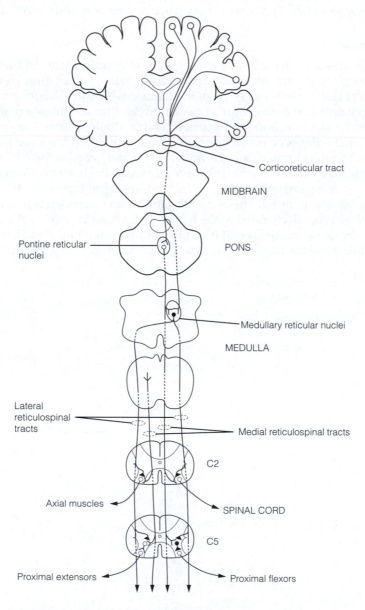

Figure 3. The reticulospinal tracts. The full extent of the medial reticulospinal tracts is shown on one side only. The action of the tracts on extensors and flexors is shown on opposite sides for clarity. Filled circles and triangles are the cell bodies and axon terminals, respectively, of inhibitory neurons.

ventral gray matter of the spinal cord. Input from the vestibular labyrinth goes to **medial and inferior vestibular nuclei,** which give rise to the bilateral **medial vestibulospinal tract**. This makes connections with neck muscle motor neurons. In general, ipsilateral motor neurons are excited whilst contralateral ones are inhibited. The pathway mediates some vestibulocollic reflexes.

Vestibular afferents to the **lateral vestibular nucleus** are implicated in the control of limb muscles. The lateral vestibular nucleus projects via the uncrossed **lateral vestibulospinal tract** to all segments of the spinal cord. Neurons in this pathway facilitate extensor motor neurons and inhibit flexor motor neurons. This pathway is responsible for some vestibulospinal reflexes.

There are two reticulospinal tracts. The **medial reticulospinal tract** originates from the **pontine reticular nuclei** and is ipsilateral. Its neurons excite axial and limb extensor motor neurons, but are inhibitory via polysynaptic pathways to limb flexors. The **medullary reticular nuclei** give rise to the **lateral reticulospinal tract** which is bilateral and produces monosynaptic inhibition of neck and axial motor neurons and polysynaptic inhibition (excitation) of proximal limb extensors (flexors). The medullary reticular nuclei get input from the mesencephalic locomotor region and project to the central pattern generators. Table 1 summarizes the effects of the descending motor pathways on motor neurons.

The reticular nuclei receive proprioceptor input from muscle spindles and vertebral joint receptors, and vestibular input, so reticulospinal tracts service both neck and vestibular reflexes. In fact, otolith signals for head forward and back pitching components of the vestibulocollic reflexes are not transmitted through vestibulospinal tracts but probably involve the reticulospinal tracts. The reticular nuclei receive input from the premotor cortex which allows postural reflexes to be modified by feedforward during intentional movement. After lesions of the medullary reticular nuclei in cats, anticipatory adjustments are compromised and the animals lose balance temporarily when attempting to move a forelimb.

Table 1. A summary of the major features of the descending motor pathways[a]

Motor system	Tract	Distribution	Principal effects on motor neurons	
			Excitatory to:	*Inhibitory to:*
Medial	Lateral vestibulospinal	Ipsilateral	Axial and proximal limb extensors	Axial and proximal limb flexors
	Medial vestibulospinal	Bilateral	Axial ipsilateral	Axial contralateral
	Pontine (medial) reticulospinal	Ipsilateral	Axial and proximal limb extensors	Proximal limb flexors
	Medullary (lateral) reticulospinal	Bilateral	Proximal limb flexors	Axial and proximal limb extensors
Lateral	Corticospinal	Largely contralateral	Distal limb flexors	Distal limb extensors
	Rubrospinal	Bilateral	Distal limb flexors	Distal limb extensors

[a] The lateral motor system is described in Section K1.

K1 Cortical control of voluntary movement

Key Notes

Intentional movement

Voluntary movements are planned and executed by commands from the motor cortex which specify the correct sequence of muscle activation. Sensory feedback can be used to optimize performance.

Lateral motor pathways

The motor cortex gives rise to two lateral pathways for the execution of intentional movements. The corticospinal tract consists of axons of pyramidal cells in layer V of the cortex which descend through the internal capsule into the brainstem where most cross the midline giving rise either to corticonuclear fibers that go to the motor nuclei of cranial nerves or to the lateral corticospinal tract. Corticospinal tract neurons are excitatory and modulate the activity of motor neurons in the ventral horn of the spinal gray. The corticospinal tract is commonly excitatory to flexors and inhibitory (via Ia inhibitory interneurons) to extensors. The rubrospinal tract from the red nucleus runs alongside the corticospinal tract.

Motor cortex

The motor cortex is divided into three reciprocally interconnected areas, the primary motor cortex (M1), and the supplementary motor area (SMA) and premotor area (PM). The SMA and PM together constitute the secondary motor cortex. M1 forms a motor loop with the cerebellum while the SMA forms a motor loop with the basal ganglia. The motor cortex has several somatotopic maps. The map in M1 has more cortical space devoted to regions such as the hands and face where the variety and complexity of movements is greatest. An M1 cell does not have exclusive control over individual muscles or movements but its activities correlate with a wide variety of movement parameters (e.g., force). Direction of limb movement is not encoded by the behavior of a single cell, but by the average firing of a population of neurons.

The secondary motor cortex is involved in complex movements. The SMA is used in tasks involving both sides of the body, usually previously learnt. The premotor area is concerned with movements that require sensory cues.

Red nucleus

The red nucleus has a motor map. Firing of rubrospinal tract neurons correlates with motor parameters in a similar manner to corticospinal tract neurons, and they are distributed to motor neuron pools in a comparable fashion. However, the two pathways have different functions. The rubrospinal tract may operate when learnt, automated movements are being executed, while the corticospinal tract is active when new motor tasks are being acquired.

Related topics	(A4) Organization of the central nervous system	(K5) Cerebellar function
	(F2) Touch	(K6) Anatomy of the basal ganglia
	(J3) Elementary motor reflexes	(K7) Basal ganglia function

Intentional movement

Voluntary movements are those made intentionally. Many neurons in the motor system fire hundreds of milliseconds before any muscle contraction occurs showing that movements are planned. This is necessary because the same motor tasks can be performed in various ways depending on the context, a property called **motor equivalence**. For example, driving a large truck needs a different motor strategy than driving a small car. The movement is executed by the output of **motor commands** which specify the correct temporal sequence of muscle activation. Sensory feedback during a movement, for example from proprioceptors such as muscle spindles and the visual system, is used to fine-tune its execution so that the performance matches the desired goal. The planning of voluntary movements and the elaboration of motor commands for their execution is done by the motor cortex which has its outputs via the lateral motor pathways.

Lateral motor pathways

There are two lateral pathways for descending control of voluntary movement. Both originate in the motor cortex which lies on the frontal lobe just anterior to the central sulcus. The **corticospinal tract** consists of the axons of about one million pyramidal cells in layer V of the cortex. Over half come from the **primary motor cortex** (**M1**, Brodmann area 4) or **secondary motor area** (**MII**, Brodmann area 6). These axons project to the ventral horns of the spinal cord to alter the activity of α and γ motor neurons. About 40% of corticospinal tract axons come from the somatosensory cortex (Brodmann areas 1, 2, and 3) or other regions of parietal cortex (Brodmann areas 5 and 7). These axons terminate in the dorsal horns of the spinal cord and regulate sensory input.

Most of the corticospinal tract consists of fine myelinated and unmyelinated axons with conduction velocities between 1 and 25 ms^{-1}. However, there are about 30 000 extremely large (20–80 μm diameter) pyramidal cells in area 4, called **Betz cells**, with big myelinated axons that conduct with velocities of 60–120 ms^{-1}. Axons of the corticospinal tract pack tightly to pass through the internal capsule which lies between the thalamus and the lentiform nucleus (Figure 1) and descend into the brainstem. Here the most medial fibers peel off and cross the midline to go to nuclei (trigeminal (V), facial (VII), hypoglossal (XII), and accessory (XI)) of the cranial nerves. These are **corticonuclear** (**corticobulbar**) **fibers** and are motor to the face, tongue, pharynx, larynx, and sternomastoid and trapezius muscles. The remaining axons descend through the medulla causing a swelling on its ventral surface, the **pyramid**, so the corticospinal tract is also called the **pyramidal**

tract. At the caudal medulla 85% of fibers cross the midline as the **pyramidal decussation**, giving rise to the **lateral corticospinal tract**. The remaining ipsilateral axons form the **anterior corticospinal tract**, which crosses over at spinal cord level.

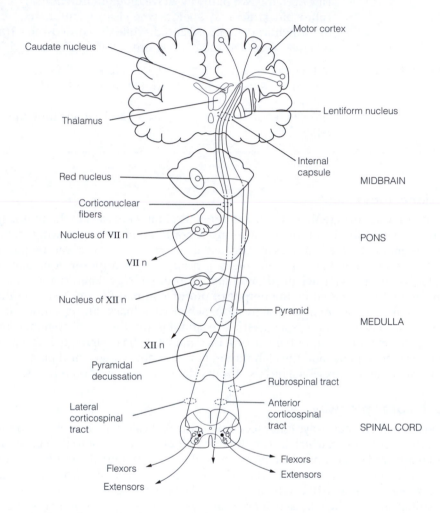

Figure 1. The lateral motor pathways. Only corticonuclear fibers in the facial (VII n) and hypoglossal (XII n) nerves are shown. Synapses of the anterior (ipsilateral) corticospinal tract with spinal cord interneurons are not shown.

Corticospinal tract neurons use glutamate as a transmitter and are excitatory. They either synapse directly with α motor neurons supplying distal limb muscles in Rexed lamina IX, or synapse with interneurons in laminae VII and VIII which make polysynaptic connections with α motor neurons of proximal limb muscles and axial muscles. Fusimotor (γ-efferent) neurons that must be coactivated with α motor neurons to override the stretch reflex during voluntary movement are excited polysynaptically. Stimulation of the corticospinal tract is predominantly excitatory to flexors but inhibitory to extensors. The corticospinal tract inhibits motor neurons disynaptically via Ia inhibitory interneurons.

The corticospinal tract axons arising from the somatosensory cortex project to cranial nerve sensory nuclei and dorsal horns and produce presynaptic inhibition on primary afferent terminals *except* for 1a spindle afferents.

Some pyramidal cells in layer V of the motor cortex send their axons in the **corticorubral tract** to the **red nucleus** in the midbrain, which also receives collaterals from the corticospinal tract. The red nucleus gives rise to the **rubrospinal tract**, the second of the lateral motor pathways. Some of its axons go to cranial nerve nuclei in the pons and medulla. In humans the rubrospinal axons descend as part of the corticospinal tract.

Motor cortex

The motor cortex is subdivided into three areas, MI, MII, which contains the **supplementary motor area** (**SMA**), and **premotor area** (**PM**). These are reciprocally connected with each other (Figure 2) and with subcortical structures which send inputs back to the motor cortex via the thalamus forming closed motor loops. Reciprocal back projections from the motor cortex to the thalamus also exist (not shown on Figure 2).

The supplementary motor area is part of a motor loop with the basal ganglia. It sends output to the striatum which projects back to the SMA via the globus pallidus and

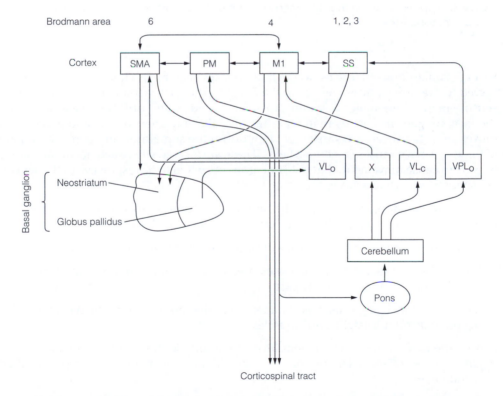

Figure 2. The connections of the motor cortex establish motor loops with the basal ganglia and the cerebellum in which the thalamus provides the input that closes the loop. SS, somatosensory cortex. Thalamic nuclei: VL_c, ventroposterior nucleus pars caudalis; VL_o, ventroposterior nucleus pars oralis; VPL_o, ventroposterolateral nucleus pars oralis; X, nucleus X.

ventrolateral (VL_O) thalamus. Many corticospinal tract axons from M1 either terminate in the pons or give off collaterals there. These make synapses with pontine neurons which project to the cerebellum. Outputs from the cerebellum to the thalamus, which in turn project back to M1 and PM motor areas, form other motor loops. These motor loops with basal ganglia and cerebellum are required for the initiation of specific motor patterns and their coordination.

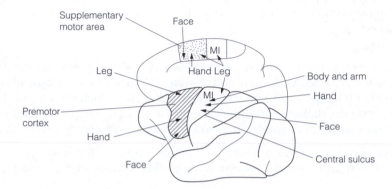

Figure 3. Approximate somatotopic mapping in the motor cortex of the macaque. M1, primary motor cortex.

Several somatotopic maps exist in the motor cortex (Figure 3). Their topography is preserved in the orderly arrangement of the fibers in the corticospinal tract. Like somatosensory maps, these motor maps are grossly distorted. More cortical space is devoted to the face, tongue, and hands than other regions, allowing great variety and precision of movements. M1 receives a substantial input from the somatosensory cortex and many M1 neurons have sensory receptive fields. These are located where activity of the neuron is likely to cause movement; that is, M1 neurons are wired to be responsive to the sensory consequences of their actions.

What the somatotopic mapping in the motor cortex represents is uncertain. It is not a one-to-one mapping to individual muscles or movements. This is shown by:

- Output from single cortical neurons diverging to several motor neuron pools

- Converging outputs from quite a wide area of M1 onto the motor neuron pools for muscles moving a specific body part

- A given muscle being subserved by a region in the motor cortex which overlaps with regions controlling neighboring muscles

During the execution of a movement involving a muscle the set of neurons within the region that are activated depends on the nature of the movement; for example, its direction and force.

Recording from single cells in the motor cortex of conscious monkeys engaged in intentional limb movements shows that M1 cell firing can correlate with force, rate of change of force, velocity, acceleration, direction of movement, or joint position. None of these parameters is mapped in an orderly way in the cortex. Firing of an M1 cell during a task is usually related to two or three of these variables so M1 cells do not exclusively encode a single movement parameter. Many M1 cells are rather broadly tuned for movement direction so a single M1 cell cannot encode direction of movement very well. However, direction of movement is very precisely coded by the *average* firing of a few hundred cells. This is an example of population coding. The population of cells encoding the direction of a given movement are quite widely distributed across the cortex.

The secondary motor cortex contains neurons that fire in a way that correlates with the direction and force of a movement. The SMA controls proximal limb muscles directly via its output to the corticospinal tract, whereas the premotor area neurons synapse with brainstem reticular neurons that go to axial and proximal limb muscles. Both control distal limb muscles via M1.

The SMA has bilateral representations of the body and is crucial for movements involving both sides of the body, particularly those that have been learnt. For this, the motor loop involving the basal ganglia is important. The premotor area consists of a number of discrete motor regions and so has several maps. It is implicated in movements in response to sensory (mostly visual) cues. The premotor area gets a large input from the **posterior parietal cortex** (Brodmann areas 5 and 7) association cortex which receives visual, somatosensory, and vestibular sensory input. The posterior parietal cortex thus provides sensory input for targeted movements. Some posterior parietal neurons are context specific, firing only during goal-directed behavior (e.g., reaching for food) but remaining silent if the limb moves in the same way in the absence of the goal.

A key role of the secondary motor cortex is in planning movements. This is shown, firstly, by the fact that neurons here fire a long time (possibly up to 800 ms) before a voluntary movement begins. Secondly, measurements of cerebral blood flow (cbf) in humans doing motor tasks show that a simple movement involves increased cbf in MI only, a more complex task is accompanied by increased cbf in secondary motor cortex as well as MI; but, remarkably, when subjects were required to mentally rehearse a complex task (but not do it) an increase in cbf was seen restricted to the secondary motor cortex.

Red nucleus

The red nucleus has a somatotopic map. Its activity precedes intentional movements and correlates with parameters such as force, velocity, and direction, much like corticospinal tract neurons. Furthermore, in many primates rubrospinal axons have the same distribution to proximal and distal limb motor neurons as the corticospinal tract and their activity moves individual digits. However, in humans the distinction between the two lateral pathways seems not to be as important as in many mammals, and the rubrospinal tract is involved in gross limb movements not fine ones.

Although the two pathways appear strikingly similar, studies in sub-human primates suggest that they operate in different contexts. While the rubrospinal tract is active when previously learnt automated movements are executed, the corticospinal tract is required when novel movements are being learnt. Another pathway acts to switch activity between the two lateral motor systems. As a new movement is successfully learnt its execution is switched from the corticospinal tract to rubrospinal tract control. The switch operates in the opposite direction if an automatic movement needs to be adapted. It is because of the switch that each lateral motor system can compensate for the loss of the other. Corticospinal tract lesions have a more severe and protracted effect than rubrospinal tract lesions because new movements cannot be executed and only the old rubrospinal tract repertoire can be called upon. The switch pathway involves the inferior olive, one function of which is to detect and correct errors in motor performance.

K2 Motor lesions

Key Notes

Lesions of spinal motor pathways	Cutting the corticospinal tract causes an ipsilateral lesion if the transection is below the pyramidal decussation, and a contralateral lesion if above it. A pure corticospinal tract lesion in nonhuman primates causes a loss of fine movements by distal muscles which eventually recovers. Cutting vestibulospinal and reticulospinal tracts causes deficits in posture and locomotion.
Brown–Sequard syndrome	Severing the spinal cord on one side causes paralysis and loss of touch and proprioceptor sensation below the lesion on the same side and loss of pain and temperature sensation on the contralateral side.
Decerebrate rigidity	Lesions which sever the brainstem between the red nucleus and the vestibular nuclei cause an increase in extensor tone known as decerebrate rigidity. It is caused by the loss of facilitation of flexor motor neurons by the rubrospinal tract.
Cerebrovascular accidents	Lower motor neurons are those that innervate skeletal muscles while upper motor neurons are cortical neurons of both lateral and medial pathways and include corticoreticular neurons. Thus, upper motor neuron lesions cause more severe deficits than corticospinal tract lesions alone. Long term symptoms are muscle weakness and spasticity on the side opposite the lesion. Spasticity is an increase in muscle tone, particularly in extensor muscles and is caused by hyperexcitability of stretch reflexes.
Related topics	(F2) Touch (F3) Pain (J5) Brainstem postural reflexes	(K1) Cortical control of voluntary movement

---

Lesions of spinal motor pathways

Experimental transection of the corticospinal tract below the pyramidal decussation in primates causes an ipsilateral motor deficit below the level of the section. Corticospinal tract lesions above the pyramidal decussation cause a contralateral deficit. The deficit seen with a *pure* corticospinal lesion is a loss of the ability to make fine movements with distal muscles. Almost complete recovery is eventually seen. Similar effects are seen if the rubrospinal tract is cut. A lesion of *both* lateral motor pathways results in permanent deficits.

In contrast, experimental transection of the vestibulospinal and reticulospinal tracts which control output to proximal limb and axial muscles produces much more extensive deficits in posture and walking, but leaves fine control of distal muscles intact.

Brown–Sequard syndrome

A classic pattern of sensory and motor deficits, the **Brown–Sequard syndrome**, is seen when the spinal cord is severed on one side. On the side of the lesion there is a motor paralysis and loss of all sensation transmitted through the dorsal columns (touch and proprioception). On the contralateral side there is loss of nociceptor and thermoreceptor sensation from a few spinal segments below the lesion. This is due to the interruption of the anterolateral columns which contain spinothalamic axons that have crossed from the opposite side (Figure 1).

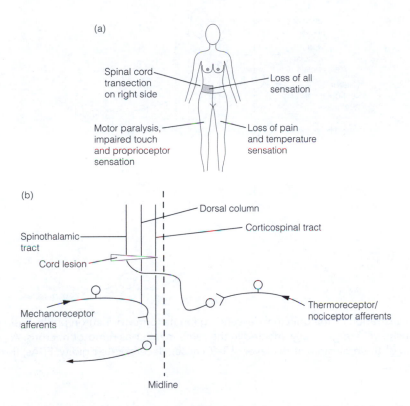

Figure 1. Brown–Sequard syndrome results from the hemisection of the spinal cord which interrupts dorsal column input and motor output on the side of the lesion and spinothalamic input from the contralateral side. (a) Signs and symptoms; (b) lesion.

Decerebrate rigidity

Patients in whom brain trauma or a tumor produces functional disconnection of the brainstem from the rest of the brain at the level between the red nucleus and the vestibular nuclei show an increase in extensor tone called **decerebrate rigidity**. It is caused by tonic activity in the vestibulospinal and reticulospinal neurons that is no longer opposed by the powerful facilitation of flexor motor neurons by the rubrospinal tract (Figure 2). The overall effect of the vestibulospinal activity is the activation of extensors. Reticulospinal inhibition and excitation of extensor motor neurons tend to cancel, but reticulospinal inhibition of interneurons mediating flexor reflexes results in net extensor activity. The high firing rates of both α and γ motor neurons facilitated by vestibulospinal and reticulospinal inputs requires on-going sensory input from muscle spindles.

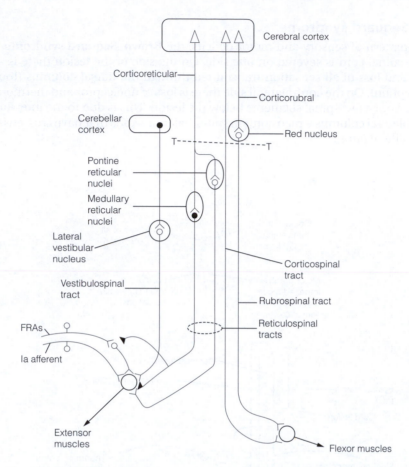

Figure 2. Illustration of the principal descending inputs to motor neuron pools. Fusimotor fibers (not shown) are generally affected in the same manner as α motor neurons. A transection of the brainstem at the level of T–T causes decerebrate rigidity. FRAs, flexor reflex afferents.

Cerebrovascular accident

Clinicians distinguish deficits due to lesions of **lower motor neurons** (brainstem and spinal cord neurons innervating skeletal muscle) from those due to lesions of **upper motor neurons**, which refers not only to the corticospinal and corticobulbar neurons of the pyramidal tract, but also to cortical cells that drive reticulospinal (medial pathway) neurons. Hence, upper motor neuron lesions produce much more severe deficits than those due to lesions of corticospinal or corticobulbar neurons alone. Strokes are the major cause of upper motor neuron lesions.

The commonest **cerebrovascular accident** (**CVA, stroke**) is caused by a thromboembolism affecting the branch of the middle cerebral artery that supplies the internal capsule. Infarction of the internal capsule produces a syndrome which *does not* resemble experimental lesions of the lateral motor pathways because the internal capsule also contains corticoreticular axons which drive lateral and medial reticulospinal tracts. After an initial period of flaccid paralysis and lack of reflexes on the side opposite the lesion two major deficits are seen:

- **Hemiparesis**. Muscle weakness on one side that is greatest in arm extensors and leg flexors, because arm flexors are stronger than extensors and in the legs the reverse is true. If the corticobulbar fibers are affected, voluntary facial movements are compromised. When the weakness is so severe that paralysis results the term **hemiplegia** is used. Weakness occurs because the loss of descending excitation means fewer motor units are recruited.

- **Spasticity**. An increase in **muscle tone** is seen in the stronger (antigravity) limb muscles. It is caused by enhanced excitability of the stretch reflex, particularly the phasic component, since attempts at rapid muscle stretch are met with much greater resistance than slow stretch. Forceful attempts to stretch a muscle are met with great resistance (caused by the stretch reflex), which fails suddenly (the **clasp knife response**) due to firing of high threshold muscle (non-spindle) afferents.

Spasticity in part results from the loss of presynaptic inhibition on Ia terminals. Normally, presynaptic inhibition is brought about by the action of the reticulospinal tracts on GABA-ergic Ia presynaptic inhibitory interneurons. GABA released from these interneurons acts on $GABA_B$ and $GABA_A$ receptors on the Ia terminals. $GABA_B$ receptors are metabotropic receptors and when stimulated act via G_i proteins to increase the K^+ conductance. The resulting hyperpolarization reduces Ca^{2+} influx into the primary afferent terminal, so curtailing the release of glutamate onto the motor neurons. In spasticity, this descending reticulospinal input is lost. This leads to failure of presynaptic inhibition and so hyperexcitability of the stretch reflex.

Baclofen is an agonist at $GABA_B$ receptors and is used orally and intrathecally in the treatment of spasticity. Because benzodiazepines (e.g., diazepam) are agonists at the $GABA_A$ receptors involved in presynaptic inhibition, they too can be used in spasticity.

K3 Anatomy of the cerebellum

Key Notes	
Gross anatomy	The cerebellum is divided into three lobes; anterior, posterior, and flocculonodular, each of which are subdivided into lobules. Longitudinally it has a central vermis and two lateral hemispheres. The cerebellum is covered with a cortex and embedded within its core of white matter are deep nuclei which, with the vestibular nuclei, provide the cerebellar output.
Cerebellar connections	Input to the cerebellum is from the spinal cord and brainstem sensory systems, cerebral cortex and inferior olive.
Proprioceptor pathways	Proprioceptor input is used by the cerebellum to provide feedback on motor performance. Collaterals of axons from the upper body ascend in the dorsal columns to the accessory cuneate nucleus which projects to the cerebellum. Proprioceptor afferents from the lower body terminate in Clark's column of the dorsal horn which gives rise to the dorsal spinocerebellar tract. Conscious proprioception (body position sense) is relayed via dorsal columns and medial lemniscus to the somatosensory cortex. A ventral spinocerebellar tract from the ventral horn conveys information from spinal circuits controlling locomotion.
Functional subdivisions of the cerebellum	The flocculonodular lobe is the vestibulocerebellum. The anterior and posterior lobes have three sagittal zones, each with distinct inputs and deep nuclei outputs. The medial zone (vermis) sends its output via the fastigial nucleus. The intermediate zone output is via the interposed nucleus. The medial and intermediate zones together make up the spinocerebellum. The lateral zone is the cerebrocerebellum, the output of which goes by way of the dentate nucleus.
Vestibulocere-bellum	The vestibulocerebellum is the flocculonodular lobe and gets input from the vestibular system. Lesions cause defects in balance.
Spinocerebellum	The medial spinocerebellum controls postural adjustments via the medial motor system, using inputs from several sensory modalities. When lesioned, animals cannot stand or walk. The intermediate spinocerebellum gets proprioceptor and sensorimotor input from the cerebral cortex via the corticopontinecerebellar pathway. It exerts control over

	limb movements via the lateral motor system and damage disrupts the ability to make accurate limb movements.	
Cerebrocerebellum	The cerebrocerebellum gets sensory, motor, and cognitive input from the cerebral cortex. Lesions have only small effects on movements involving single joints but severe effects on more complex multijoint movements, and on language.	
Related topics	(F2) Touch (G7) Oculomotor control and visual attention (J4) Spinal motor function	(K1) Cortical control of voluntary movement (K4) Cerebellar cortical circuitry (K5) Cerebellar function

Gross anatomy

The human cerebellum is about one quarter of the mass of the brain and is divided into three lobes, the **anterior lobe** and **posterior lobe**, separated by the primary fissure, and the **flocculonodular lobe**, separated from the posterior lobe by the posterolateral fissure (Figure 1). The lobes are further subdivided into **lobules** which are differently named in humans and other mammals. Longitudinally the cerebellum has a central **vermis** and two **lateral hemispheres**.

Over the surface of the cerebellum lies the cerebellar cortex which is folded into coronal strips called **folia** (singular; **folium**). The cerebellum has an internal core of white matter containing **deep** (**intracerebellar**) **nuclei**. These, together with the vestibular nuclei, provide the output of the cerebellum. Afferent and efferent connections of the cerebellum go by way of three pairs of **cerebellar peduncles**.

Cerebellar connections

Input to the cerebellum includes:

- Skin, proprioceptor, and vestibular input from brainstem and spinal cord

- Cognitive, motor, and sensory signals from the cerebral cortex via pontine nuclei in the huge corticopontinecerebellar tract

- Motor information from the inferior olivary nucleus

Details of these inputs are given in Table 1.

Cerebellar output goes to three principal destinations:

- The ventrobasal thalamus which projects to the motor cortex, to modify corticospinal tract outflow

- The red nucleus, to modify the rubrospinal tract output

- Vestibular and reticular nuclei to modulate the activity of medial motor pathways

These general input–output relations are depicted in Figure 2, although there are variations on this arrangement for different parts of the cerebellum.

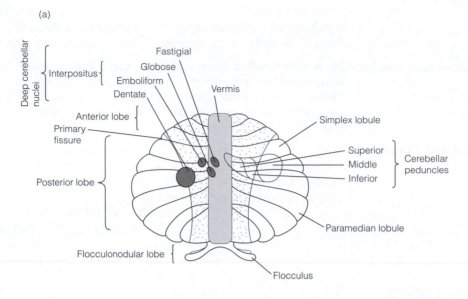

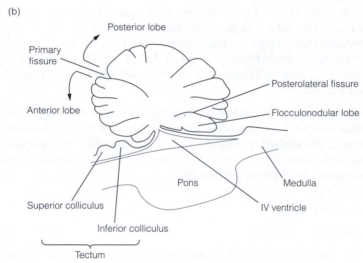

Figure 1. Cerebellar anatomy. (a) A diagram of the cerebellum, unfurled and viewed from above. Locations of the deep cerebellar nuclei are shown on the left and of the cerebellar peduncles on the right. The medial zone is pale gray, the intermediate zone stippled, and the lateral zone clear. (b) Sagittal section through the brainstem and cerebellum.

Proprioceptor pathways

Proprioceptor information coming from muscle spindles, Golgi tendon organs, and receptors in joints provides for conscious awareness of body position and movement. It is also used by the cerebellum to guide motor performance. Proprioceptor input from the neck, arms, and upper trunk is relayed in the dorsal columns to the cuneate nucleus, from where it follows exactly the same path as touch sensation from the same areas. This is the route for conscious upper body proprioception. Axon collaterals from the dorsal columns go to the **accessory cuneate** (**external arcuate**) **nucleus**, the origin of the **cuneocerebellar tract** that supplies upper body proprioceptor input to the cerebellum.

Table 1. Principle inputs to the cerebellum

Tract	Origin	Peduncle	Distribution	Modality
Vestibulocerebellar	Vestibular nuclei	ICP	Crossed and uncrossed to flocculonodular lobe cortex and fastigeal nucleus	Vestibular
Trigeminocerebellar	Secondary afferents in trigeminal nerve (V nerve) nuclei	ICP	Crossed and uncrossed	Proprioceptive and cutaneous somatosensory from jaw and face
Cuneocerebellar	Accessory cuneate nucleus	ICP	Uncrossed	Proprioceptive from arm and neck
Dorsal spinocerebellar	Clark's column	ICP	Uncrossed	Proprioceptive and cutaneous somatosensory from trunk and leg
Ventral spinocerebellar	Ventral horn	SCP	Crossed and uncrossed	Proprioceptive and cutaneous somatosensory from all parts of the body
Tectocerebellar	Superior colliculi, inferior colliculi	SCP	Crossed	Visual and auditory
Pontocerebellar	Pontine nuclei	MCP	Crossed	Cognitive, motor, somatosensory, and visual from cerebral cortex
Olivocerebellar[a]	Inferior olivary nucleus	ICP	Crossed to all deep cerebellar nuclei	Motor error signals

[a] The olivocerebellar input is via climbing fibers, all other afferents are mossy fibers. ICP, Inferior cerebellar peduncle; SCP, superior cerebellar peduncle; MCP, middle cerebellar peduncle.

A separate proprioceptor pathway serves the lower trunk and legs. Proprioceptor afferents from the lower body enter the **nucleus dorsalis (Clark's column)**, located in lamina VII between spinal segments C8 to L3. Axons of Clarke's column neurons ascend on the same side as the **dorsal spinocerebellar tract (DST)** to the cerebellum. Collaterals of the DST synapse with neurons in **nucleus Z** which project to the medial lemniscus to provide input for conscious lower body proprioception.

The **ventral spinocerebellar tract (VST)** arises from the ventral horn and transmits signals reflecting the current state of spinal cord central pattern generators involved in locomotion.

Functional subdivisions of the cerebellum

The inputs and outputs of the cerebellum are segregated giving rise to three functional subdivisions. The vestibulocerebellum corresponds to the flocculonodular lobe. The anterior and posterior lobes are organized into three parallel sagittal zones in humans

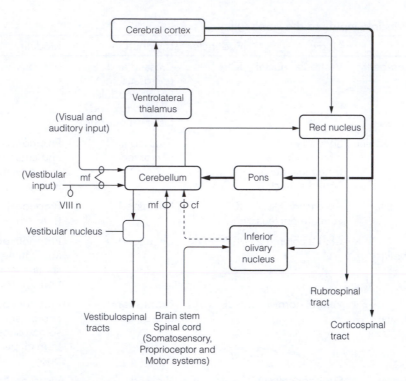

Figure 2. Major connections of the cerebellum. cf, climbing fibers; mf, mossy fibers;
| , corticopontine cerebellar tract.

(Figure 1a). The medial zone occupies the vermis and sends its output to the **fastigial nucleus**. The intermediate zone projects to the **interpositus nucleus** (separate **emboliform** and **globose** nuclei in humans). Together the medial and intermediate zones constitute the spinocerebellum. The lateral zone of the posterior lobe corresponds to the cerebrocerebellum the cortex of which sends its output to the **dentate nucleus**.

Vestibulocerebellum

The **vestibulocerebellum** corresponds to the flocculonodular lobe. It gets input from the ipsilateral vestibular labyrinth via the vestibulocochlear (VIII) nerve, and projects directly to the vestibular nuclei. These bring about postural adjustments via the lateral and medial vestibulospinal tracts. Lesions of the vestibulocerebellum in primates causes swaying and **truncal ataxia** (staggering gait). If the lesion is unilateral the head is tilted to the side of the injury and **nystagmus** is seen.

Spinocerebellum

The **spinocerebellum** consists of the anterior lobe and part of the posterior lobe. Its medial zone receives sensory input from vestibular, proprioceptor, and cutaneous somatosensory input from the trunk, together with visual and auditory input. The output from this part of the spinocerebellum goes by way of the fastigial nucleus to the vestibular nuclei. It controls postural adjustments in response to sensory input by signaling to axial muscles via the medial motor systems. In primates, inactivation of the fastigial nucleus causes animals to fall towards the side of the lesion.

The intermediate zone spinocerebellum gets input from the cuneocerebellar and dorsal spinocerebellar tracts conveying proprioceptor and cutaneous somatosensory data and from the ventral spinocerebellar tract which imparts information about the activity of spinal motor circuits. In addition, it receives inputs from the somatosensory and motor cortex via axon collaterals of the corticospinal tract that synapse with nuclei in the pons. These pontine nuclei give rise to fibers that cross the midline to enter the contralateral cerebellar cortex by way of the middle cerebellar peduncle. This **corticopontinecerebellar pathway**, containing 20 million axons, is one of the largest tracts in the CNS, and also goes to the cerebrocerebellum.

The output of the intermediate zone spinocerebellum is via the interpositus nucleus which projects to the ventrolateral thalamus and the red nucleus. By this route the spinocerebellum controls the lateral motor pathways to the limbs. Inactivation of the interpositus nuclei has little effect on standing or walking but results in a large amplitude tremor of the limbs when an animal attempts to reach for an object. This is known as an **intention tremor** and is often seen in humans with cerebellar damage.

Cerebrocerebellum

The **cerebrocerebellum** receives inputs from frontal, parietal, and occipital cerebral cortex relaying sensory, motor, and visual information by way of the corticopontinecerebellar tract. However, it also gets input from prefrontal cortex concerned with cognitive *not* motor functions. The cerebrocerebellar outflow via the dentate nucleus goes to the ventrolateral thalamus which in turn projects to frontal cortex motor and prefrontal areas. Additionally, the dentate nucleus has reciprocal connections with the red nucleus. Lesions of the cerebrocerebellar cortex or dentate nucleus cause slight delays and modest overshooting of movements involving single joints (**dysmetria**). However, for multijointed movements the deficits are much more severe, so animals and patients have difficulty using their fingers (**dysdiachokinesia**). Cognitive, including language, deficits are also seen in patients with cerebrocerebellar damage.

K4 Cerebellar cortex circuitry

Key Notes

Inputs to the cerebellar cortex	Axons of spinal cord and brainstem neurons conveying sensory and motor information enter the cerebellum as mossy fibers to synapse with granule cells in multisynaptic complexes called glomeruli. Granule cell axons bifurcate into parallel fibers which form synapses with thousands of Purkinje cells aligned in a row. Each parallel fiber makes just one synapse with each Purkinje cell but every Purkinje cell is contacted by 200 000 parallel fibers. Parallel fibers excite Purkinje cells to fire simple spikes. Climbing fibers from the inferior olive wrap themselves around 10 Purkinje cells making very powerful excitatory synapses with each one. Climbing fiber stimulation causes Purkinje cells to fire complex spikes.
Cerebellar cortex output	The output of the cerebellar cortex is provided exclusively by large GABAergic inhibitory Purkinje cells and goes to the deep cerebellar nuclei.
Cerebellar cortical interneurons	There are three types of GABAergic inhibitory neuron in the cerebellar cortex. Basket and stellate cells produce lateral inhibition by inhibiting those Purkinje cells that are the immediate neighbors of those activated by parallel fibers. Golgi II cells terminate the effect of parallel fibers on Purkinje cells. Hence inhibitory interneurons sculpt Purkinje cell output in space and time.
Somatotopic mapping	Inputs to the cerebellum are organized topographically and this gives rise to somatotopic maps in the cerebellar cortex, the deep cerebellar nuclei, and in their outputs to the red nucleus and thalamus. Each map represents not only sensory input but is also a motor output map.
Related topics	(K3) Anatomy of the cerebellum (K5) Cerebellar function

Inputs to the cerebellar cortex

The cerebellar cortex has three layers and contains five cell types that are organized into a simple circuit repeated millions of times (Figure 1).

The major input to the cerebellum is **mossy fibers** (**mf**), axons of second-order neurons from the spinal cord and brainstem conveying proprioceptor input, or the pontine–cerebellar relay from the cerebral cortex conveying sensory and motor signals. Each mossy fiber terminates in a discrete patch of cortex. Mossy fibers are glutamatergic and

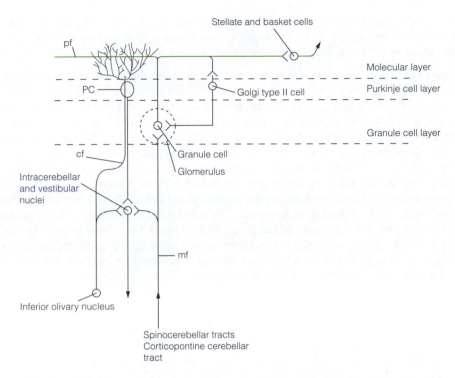

Figure 1. The basic circuitry of the cerebellar cortex. Mossy fibers (mf) and climbing fibers (cf) are excitatory, as are granule cells and the intracerebellar nuclei cells. All other cell types are inhibitory. Stellate and basket cells inhibit adjacent Purkinje cells (PCs). pf, parallel fiber.

excitatory, and after giving off an axon collateral which goes to the appropriate deep cerebellar (or vestibular) nuclei, they synapse with granule cells in synaptic complexes called **glomeruli** (Figure 2). Each glomerulus consists of the swollen terminal of a single mossy fiber which forms 15–20 synapses with the surrounding dendrites of four to five

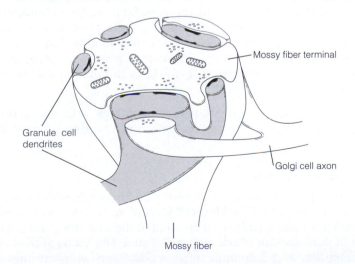

Figure 2. The structure of a cerebellar glomerulus. All the synapses are axodendritic.

granule cells. Mossy fibers branch, so each one can excite about 30 granule cells. Every granule cell is contacted by five to eight mossy fibers. Glomeruli also contain axodendritic synapses between Golgi cells (see below) and granule cells.

Granule cells in the **granule cell layer** are small (5–8 μm diameter) and one of the most numerous neurons types in the brain. Granule cell axons ascend to the most superficial layer of the cerebellar cortex, the **molecular layer**, where they bifurcate into **parallel fibers** that in primates extend for 6 mm or so in each direction along the long axis of a folium. Parallel fibers intersect with the perpendicularly oriented planar dendritic trees of **Purkinje neurons**. Because of this arrangement every parallel fiber excites a longitudinal beam of 2000–3000 Purkinje cells, making just a single synapse with each one. Every Purkinje cell is contacted by about 200 000 parallel fibers. Mossy fibers, and the granule cells driven by them, have high background firing rates (50–100 Hz) that is changed by sensory input and during movements. The effect of parallel fiber activity is to cause the Purkinje cell to fire **simple spikes** repetitively (Figure 3a). Purkinje cells fire at background rates between 20 and 50 Hz and even weak mossy fiber input produces increased Purkinje cell firing. Moreover, since Purkinje cells can fire in excess of 400 Hz they can follow a wide range in firing of mossy fiber inputs.

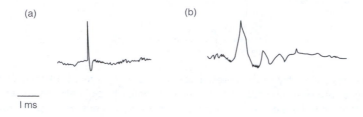

(a)   (b)

I ms

Figure 3. Simple (a) and complex (b) spikes of Purkinje cells produced by mossy fiber or climbing fiber activation respectively.

The second input to the cerebellum is **climbing fibers** that come exclusively from the **inferior olivary nucleus** (**ION**) via the **olivocerebellar tract**. Each climbing fiber, of which there are about 15 million in humans, establishes contact with around 10 Purkinje cells; each Purkinje cell getting input from just one climbing fiber that winds its way round the soma and dendrites making about 300 powerful glutamatergic, excitatory synapses. Climbing fibers fire with a frequency of 1–10 Hz, each time causing the Purkinje cell to discharge a **complex spike** (Figure 3b).

A third diffuse set of inputs come from monoaminergic cells in the brainstem. They establish sparse connections with deep cerebellar nuclei and cortex to produce modulating effects.

Cerebellar cortex output

The sole output of the cerebellar cortex is via the Purkinje cells, which have their large cell bodies (50 μm diameter) in the **Purkinje cell layer** of the cortex. Their extensive dendrites are aligned into a flat plane and are all oriented in the same direction, at right angles to the long axis of the folium in which they are located. The axons of the Purkinje cells go to the deep cerebellar nuclei. Purkinje cells use GABA as a neurotransmitter so the entire output of the cerebellar cortex is inhibitory.

Cerebellar cortical interneurons

The cerebellar cortex has three types of GABAergic inhibitory interneuron. **Basket cells** and **stellate cells** in the molecular layer receive input from parallel fibers and send axons at right angles to them to synapse with proximal or distal dendrites, respectively, of neighboring Purkinje cells. Activation of a mossy fiber excites a cluster of granule cells and hence stimulates linear arrays of **on-beam** Purkinje cells via the parallel fibers. However, the basket and stellate cells inhibit surrounding **off-beam** Purkinje cells. This is a surround antagonism mechanism that produces spatial focusing of cerebellar cortex output. It is akin to the lateral inhibition seen in sensory systems.

Golgi cells get input from parallel fibers and synapse with granule cells to produce feedback inhibition. In this way the Golgi cell brings about temporal focusing so that the net effect of mossy fiber input is *brief* firing of Purkinje cells.

In summary, the inhibitory interneurons constrain Purkinje cell output both in space and time.

Somatotopic mapping

Mossy fiber input and climbing fiber input are organized topographically, giving somatotopic maps in the cerebellar cortex that are retained in the deep cerebellar nuclei, and in their output to the thalamus and red nucleus. The cerebellar cortex has several maps which exhibit fractured somatotopy (Figure 4) and encode visual and auditory input from the tectum as well as somatosensory input. Each of these representations is actually three maps in register. One is formed by mossy fiber input, a second is corticopontine input and the third is an output map that preserves a somatotopic projection of movements.

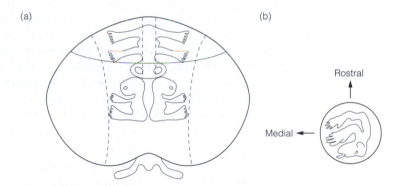

Figure 4. Somatotopic maps in: (a) cerebellar cortex based on inputs and clinical lesions, the fractured nature of these inputs revealed by detailed studies is not shown; (b) deep cerebellar nucleus.

K5 Cerebellar function

Key Notes

General principles	The cerebellum coordinates precise timing of movements initiated from the motor cortex, can initiate movements itself, and learn new motor tasks. The cerebellum is important for complicated multijoint movements. It can operate in feedback or feedforward mode.
Feedback mode	Here the cerebellum compares motor intentions with motor performance. A discrepancy between these generates an error signal which is used to make the mismatch smaller. Sensory signals caused by movement errors activate mossy fibers which excite Purkinje cells. These inhibit the deep cerebellar nuclei which drives the red nucleus and thalamus to prevent the erroneous movement.
Feedforward mode	For movements that are too fast for feedback to operate, the cerebellum runs pre-programmed sequences that have predictable effects on motor function. This feedforward works well provided nothing unexpected happens.
Motor learning	In humans most voluntary movements must be learnt. In motor learning the cerebellum acquires a program which specifies the motor commands needed for a given movement by trial and error. Sensory errors are translated into motor errors by the inferior olive and sent to the cerebellum via climbing fibers. These motor error signals cause the Purkinje cells to become less responsive to the mossy fiber input occurring at the same time. Whenever the same input recurs the Purkinje cells are excited less than before the learning.
Related topics	(J4) Spinal motor function (K4) Cerebellar cortical (K1) Cortical control of circuitry voluntary movement (N5) Motor learning in the (K3) Anatomy of the cerebellum: LTD cerebellum

General principles

Despite the functional subdivisions, the same circuit is repeated across the entire cerebellum, so it is likely that the same computations are performed by all parts of the cerebellum. The cerebellum seems to be concerned with high precision timing of movements initiated by the motor cortex. In addition the cerebrocerebellum can *initiate* movements, particularly in response to visual and auditory stimulation. In movements triggered in this way the order of activation is: dentate nucleus–motor cortex–interpositus nucleus–muscle. Furthermore, central to the operation of the cerebellum is motor learning, the acquisition of new motor skills.

In primates, parallel fibers average 6 mm in length and so affect a comparable length array of Purkinje cells (PCs) that lie across the cerebellum. This is sufficiently long to span an entire deep cerebellar nucleus or bridge adjacent nuclei. PC arrays coupling both fastigial nuclei, for example, would ensure coordination of postural muscles across the midline, which is important in gait. The PC arrays influenced by a given set of parallel fibers span muscles over several joints. This anatomical configuration supports the results of recording and lesion studies showing that the cerebellum is much more concerned with control of movements involving many joints rather than single joints.

The cerebellum is thought to operate in one of two modes, feedback or feedforward, depending on the circumstances.

Feedback mode

During the execution of well-rehearsed movements that are not too fast the cerebellum acts as a feedback device to compare motor intentions with motor performance, and works to reduce any mismatch between them. For the spinocerebellum, the motor intentions are the signals relayed by the corticopontinecerebellar tract. Motor performance is monitored by proprioceptor (and other sensory) input, and by the ventral spinocerebellar signals reporting on the activity of spinal cord and brainstem motor circuits. Similarly, the cerebrocerebellum compares inputs from the supplementary motor cortex and the primary motor cortex to produce error signals that reflect a discrepancy between motor planning and motor commands. In each case the error signals are used to correct the mismatch.

The intermediate spinocerebellum seems to be involved in correcting errors in limb movements such as when a limb is perturbed by unexpected force. The order in which various neural elements fire in this case is: muscle afferents–interpositus nucleus–motor cortex–dentate nucleus.

Feedback error correction probably works as follows. An error means that the actual position of a limb is not the intended one. This produces unpredicted muscle stretch because the appropriate coactivation of α and γ efferents to the muscle spindle has not happened. Precisely the same thing happens if an unexpected force is applied to a limb. In either case the muscle stretch will excite Ia and Ib afferents in the loaded muscles. These proprioceptor signals are relayed to the cerebellum by mossy fibers. The stimulated mossy fibers deliver tonic excitation to the intracerebellar (interpositus) nucleus via axon collaterals and stimulate a group of granule cells. The granule cell parallel fibers activate several arrays of on-beam Purkinje cells (Figure1) which strongly inhibit their target neurons in the interpositus nucleus. Normally, when mossy fibers are firing at background rates their tonic facilitation of the interpositus dominates the inhibition by Purkinje cells. Consequently, interpositus neurons maintain excitation of the red nucleus and the ventrolateral thalamus. When the mossy fibers are activated during a movement, however, Purkinje cell inhibition **disfacilitates** the interpositus neurons and this inhibition is transmitted downstream to the red nucleus and thalamus. In contrast, neighboring off-beam PC arrays, inhibited by the GABAergic interneurons in the cortex allow *their* interpositus cells to fire at higher than background rates. Thus, the pattern of activation of the deep cerebellar nucleus is a negative image of the input activation.

The overall effect, mediated by the rubrospinal and corticospinal tracts, is to correct the movement error by activating spinal reflexes that defend the correct limb position and dampen those that do not. This is probably done by adjusting the firing of γ efferents to muscle spindles.

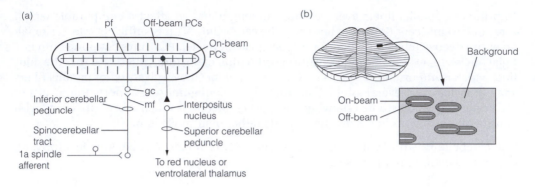

Figure 1. (a) Mossy fiber (mf) input activates an array on on-beam Purkinje cells (PCs). Each short vertical line represents the planar dendritic field of a PC, viewed from above. The output of only one PC is shown. (b) Pattern of activation produced by single mossy fiber input. gc, granule cell; pf, parallel fiber.

The intermediate spinocerebellum seems to control the precise timing of the contraction of agonist and antagonist muscles during a movement. During reciprocal activation of agonist and antagonist muscles, Purkinje cells responsible for controlling these muscles fire alternately, driving interpositus neurons to do the same. During co-contraction, however, the Purkinje cells are silent. A role for the cerebellum in organizing these patterns of muscle activity is supported by the fact that the action tremor resulting from lesions of the interpositus nucleus or intermediate cerebellar cortex appears to be due to derangement in the timing of agonist contraction. In normal humans, a rapid wrist movement involves an initial burst of activity in the agonist, followed by a burst in the antagonist to produce braking, and finally a second agonist burst to stabilize the joint at the desired end point (refer to Figure 2 in Section J4 for an illustration of this pattern of activity). In cerebellar tremor the start of the movement is normal but the second agonist burst is late. Consequently the antagonist burst moves the wrist beyond the end point. This sets up the tremor.

Feedforward mode

For the execution of well-practiced, very rapid (**ballistic**) movements (e.g., playing fast passages on a musical instrument or a tennis serve), there is insufficient time for feedback correction of errors. For these movements the cerebellum operates in a feedforward mode in which it runs a program that predicts the motor consequences of its own action. *Unexpected* perturbations that occur when the cerebellum is in this mode cannot be corrected for in time and so performance will be degraded.

Motor learning

The predictions inherent to feedforward operation must be learnt during numerous trials attempting to perform the task. This is motor learning and is probably important in acquiring skill in all voluntary motor tasks, including (in humans) learning to walk.

In one model of motor learning (the **cerebellar feedback-error-learning model**) the cerebellum acquires a program called an **inverse model** of the motor task, which is the transformation from the desired trajectory to the motor commands needed to bring about the movement (Figure 2). In this model movement errors are initially represented

Figure 2. The operation of the inverse model on the controlled system (motor neurons and muscles) turns a desired into an actual trajectory. The better the inverse model the closer the actual is to the desired trajectory.

as sensory errors. For example, during a tennis match an incorrect arm movement will be visually obvious from the direction of the ball, and an error in playing a musical instrument will sound wrong. Sensory errors are converted into a neural firing pattern that specifies errors in motor performance, an operation of the **inferior olivary nucleus (inferior olive)**. The inferior olive sends these motor error signals to the cerebellum via the olivocerebellar tract climbing fibers (Figure 3). Hence, the cerebellum gets mossy fiber input that represents the desired trajectory (from motor cortex), sensory input (e.g., from visual cortex or proprioceptor pathways), and feedback from the inferior olive that represents errors in motor performance.

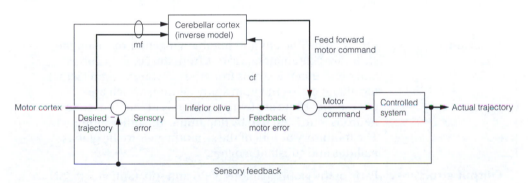

Figure 3. A model for motor learning by the cerebellum. When the cerebellum is operating in feedforward mode the bold pathway is activated. cf, climbing fibers; mf, mossy fibers.

The error signals carried by the climbing fibers causes concurrent mossy fiber input to become less effective in activating Purkinje cells. Whenever the same pattern of mossy fiber input occurs subsequently, the Purkinje cells fire fewer simple spikes and cause less inhibition on downstream motor pathways. This corrects the motor output. The cellular mechanism that alters the responsiveness of the Purkinje cells is long term depression.

K6 Anatomy of the basal ganglia

Key Notes	
Overview	The basal ganglia are important in selecting appropriate learned motor sequences movement and stimulus–reward learning to form novel motor strategies. Several interconnected structures make up the basal ganglia; striatum, globus pallidus, substantia nigra, and subthalamus. Cerebral cortical input to the basal ganglia goes to the striatum. The basal ganglia output goes from the globus pallidus and substantia nigra to the cortex via the thalamus. The basal ganglia are responsible for producing motor sequences during voluntary movement.
Striatum	The caudate nucleus and putamen together make up the striatum. Glutamatergic axons from the cerebral cortex and dopaminergic axons from the substantia nigra (SNpc) terminate on the medium spiny neurons which are GABAergic. There are two populations of medium spiny neurons, one is excited by dopamine, the other inhibited. The inhibitory output of the striatum goes to the globus pallidus and substantia nigra.
Output structures of the basal ganglia	Parts of the globus pallidus (GPi) and substantia nigra (SNpr) send axons to specific thalamic nuclei which in turn project to particular areas of the cerebral cortex. This circuitry provides basal ganglia control of limb, facial, and eye movements. Part of the globus pallidus (GPe) projects to the subthalamic nucleus.
Subthalamic nucleus	Excitatory neurons of this nucleus are excited by the motor cortex, inhibited by the globus pallidus (GPe) and send their axons to the globus pallidus and substantia nigra.
Parallel processing in the basal ganglia	Somatotopic maps are found in all basal ganglia nuclei except the compact part of the substantia nigra. Five circuits form loops between specific regions of the cortex and basal ganglia and each circuit seems to have a distinct functional role. One of the circuits is concerned with motor output, the others with eye movements, executive functions, limbic cortex, and lateral orbitofrontal cortex.
Related topics	(D4) Dopamine (K1) Cortical control of voluntary movement

Overview

The classical role of the basal ganglia is in selecting appropriate learned motor sequences, and in stimulus–reward learning to form novel motor strategies. They consist of several extensively interconnected structures, the striatum (the caudate and putamen), the globus pallidus (pars interna and pars externa), the substantia nigra (pars compacta and pars reticulata), and the subthalamic nucleus. Most inputs to the basal ganglia are from the cerebral cortex and enter the striatum. The output of the basal ganglia emerges from the pars interna (internal part) of the globus pallidus, and the substantia nigra pars reticulata, to go to the thalamus. The thalamus projects back to the cortex thus closing a loop. The thalamocortical axons return to the same region of cortex which gave rise to the striatal inputs (Figure 1).

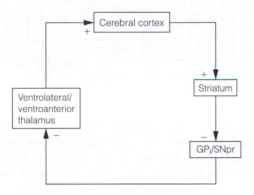

Figure 1. Block diagram of the interface between basal ganglia and cerebral cortex. GPi, globus pallidus pars interna; SNpr, substantia nigra pars reticulata.

Motor basal ganglia circuitry is responsible for the execution of appropriate preprogrammed motor sequences during voluntary movements. Classically the basal ganglia constitute part of the **extrapyramidal system**, on the basis that lesions of the basal ganglia produce quite different symptoms from lesions of the corticospinal tract.

Striatum

The caudate nucleus and putamen are functionally a single unit, the **dorsal striatum** (neostriatum) but are split anatomically by the internal capsule. The adjacent ventral striatum (nucleus accumbens), which is part of the limbic system, has similar circuitry.

The striatum receives excitatory input from the cortex via the glutamatergic corticostriate pathway (Figure 2). The input is organized topographically so that somatotopy is preserved in the projections from the somatosensory cortex and motor cortex. Corticostriate axons terminate on the major neuron type in the striatum, the **medium spiny neuron**. These make up 95% of striatal neurons, use GABA as their transmitter, and provide the inhibitory output of the striatum. There are two populations of medium spiny neurons with different connections and neurochemistry. One type has **substance P** (**SP**) and **dynorphin** (**DYN**) as co-transmitters, expresses dopamine D_1 receptors, and projects to the globus pallidus pars interna (GPi) and **substantia nigra pars reticulata** (**SNpr**). The second uses **enkephalin** (**ENK**) as a co-transmitter, expresses D_2 dopamine receptors, and projects to the globus pallidus pars externa (GPe).

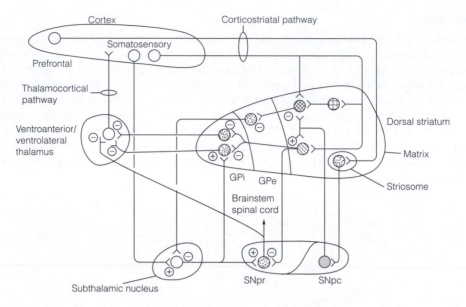

Figure 2. Connections of the basal ganglia. GABAergic neurons in GPe also project to GPi and SNpr (not shown). GPe, globus pallidus pars externa; GPi, globus pallidus pars interna; SNpc, substantia nigra pars compacta; SNpr, substantia nigra pars reticulata. ⊕, excitatory synapse; ⊖, inhibitory synapse. The neurons are coded by their neurotransmitter: ○, glutamate; ⊕, acetylcholine; ○, dopamine; ⊗, GABA; ◐, GABA/substance P/dynorphin; ⊛, GABA/enkephalin.

The medium spiny neurons receive projections via the nigrostriatal pathway from the **substantia nigra pars compacta** (**SNpc**). The SNpc uses dopamine as a transmitter. Because the two types of medium spiny neuron express different dopamine receptors they are differently modulated by this input. At the GABA/SP/DYN cells, dopamine acting on D_1 receptors enhances the effect of excitatory cortical input. In contrast, the action of dopamine on D_2 receptors on GABA/ENK cells is to reduce the effect of cortical excitation. Medium spiny neurons have a third input from large **aspiny interneurons** that constitute about 2% of striatal neurons. These cells use ACh as a transmitter, are excitatory, and driven by cortical inputs.

Staining of the striatum for acetylcholinesterase shows it to be compartmented into a heavily stained **matrix** and a lightly stained three-dimensional labyrinth, the **striosomes**, that are about 10–20% of the striatal bulk. The connectivity of cells in these two compartments is different. The matrix gets inputs from throughout the cerebral cortex and sends outputs to the entire globus pallidus and SNpr, whereas the striosomes get restricted input from the prefrontal cortex and project to the SNpc. The matrix is concerned with sensorimotor function, while striosomes are associated with the limbic system, and may control the dopaminergic pathway from SNpc to striatum.

Output structures of the basal ganglia

Both the globus pallidus and substantia nigra are divided into two parts. The **globus pallidus pars interna** (**GPi**) of primates (equivalent to the **entopeduncular nucleus** in rodents) and the substantia nigra pars reticulata (SNpr) have very similar structures and are functionally equivalent. Both get inhibitory connections from the GABA/SP/DYN population of striatal neurons and excitatory inputs from the subthalamic nucleus, and both send GABAergic inhibitory outputs to the thalamus. The thalamus in turn projects to specific locations in the cerebral cortex. The GPi and SNpr make connections with several thalamic nuclei that project to motor cortex, providing basal ganglia output for limb and facial movements. Part of the SNpr is concerned with eye movements.

The **globus pallidus pars externa** (**GPe**) receives its striatal connections from the GABA/ENK medium spiny neurons. The GPe neurons are GABAergic and project mostly to the subthalamic nucleus.

Subthalamic nucleus

The subthalamic nucleus (**STN**) lies at the junction between midbrain and diencephalon, and is particularly well developed in primates. It gets excitatory input from the motor cortex and GABAergic input from the GPe. The STN neurons use glutamate as a transmitter, are excitatory, and send their axons principally to the GPi and SNpr.

Parallel processing in the basal ganglia

Somatotopic representations exist in all structures of the basal ganglia, except the SNpc. The output of the basal ganglia to the thalamus establishes somatotopic maps there. Hence, the connectivity of basal ganglia circuits allows highly specific and well-focused behavioral outcomes. The somatotopic maps in the thalamus produced by the basal ganglia are separate from those produced by the cerebellum.

The basal ganglia are important in many functions besides the classic motor activities. There are five parallel circuits broadly like that in Figure 1, each getting corticostriatal inputs from an area of frontal cortex and projecting back to the same area by way of specific basal ganglia regions and thalamic nuclei. These are referred to as **basal ganglia-thalamocortical circuits** (Table 1). Apart from the motor circuit there are circuits associated with eye movement, executive functions (problem solving, working memory), the limbic system, and lateral orbitofrontal cortex. The limbic circuit, unlike the others, goes by way of the ventral striatum (nucleus accumbens), which is part of the dopamine motivation system, rather than the caudate or putamen. The limbic circuit contains the medial orbitofrontal cortex. The orbitofrontal cortex (OFC) seems to be implicated in rapid stimulus-reinforcement learning and its reversal, with the medial OFC being more concerned with positive reinforcers and the lateral OFC with negative reinforcers. In one model the OFC is the brain structure responsible for measuring the hedonic quality of stimuli which is crucial for goal-directed behaviors such as eating and drinking. Both the limbic and lateral orbitofrontal circuits are important in emotion.

Since the circuits are all wired in much the same way it is likely that they all perform the same computations. The different outcomes of the operation of each of the circuits depends on the cortical regions they are connected to, and the contexts in which they are activated. Interestingly basal ganglia disorders can produce deficits in thought processes not unlike those that affect movement.

Table 1. Parallel basal ganglia-thalamocortical circuits

Component	Circuit				
	Motor	Oculomotor	Executive	Limbic	Association
Cortical area	MI (4), MII(6), somatosensory (1, 2, 3)	Frontal eye fields (8)	Dorsolateral prefrontal (DLPFC, 9, 46)	Anterior cingulate (25), medial orbitofrontal (10, 11)	Lateral orbitofrontal (12, 47)
Striatal target	Putamen	Caudate	Caudate	Nucleus accumbens	Caudate
Striatal output	GPi/SNpr[a]	GPi/SNpr	GPi/SNpr	Ventral pallidum	GPi/SNpr
Thalamic nuclei	Ventrolateral	Mediodorsal	Ventroanterior, medial dorsal	Medial dorsal	Ventroanterior, medial dorsal
Functions	Motor	Eye movements	Planning, problem solving, working memory, spatial working memory, long-term reward learning, self control. Inactivated in REM sleep	Evaluating errors and conflict, motor expression of emotion, short-term reward learning (positive reinforcers; e.g., attractive faces, financial reward)	Short-term reward learning (negative reinforcers), suppression of previously learnt associations
Lesions	Hypo- and hyperkinetic disorders	Oculomotor hypokinesia	Deficits in planning, working memory, self control. Craving in addictions	Akinetic mutism	Defects in social behavior, risk taking, obsessive-compulsive disorder

[a] GPi, globus pallidus pars interna; SNpr, substantia nigra pars reticulata.

K7 Basal ganglia function

Key Notes

Direct and indirect pathways	There are two pathways in basal ganglia circuits with opposing effects on firing of thalamic and cortical neurons. The direct pathway activates thalamic neurons and this allows movement sequences to occur. The indirect pathway inhibits thalamic neurons and suppresses unwanted movement. Both of these pathways are activated when the motor cortex initiates a specific movement.
Dopamine modulation	These two pathways are modulated by dopaminergic axons running from the substantia nigra to the striatum. Activity in this nigrostriatal tract enhances the direct pathway but suppresses the indirect pathway. Hence dopamine neurotransmission enables movements to occur.
Basal ganglia operation	The group of cells that represents a given movement sequence are normally inhibited by tonic discharge from the globus pallidus (GPe) and substantia nigra (SNpr). Executing a given movement needs activation of the direct pathway which inhibits the GPe and SNpr. The basic principle of basal ganglia action is to select the appropriate action from a repertoire on the basis of the most salient current inputs.
Basal ganglia disorders	Motor disorders due to malfunction of the basal ganglia are of two types. Hyperkinesias feature motor overactivity and include Huntington's disease, a genetic disorder in which the GABAergic medium spiny neurons of the striatum controlling the indirect pathway die. Hypokinesias are conditions in which motor activity is reduced. The most common, characterized by rigidity, slowness of movement, and tremor is Parkinson's disease. Obsessive compulsive disorder is characterized by an inability to stop repeating the same actions or thoughts, even when they are recognized to be unnecessary. Its cause may lie with reduced activity in the orbitofrontal basal ganglia circuit.
Related topics	(K6) Anatomy of the basal ganglia

Direct and indirect pathways

There are two routes through the basal ganglia circuitry with opposite effects on firing of thalamic, and hence cortical, neurons (Figure 1). The **direct pathway** uses the GABA/ SP/DYN medium spiny striatal neurons which inhibits GABAergic outflow of the GPi and SNpr to the thalamus. Cortical activation of this pathway *increases* the firing of thalamic neurons (since inhibiting an inhibition is excitation).

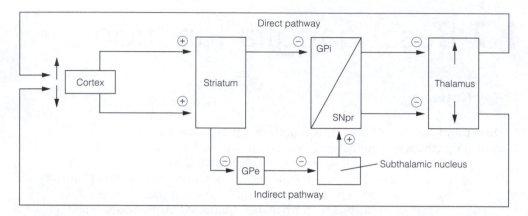

Figure 1. Direct and indirect pathway activation increased (↑) or decreased (↓) firing of thalamic (and cortical) neurons respectively. GPe, globus pallidus pars externa; GPi, globus pallidus pars interna; SNpr, substantia nigra pars reticulata.

The **indirect pathway** starts with the GABA/ENK medium spiny neuron output to the GPe, inhibitory neurons from which go to the STN. The STN excites inhibitory neurons in the GPi and SNpr that go to the thalamus. Corticostriate activation of the indirect pathway results in *decreased* firing of thalamic neurons.

This dual circuitry allows the possibility that given movement sequences may be triggered or suppressed by differential activation of direct or indirect pathways respectively.

Dopamine modulation

The dopaminergic neurons of the substantia nigra pars compacta (SNpc) alter their firing pattern in response to stimuli that reward a movement. They modulate the response of medium spiny neurons in the striatum to corticostriate inputs but have opposite effects on the two populations. While the GABA/SP/DYN neurons are made more excitable, the GABA/ENK cells become less excitable in the face of SNpc inputs. In summary, the direct pathway is enhanced by, and the indirect pathway suppressed by, the nigrostriatal tract from the SNpc.

Basal ganglia operation

One function of the basal ganglia is to enable the execution of motor sequences. Each sequence is represented by an array of cells, a micro-loop, within the basal ganglia-thalamocortical motor or oculomotor circuit and can be either activated or (if unwanted) inhibited. While some sequences are stereotyped movements, the circuitry for which is genetically specified, many sequences are learnt; that is, many micro-loops are entrained by experience.

At rest, most medium striatal neurons fire at low frequencies (0.1–1 Hz), while GPi and SNpr neurons have high background firing rates (about 100 Hz).

The current model is that movements are initiated by activity in the motor cortex which is relayed to the striatum. During a movement, striatal neurons increase firing, as a result of elevated activity of the corticostriatal neurons that drive them. Tonic inhibitory output of the GPi and SNpr at rest, which is increased about 50 ms before a movement by excitatory drive from the subthalamus, is due to the operation of the indirect pathway and results in widespread, and complete, suppression of unwanted movement sequences. Making a particular movement requires that the direct striatopallidal pathway to the GPi/SNpr cells that enable the movement (those belonging to the correct micro-loop) become activated. These GPi/SNpr cells reduce their firing, releasing their corresponding thalamocortical cells from inhibition. The nigrostriatal dopamine system acts to raise the likelihood that a movement sequence is actually made.

Current thinking is that the essential operating principle of the basal ganglia is to select the most appropriate behavior from among the repertoire of possibilities on the basis of the relative salience (importance) of prevailing inputs, each of which generates a phasic dopamine signal that corresponds to its salience. The role of dopamine is to promote reselection of behaviors and context that immediately precede unpredicted sensory events/rewards. In this way, over a series of trials, the basal ganglia learns an association between an event and the behavioral and contextual elements that elicited it, generating new memories.

The basal ganglia-thalamocortical circuits can switch between a top-down mode, in which behavior is triggered from the appropriate cortical region, to a bottom-up sensory input mode driven by attention and by unpredictable events. The switch is provided by excitatory input from the intralaminar nuclei of the thalamus to the striatum.

The basal ganglia make extensive connections with subcortical structures involved in sensorimotor function and motivation (e.g., superior colliculus, periaqueductal gray, parabrachial nucleus) forming circuits in which the thalamus sends input into the striatum rather than being on the output side of the basal ganglia (Figure 2). It is possible that these form extensions to the basal ganglia-thalamocortical circuits.

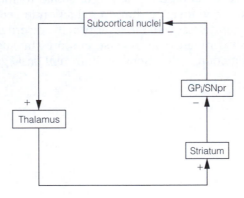

Figure 2. Subcortical-basal ganglia loop.

Basal ganglia disorders

Motor disorders arising from dysfunction of the basal ganglia, whether caused by disease, or by lesions in animal studies, fall into two distinct categories, hyperkinesias and hypokinesias:

- **Hyperkinesias**. These are disorders in which motor overactivity occurs. Characteristically these disorders consist of frequent, random, twitch-like or writhing movements, resembling fragments of normal movements, termed **choreoathetosis**. It is the principal symptom of Huntington's disease, and of **tardive dyskinesia**, an unwanted effect of the treatment of Parkinson's disease with L-DOPA, or infarcts of the subthalamic nucleus. **Huntington's disease** (**HD**) is a progressive neurodegenerative disorder in which motor and cognitive symptoms begin between 40 and 50 years of age. It is an autosomal dominant disease caused by an excessive number (> 36) of CAG repeats in the coding region of the gene for **huntingtin** (**Htt**) which consequently has a long string of glutamine residues near its N terminus. The normal functions of Htt and how the abnormal protein kills cells are unclear, but particularly afflicted are the GABA/ ENK medium spiny neurons of the striatum. Their death causes excessive inhibition of the subthalamic nucleus, so increased and inappropriate firing of thalamocortical neurons (Figure 2 in Section K6). In summary, the chorea is a failure of the indirect pathway to block unwanted movement sequences.

- **Hypokinesias**. The prototypical hypokinetic disorder is **Parkinson's disease**, characterized by rigidity, bradykinesia, and tremor. Here the primary lesion is the massive loss of dopaminergic cells in the SNpc. Since the normal situation is that dopamine tonically activates the direct pathway via D_1 receptors but inhibits the indirect pathway via D_2 receptors, the loss of the nigrostriatal pathway leads to an increase in the activity of GABAergic striatal neurons in the indirect pathway and a decrease in the GABAergic striatal neurons in the direct pathway. The overall effect is increased inhibition on the thalamocortical connections that are required for movement. Although this model accounts for the bradykinesia it does not explain the tremor.

Disorders of other basal ganglia-thalamocortical circuits seem to include **obsessive-compulsive disorder** (**OCD**). This is a chronic psychiatric disorder in which a person is unable to prevent themselves endlessly repeating the same actions or thoughts. The afflicted individual may spend many hours each day acting out pointless rituals, such as hand washing because of an obsessional fear of contamination. OCD appears to be excessive perseveration. Brain imaging shows a reduction in cerebral blood flow in the orbitofrontal cortex that correlates with the severity of the disorder in patients. Lesions of the orbitofrontal cortex in primates causes perseveration. This suggests that the cause of OCD may lie with dysfunction of the lateral orbitofrontal basal ganglia-thalamocortical circuit.

L1 Anatomy and connections of the hypothalamus

<table>
<tr><td colspan="2">Key Notes</td></tr>
<tr>
<td>Hypothalamus</td>
<td>The hypothalamus is a collection of nuclei clustered around the third ventricle implicated in sleep, appetitive behaviors, and the control of autonomic and endocrine functions.</td>
</tr>
<tr>
<td>Limbic connections of the hypothalamus</td>
<td>The hypothalamus is part of the limbic system. Input from the hippocampus enters the hypothalamus by way of the fornix. The mammillary bodies of the hypothalamus project (via the fornix) to the anterior thalamic nuclei and thus to the cingulate cortex which, in turn, sends output back to the hippocampus. This circuit is part of a larger network (including the amygdala and prefrontal cortex) which is implicated in emotion and memory.</td>
</tr>
<tr>
<td>Hypothalamic–pituitary connections</td>
<td>The pituitary gland consists of a posterior lobe and an anterior lobe. It is connected to the hypothalamus by a pituitary stalk. The large paraventricular nucleus neurons release their hormones into the posterior lobe. Hormones secreted into the median eminence by small hypothalamic cells are, in contrast, carried in a vascular network, the portal system, to the anterior lobe.</td>
</tr>
<tr>
<td>Related topics</td>
<td>
(A4) Organization of the central nervous system
(L2) Posterior pituitary function
(L3) Neuroendocrine control of metabolism and growth

(L4) Neuroendocrine control of reproduction
(M3) Control of feeding
(M5) Sleep
</td>
</tr>
</table>

Hypothalamus

The hypothalamus is involved in the control of a variety of functions: sleep–wakefulness, thermoregulation, feeding, metabolic energy expenditure, drinking and fluid homeostasis, growth and reproduction. Much of this is done by hypothalamic regulation of the autonomic nervous system and the pituitary gland.

The hypothalamus consists of many nuclei clustered around the third ventricle (Figure 1). At its anterior end the floor of the ventricle thickens to become the **median eminence** which projects as the **infundibulum** (part of the pituitary stalk) to the posterior pituitary gland. The hypothalamus is divided into three longitudinal **zones**—the innermost of which, the periventricular zone, surrounds the third ventricle—and four **subdivisions** along its rostro-caudal axis (Table 1).

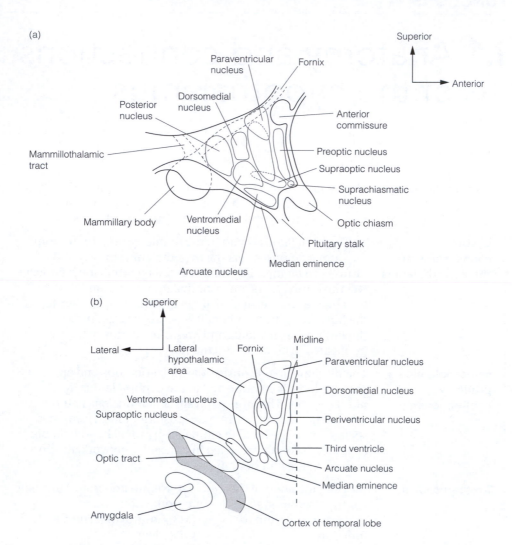

Figure 1. Human hypothalamus of the left cerebral hemisphere shown diagrammatically: (a) midsagittal section; (b) coronal section.

Limbic connections of the hypothalamus

The hypothalamus is connected with limbic structures concerned with emotion and its expression (Figure 2). It gets input from the hippocampus:

- By way of the **subiculum**, part of the hippocampal formation, which projects via the **fornix** to the mammillary bodies

- By way of the **septum**, also via the **fornix**, to connect with all three zones of the hypothalamus

Input from the amygdala arrives at the hypothalamus via the **stria terminalis**, a loop that follows a similar course to the fornix, and the **amygdalofugal pathway**.

Hypothalamic output from **mammillary bodies** (MB) via the **mammillothalamic tract** goes to the **anterior thalamic nuclei** (ATN) that are connected to the **cingulate cortex**

Table 1. Location of some hypothalamic nuclei

Subdivision	Zone		
	Periventricular	Medial	Lateral
Preoptic		Medial preoptic nucleus	Lateral preoptic nucleus
Anterior	Suprachiasmatic nucleus	Anterior nucleus	Supraoptic nucleus
	Paraventricular nucleus		
	Anterior periventricular nucleus		
Tuberal	Arcuate nucleus	Ventromedial nucleus	Lateral hypothalamic area
		Dorsomedial nucleus	
Mammillary	Posterior hypothalamic nucleus	Medial mammillary nuclei[a]	Lateral hypothalamic area
			Lateral mammillary nuclei[a]

[a] Mammillary bodies.

(**CC**). The cingulate cortex projects back to the hippocampal formation, so closing a loop (hypothalamus–ATN–CC–hippocampus–hypothalamus) termed the **Papez circuit**. This is part of a larger network that includes the septum, amygdala, and prefrontal cortex which is concerned with emotion and memory. The mammillary bodies make reciprocal connections via the **mammillotegmental tract** with the ventral tegmental nuclei in the midbrain; these have been implicated in memory.

The **medial forebrain bundle** passes through the lateral hypothalamic zone. It consists of monoaminergic axons ascending from brainstem nuclei. Many noradrenergic and serotonergic, but not dopaminergic, axons synapse with hypothalamic neurons.

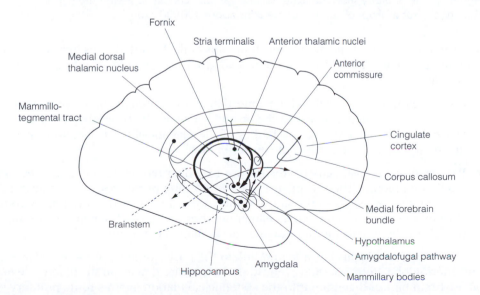

Figure 2. Major connections of the human hypothalamus.

The paraventricular hypothalamus and the lateral hypothalamic area receive visceral sensory input from the nucleus of the solitary tract (NST) which is important for hypothalamic control of the ANS.

Hypothalamic–pituitary connections

The **pituitary gland** is divided into the **neurohypophysis** and the **adenohypophysis**. The neurohypophysis, which is a direct outgrowth of the hypothalamus, consists of the **posterior lobe**, the median eminence, and the infundibulum. The adenohypophysis consists of the **anterior lobe**, an intermediate lobe (poorly developed in humans), and the pars tuberalis (an extension surrounding the infundibulum). The pars tuberalis and the infundibulum together make up the **pituitary stalk** (Figure 3).

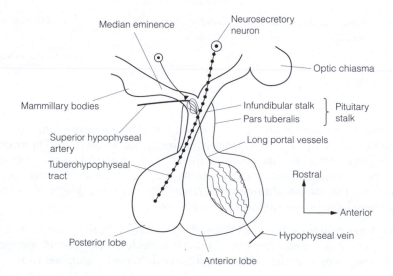

Figure 3. The pituitary showing its neural and vascular connections with the hypothalamus. In humans the tuberohypophyseal tract contains about 100 000 axons.

Many of the endocrine and autonomic functions of the hypothalamus involve the **paraventricular nuclei** (**PVN**). These contain two groups of peptide-secreting neuroendocrine cells:

1. **Magnocellular** (large) **cells** which send their axons through the median eminence down the infundibular stalk into the posterior lobe as the **tuberohypophyseal tract**. Hormones are made in the cell bodies of the magnocellular cells and transported down their axons for release in the posterior lobe.

2. **Parvocellular** (small) **cells** have short axons which terminate on capillaries in the median eminence. These capillaries drain into **long portal vessels** that descend to form venous sinusoids in the anterior lobe; this vascular bed is the **portal system**. Hormones secreted by the parvocellular neurons into the median eminence are carried via the hypothalamic-pituitary portal circulation to the anterior lobe.

Hence the hypothalamus has a *neural* link to the posterior lobe, but a *vascular* link to the anterior lobe. Both lobes have fenestrated capillaries that lie on the blood side of the blood–brain barrier that drain into the systemic circulation. By this route pituitary hormones are delivered to the body.

L2 Posterior pituitary function

Key Notes

Posterior lobe hormones	Neurons in the supraoptic (SON) and paraventricular (PVN) nuclei release two peptide hormones, arginine vasopressin (AVP) and oxytocin.
Arginine vasopressin (AVP)	AVP is secreted from the posterior pituitary in response to a rise in the osmolality of the extracellular fluid or a fall in blood volume. It acts to restore these by increasing the reabsorption of water by the kidney. Changes in osmolality are detected by neurons in a circumventricular organ which synapses on the PVN and SON. Alterations in blood volume are detected in two ways. Firstly, as changes in mean arterial blood pressure signaled by baroreceptors; reduced blood pressure causes a rise in AVP secretion. Secondly, a fall in blood volume detected by the kidney, which responds by secreting renin. This enzyme triggers a cascade that generates angiotensin II (AII), a peptide that stimulates AVP secretion and drinking.
Oxytocin	Oxytocin, released from cells in the PVN and SON different from those that secrete AVP, stimulates contraction of smooth muscle. Suckling stimulates reflex milk ejection by release of oxytocin, and uterine contractions during childbirth occur in response to oxytocin release triggered reflexly by the pressure of the fetus on the neck of the uterus.
Oxytocin and behavior	Oxytocin neurons in the hypothalamus project to the limbic (emotion) and brain reward systems. Oxytocin is implicated in pro-social behaviors such as pair bonding, maternal behavior, and trust between individuals.
Related topics	(A6) Blood–brain barrier (M1) Emotion (L5) Autonomic nervous system (ANS) function

Posterior lobe hormones

Magnocellular cells in the **supraoptic** (**SON**) and **paraventricular** (**PVN**) nuclei manufacture and secrete the nonapeptides **arginine vasopressin** and **oxytocin**.

Arginine vasopressin (AVP)

AVP, also known as **antidiuretic hormone** (**ADH**), is secreted in response to an increase in extracellular fluid osmolality or reduced blood volume. It increases the water permeability of nephron collecting ducts, thereby promoting water reabsorption. This reduces extracellular fluid osmolality and urine output (an antidiuretic effect) so restoring blood volume. Thus, AVP acts as a negative feedback regulator, defending set points in body fluid osmolality and blood volume.

The stores of AVP in the posterior lobe are large, sufficient to maintain maximum antidiuresis during several days of dehydration. The osmoreceptors which respond to changes in osmolality are in the **vascular organ of the lamina terminalis** (**OVLT**), one of the **circumventricular organs** of the brain which lie on the blood side of the blood–brain barrier. Osmolality-sensitive neurons in the OVLT synapse with the PVN and SON cells (Figure 1), and increase their firing rate as osmolality rises.

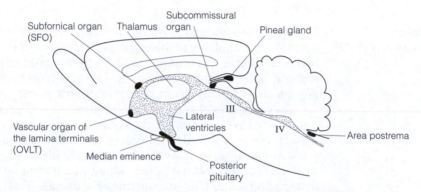

Figure 1. Location of the circumventricular organs (CVOs, shaded) in the rat brain (midsagittal section); the ventricles are stippled.

A reduction in blood volume (hypovolemia) greater than about 10% (caused by dehydration or hemorrhage) stimulates AVP secretion. Two mechanisms operate:

- Hypovolemia lowers mean arterial blood pressure. This is detected by stretch receptors (**baroreceptors**) in the walls of the carotid sinus and aorta. The afferents of these

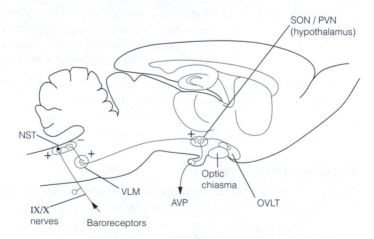

Figure 2. A model for neural control of arginine vasopressin (AVP) secretion. Increased osmolality detected by the vascular organ of the lamina terminalis (OVLT) stimulates supraoptic and paraventricular nuclei (SON and PVN) cells to secrete AVP. Reduced arterial blood pressure is signaled via the nucleus of the solitary tract (NST) and the ventrolateral medulla (VLM) to the SON/PVN.

pressure sensors run in the glossopharyngeal (IX) and vagus (X) cranial nerves to the **nucleus of the solitary tract** (**NST**) in the medulla. The NST activates noradrenergic neurons in the ventrolateral medulla which project to the PVN and SON to bring about AVP release. A reduced blood pressure causes *decreased* firing of the baroreceptor afferents and hence disinhibition of the circuitry triggering AVP secretion. This is shown in Figure 2.

● Activation of the renin–angiotensin cascade (Figure 3). Renin is secreted by the **juxta-glomerular apparatus** (**JGA**) of the kidney in response to several factors contingent on a fall in blood volume.

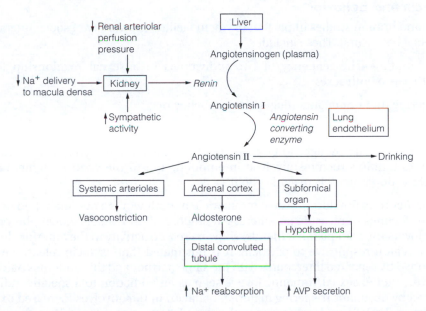

Figure 3. The renin–angiotensin cascade helps to maintain body fluid osmolality and blood volume.

Renin is a proteolytic enzyme which cleaves a plasma protein, **angiotensinogen**, to yield a decapeptide, **angiotensin I**. This is further cleaved by angiotensin converting enzyme (ACE), expressed on pulmonary endothelial cells, to the octapeptide, **angiotensin II** (**AII**). Angiotensin II stimulates the **subfornical organ** (a circumventricular organ), neurons of which stimulate AVP secretion. In addition, AII stimulates drinking, vasoconstriction, and the secretion of aldosterone; all actions which help restore blood volume and pressure.

Oxytocin

Oxytocin is implicated in several aspects of reproductive function. Its stimulation of smooth muscle contraction underlies the **milk ejection** reflex in lactating females, and maintenance of uterine contractions during **parturition** (birth).

Suckling is the most potent stimulus for milk ejection. Primary afferents from the areolar and nipple skin relay with spinothalamic tract neurons in the dorsal horn of the spinal cord. Spinothalamic input causes oxytocin secretion via an undefined neural pathway from midbrain to the PVN and SON. Neurons that secrete oxytocin are distinct from those that secrete AVP.

Oxytocin is not the trigger for parturition. However, at term, a rise in maternal estradiol/progesterone ratio upregulates oxytocin receptors in uterine smooth muscle which consequently becomes very sensitive to oxytocin. Once parturition is established, pressure of the fetal head on the cervix evokes the secretion of oxytocin from the posterior pituitary via a reflex pathway similar to that for milk ejection. The oxytocin stimulates contractions of the uterine smooth muscle, further increasing the pressure of the fetus on the cervix. This positive feedback mechanism is the **Ferguson reflex**. However, since parturition can be normal in spinally transected women, or those deficient in oxytocin, additional mechanisms must also be important.

Oxytocin and behavior

Rodent and human studies imply that oxytocin facilitates a variety of social interactions (pro-social behaviors). These include:

- Pair bonding—the tendency of two individuals in a sexual relationship to stay together—in both sexes

- Triggering (but not maintaining) maternal behavior

- Trust between individuals

These behaviors are thought to be mediated by projections of hypothalamic oxytocin neurons to limbic structures involved in emotion, and to the **ventral tegmental area** (**VTA**) of the dopaminergic reward system.

Oxytocin neurons fire during sexual intercourse in both women and men, and oxytocin produces feelings of pleasure and increased libido. MRI of women shown photographs of their lovers (but not platonic friends) show increased activity in the anterior cingulate cortex, a region responsive to oxytocin. It is postulated that oxytocin release becomes conditioned by repeated intercourse with the same partner and this underlies pair bonding. In this model sexual/romantic love is seen as an addiction to a specific individual organized by oxytocin. Triggering maternal behavior in rats involves increased oxytocin signaling to the VTA and consequently increased dopamine release from the nucleus accumbens. Human MRI suggests that facilitation of trust by oxytocin could operate by suppressing fear circuitry of the amygdala and brainstem.

The feelings of love and empathy felt by individuals dosed with 3,4-methylenedioxy-methamphetamine (MDMA, "ecstasy") could be because MDMA stimulates oxytocin secretion by activation of 5-HT_{1A} serotonin receptors.

Acute stress inhibits oxytocin release via actions of noradrenaline (norepinephrine) and adrenaline (epinephrine). Hence stress reduces milk let-down, prolongs labor, and has adverse effects on pro-social behaviors.

L3 Neuroendocrine control of metabolism and growth

Key Notes

Hypothalamic–anterior pituitary axes

The hypothalamus and anterior pituitary, acting in concert, control five endocrine axes that regulate aspects of metabolism, reproduction, development, and growth. Hypothalamic neurons secrete hormones that either stimulate or inhibit the anterior pituitary secretion of trophic hormones. Trophic hormones released into the circulation in turn stimulate target tissues (e.g., adrenals, thyroid, and gonads) to secrete their hormones. Secretion from the endocrine axes is under negative feedback regulation which maintains set point concentrations of hormones. By varying the set point the endocrine axes can alter their hormone output.

Hypothalamic–pituitary–adrenal (HPA) axis

The HPA axis controls the secretion of glucocorticoids by the adrenal cortex. Paraventricular neurons secrete corticotrophin releasing hormone (CRH) which causes a population of anterior pituitary cells to release adrenocorticotrophic hormone (ACTH) into the circulation. ACTH is the trophic hormone that stimulates the adrenals to release cortisol. This acts at two types of steroid receptor. On binding, these intracellular receptors translocate from cytoplasm to nucleus where they alter gene transcription.

Stress

Stress can be defined as a state in which there is high/prolonged elevation in ACTH and cortisol concentrations. Cortisol is adaptive in stress by promoting the synthesis of glucose and glycogen. The HPA is activated in stress by catecholaminergic neurons concerned with arousal, or hunger and thirst sensations, by brainstem cholinergic neurons conveying sensory input associated with the stressor, and from other hypothalamic nuclei relaying limbic system information.

Hypothalamic–pituitary–thyroid (HPT) axis

Paraventricular neurons release thyrotrophin releasing hormone (TRH) which causes anterior pituitary cells to secrete thyroid stimulating hormone (TSH), a trophic hormone which stimulates growth of the thyroid gland and release of the thyroid hormones (T3 and T4). Thyroid hormone receptors are intracellular receptors that are bound to nuclear DNA in the absence of hormone. On binding T3 the receptor activates gene transcription. Cold exposure excites neurons in the preoptic hypothalamus that activates

	the HPT axis. Increased secretion of thyroid hormones raises the metabolic rate.
Growth hormone (GH)	GH released from the anterior pituitary stimulates cell division and growth of many tissues, and mobilizes fatty acids as energy substrates. It is secreted in increased amounts during exercise, stress, and fasting. GH secretion is stimulated by growth hormone releasing hormone (GHRH), and inhibited by somatostatin. GH stimulates the production of insulin-like growth factor, either in the brain or by peripheral tissues which exerts negative feedback control on growth hormone release. Most GH release occurs at night. Several neurotransmitter systems and hormones modify GH output; sex steroids particularly stimulate the high GH output responsible for the growth spurt.
Related topics	(L1) Anatomy and connections of the hypothalamus (M1) Emotion (L4) Neuroendocrine control of reproduction (M4) Brain biological clocks

Hypothalamic–anterior pituitary axes

Acting through the anterior lobe of the pituitary gland the hypothalamus is the hub of five neuroendocrine **axes** that regulate aspects of metabolism, reproduction, development, and growth. Neurons in several hypothalamic nuclei send their axons to the external zone of the median eminence and the **tuberoinfundibular tract**. These axons secrete **hypophysiotropic hormones** into the hypothalamic–pituitary portal system which carries them into the anterior lobe. Each hypophysiotropic hormone acts on a particular population of cells in the anterior lobe, either exciting or inhibiting their secretion of a specific **stimulating** (**trophic**) hormone. Hypophysiotropic hormones that excite secretion are termed **releasing** hormones, those that inhibit are called **release-inhibiting** hormones. Trophic hormones of the anterior pituitary are secreted into the systemic circulation and have endocrine effects on target tissues, particularly endocrine glands (Table 1).

Secretion from the neuroendocrine axes is modulated by negative feedback, defined as a mechanism that acts to hold some variable at a set point. In endocrinology the set point is generally the blood concentration of a hormone and the negative feedback operates at several levels of the neuroendocrine axis.

In Figure 1, if the concentration of the end product hormone exceeds the set point, more receptors are activated in the hypothalamus and anterior pituitary which consequently *reduce* their output of hormones. The effect is that, after some delay, the concentration of end product hormone falls. If it falls below the set point the hypothalamus and pituitary secrete more of their hypophysiotropic and trophic hormones, provoking an increase in concentration of the end product hormone. Autofeedback inhibition is a special case of negative feedback in which a substance directly inhibits its own synthesis. Several mechanisms exist to alter the set points of physiological systems so that hormone concentrations can be varied as circumstances change. Thus, secretion of hypothalamic hormones is pulsatile with a period of 60–180 min. This drives pulsatile release of anterior pituitary

Table 1. Five hypothalamic–anterior pituitary neuroendocrine axes

Hypophysiotropic hormone		Anterior pituitary stimulating (trophic) hormone [anterior pituitary cell type]	Target tissue for trophic hormone	Secreted hormone
Releasing hormone	Release-inhibiting hormone			
Corticotrophin releasing hormone	–	Adrenocorticotrophic hormone [corticotroph]	Adrenal cortex	Glucocorticoids
Thyrotrophin releasing hormone	–	Thyroid stimulating hormone [thyrotroph]	Thyroid	Triiodothyronine (T3) and thyroxine (T4)
Gonadotrophin releasing hormone	–	Follicle stimulating hormone Luteinizing hormone [gonadotroph]	Gonads	Sex steroids: estrogens, progestogens, and androgens
Growth hormone releasing hormone	Somatostatin	Growth hormone (GH, somatotrophin) [somatotroph]	Liver, fibroblasts, myoblasts, chondrocytes, osteoblasts, and others	Somatomedins (insulin-like growth factors)
Prolactin releasing factors	Dopamine (acting at D_2 receptors)	Prolactin [lactotroph]	Mammary glands	

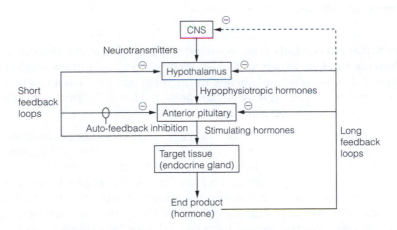

Figure 1. Negative feedback loops that control neuroendocrine secretion.

hormones. The amplitude and period of the pulses varies on a **circadian** (about a day) basis and in some instances on longer time scales.

Hypothalamic–pituitary–adrenal (HPA) axis

The HPA axis regulates the synthesis and secretion of **glucocorticoids**, a group of steroid hormones which influence energy substrate metabolism. The most important glucocorticoid in humans is **cortisol**. Cells in the **paraventricular nucleus** of the hypothalamus secrete **corticotrophin releasing hormone** (**CRH**), a peptide which acts synergistically with arginine vasopressin to stimulate release of **adrenocorticotrophic hormone** (**ACTH**) from corticotrophs. This is cleaved from a large precursor, **pro-opiomelanocortin**. In response to ACTH, cells in the adrenal cortex synthesize and secrete glucocorticoids. Negative feedback by cortisol operates at the hippocampus, hypothalamus and pituitary (Figure 2).

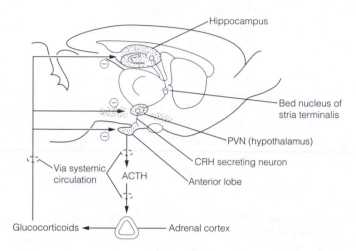

Figure 2. Feedback in the hypothalamic–pituitary–adrenal axis. Stippled regions harbor glucocorticoid receptors. ACTH, adrenocorticotrophic hormone; CRH, corticotrophin releasing hormone; PVN, paraventricular nucleus.

A circadian rhythm in cortisol output is driven by a brain biological clock located in the **suprachiasmatic nucleus** acting on CRH-secreting cells. In humans ACTH pulses are greatest early in the morning and decline through the day to reach a low point around midnight. Cortisol secretion follows a similar pattern. This daily rhythm is influenced by the timing of light and dark, sleep and meals.

The effects of cortisol are mediated by two receptors, the **mineralocorticoid receptor** (**MR**) and the **glucocorticoid receptor** (**GR**). These are members of a nuclear receptor superfamily that includes receptors for other steroids and for thyroid hormones. When unbound, MRs and GRs are present in the cytoplasm, complexed with **heat shock proteins**, molecular chaperones that stabilize the receptors into their functional configuration. Glucocorticoids such as cortisol diffuse readily across cell membranes. Binding of cortisol causes the receptor to translocate into the nucleus where it binds to specific sequences of DNA, **hormone responsive elements**, thereby increasing or decreasing the transcription of specific genes (Figure 3).

Mineralocorticoid receptors are in greatest numbers in limbic structures. Glucocorticoid receptors are more widespread, and expressed in glia as well as neurons. Because MRs have a high cortisol affinity they are mostly occupied at basal concentrations of the steroid. By contrast, GRs have a low affinity so are only occupied when the cortisol concentration is high, such as early morning. This means that cortisol affects different target tissues depending on its concentration. The corticotrophs of the anterior pituitary, CRH-secreting neurons of the hypothalamus, and hippocampal neurons express GRs. When the concentration of cortisol is high these GRs are activated and they inhibit transcription

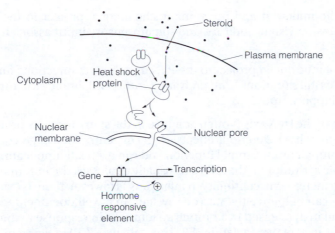

Figure 3. Model of steroid hormone receptor action to modulate gene transcription.

of the genes for CRH and arginine vasopressin. This is one mechanism for negative feedback control of glucocorticoid concentrations.

Stress

The hypothalamic–pituitary–adrenal axis is activated in **stress**. No *comprehensive* definition of stress exists, but generally it is a state seen in situations that derange homeostasis, or in which there is actual or perceived harm, loss, or challenge. **Stressors**, agents that cause stress, can be physiological (e.g., dehydration, starvation, infection, trauma), or psychosocial, often arising in situations over which an individual has little control (e.g., unemployment, bereavement). Responses to stress can be adaptive, often the case in acute physiological stress, but psychosocial stress can give rise to counterproductive emotional states (e.g., depression). There are large individual differences in how stressful a situation is perceived to be, which partly depends on early learning. A frequently adopted *operational* definition of stress is any state in which there is a large and/or prolonged rise in ACTH and glucocorticoid concentrations.

The increased secretion of cortisol is useful in stress because its overall effect is to harness long-term energy substrates and convert them to readily available substrates, glycogen and glucose. The early morning peak is timed to correspond with what is generally the longest interval each day without food. In addition, cortisol potentiates the effects of catecholamines.

Activation of the HPA in stress occurs through inputs from a variety of sources converging on the CRH-secreting cells of the paraventricular nucleus (PVN):

● Stress-evoked arousal activates noradrenergic neurons in the locus coeruleus which project to the PVN.

● Visceral sensations associated with thirst and hunger are transmitted via the glossopharyngeal (IX) and vagus (X) nerves to the nucleus of the solitary tract and adjacent regions of the medulla. These structures project catecholaminergic axons to activate the PVN.

● Inputs from the circumventricular organs which monitor osmolality or release arginine vasopressin go to the PVN to activate the HPA during dehydration.

- Neurons in the midbrain and pons, many cholinergic, project to the PVN and are thought to transmit visual, auditory, and somatosensory input associated with stressful situations.

- Most hypothalamic nuclei project to the PVN and these connections funnel information about stressful situations from prefrontal cortex and limbic structures such as the amygdala or hippocampus.

While activation of the HPA axis is often adaptive in the short-term, its protracted activation in *chronic* stress has deleterious effects. High concentrations of glucocorticoids acting via GRs can impair hippocampal function and can even kill hippocampal cells. More generally, chronic activation of the HPA axis usually depresses the immune system. (One effect of this is to increase susceptibility to infections.) However, there is evidence that the immune system can influence the nervous system. In a well documented example, the cytokine interleukin-1, released by immune system cells in response to infection, stimulates the OVLT to secrete prostaglandin E2. This enhances CRH release by the PVN. The study of the reciprocal interactions between nervous and immune systems is **psychoneuroimmunology**.

Hypothalamic–pituitary–thyroid (HPT) axis

Thyroid hormones, amongst other functions, regulate basal metabolic rate by increasing metabolic heat production. Thyroid hormone secretion is regulated by the hypothalamus and pituitary, and is influenced by several factors, such as ambient temperature.

Thyrotrophin releasing hormone (**TRH**) is a tripeptide synthesized in PVN neurons. TRH secreted by axon terminals of these cells in the median eminence is carried by the hypothalamic–pituitary portal system to the anterior pituitary, stimulating thyrotrophs to secrete **thyroid stimulating hormone** (**TSH**). TSH is a glycoprotein consisting of two chains, α and β. It is liberated into the systemic circulation and stimulates division and growth of cells in the thyroid gland, and the synthesis and secretion of the thyroid hormones. There are two thyroid hormones, **thyroxine** (**T4**) and **triidothyronine** (**T3**), named for the number of iodine atoms they contain.

Thyroid hormone receptors (**TRs**) are members of the steroid receptor superfamily. They form heterodimers with retinoid X receptors. They differ from glucocorticoid receptors in that the heterodimer binds to hormone-responsive elements in the DNA in the *absence* of ligand. These receptors have a higher affinity for T3 than T4. T4, which forms the bulk of the secreted hormones, is a prohormone which is converted to T3 by the neuronal cytosolic enzyme, 5′-deiodinase II. On binding of T3 the thyroid hormone receptor activates gene transcription.

Thyroid hormone output is controlled by negative feedback acting at several levels of the HPT axis. A drop in thyroid hormone concentration causes the increased secretion of TSH by thyrotrophs of the anterior pituitary. TRH secretion from the hypothalamus is also subject to feedback inhibition by both T4 and T3.

Pulses of TRH secretion drives pulsatile TSH output. The frequency and amplitude of the pulses is entrained into a circadian rhythm by the suprachiasmatic nucleus, rising throughout the night, falling during the morning, and remaining low throughout the afternoon. This circadian rhythm is sensitive to light and dark, but unaffected by sleep patterns.

Thyroid hormone secretion is increased in cold exposure. Temperature-sensitive neurons in the preoptic hypothalamus which get input from skin thermoreceptors project to brainstem noradrenergic neurons. These, in turn, synapse with the TRH-secreting cells of the PVN. Cold exposure activates the noradrenergic neurons, provoking a rise in TRH secretion within a few minutes. The resulting increase in the concentrations of thyroid hormones enhances metabolic rate, helping to maintain core temperature.

Growth hormone (GH)

Growth hormone (somatotrophin) stimulates cell division and growth of many tissues, particularly during the perinatal period and the growth spurt that heralds puberty, enhancing protein synthesis by increasing transcription and translation. GH mobilizes fatty acids as energy substrates. This is adaptive during exercise, stress, and fasting, three major physiological variables which increase GH secretion.

GH secretion from somatotrophs of the anterior pituitary is regulated by two peptide hormones, **growth hormone releasing hormone (GHRH)** and **somatostatin**. GHRH is synthesized by neurons in the **arcuate nucleus** of the hypothalamus. Somatostatin is an important transmitter throughout the CNS, but the somatostatin-containing cells responsible for inhibiting GH secretion are restricted to the hypothalamic **periventricular nucleus**. GHRH and somatostatin exert their opposing effects on GH secretion via GPCRs linked to the cAMP second messenger system; GHRH receptors enhance, while somatostatin receptors reduce, cAMP.

The secretion of GH is circadian and pulsatile, driven mostly by pulses of GHRH from the hypothalamus. The pulses are much bigger at night and triggered by deep slow wave sleep. This nocturnal GH secretion is greatest in children and declines with age. It is brought about by a serotonergic pathway from the brainstem to the hypothalamus. GHRH secretion is also stimulated by dopaminergic, noradrenergic, and enkephalinergic pathways in the brain. Thyroid hormones are required for normal levels of GH synthesis and secretion.

Negative feedback control of GH secretion occurs at the pituitary by suppression of the synthesis and secretion of GH, and at the hypothalamus by reduction of GHRH secretion. GH also stimulates the secretion of somatostatin. This negative feedback is exerted by **insulin-like growth factor (IGF-1)**, one of a group of peptides called **somatomedins** which mediate the effects of GH. IGF-1 is produced either in the brain, or peripherally, in response to GH.

A rapid rise in the rate of growth due to high levels of GH secretion, the **growth spurt**, occurs during puberty. During this time gonadal secretion of androgens and estrogens rises, stimulating GH secretion.

L4 Neuroendocrine control of reproduction

Key Notes	
The hypothalamic– pituitary–gonadal (HPG) axis	Pulsatile secretion of gonadotrophin releasing hormone by hypothalamic neurons stimulates the anterior pituitary to release two gonadotrophins, follicle stimulating hormone (FSH) and luteinizing hormone (LH). These stimulate the gonads to produce sex steroids.
Control in males	Gonadotrophins stimulate the testis to produce testosterone (which drives sperm production) and inhibin, both of which produce negative feedback suppression of the HPG axis. Testosterone acts at both the hypothalamus and the anterior pituitary but the effects of inhibin are confined to suppression of anterior lobe secretion of FSH.
Control in females	In women, gonadotrophins stimulate the ovarian follicles to grow, producing estradiol and inhibin during the first half of the cycle (the follicular phase). Ovulation is triggered by a mid-cycle surge in luteinizing hormone (LH) which then drives the follicle to become a corpus luteum. This secretes progesterone through the second half of the cycle (the luteal phase). During most of a cycle sex steroids exert a negative feedback suppression of gonadotrophin output. In the follicular phase it is mediated by estradiol (and inhibin) whereas in the luteal phase it results from both estradiol and progesterone. However, just before ovulation the high concentrations of estradiol produced by the mature follicle cause the HPG to switch briefly into a positive feedback mode. This is what stimulates the midcycle surge of LH.
How steroid feedback works	Negative feedback by estradiol is due to the anterior pituitary having low sensitivity to GnRH. High estradiol concentration near mid-cycle causes GnRH release to occur in high frequency, low amplitude pulses, allowing upregulation of GnRH receptors in the anterior pituitary so it becomes very sensitive to GnRH. This results in the positive feedback mid-cycle surge in LH. Rising progesterone concentrations after ovulation act on these to switch the secretion of GnRH into a low-frequency, high-amplitude pattern of release which downregulates GnRH receptors, re-establishing negative feedback.
Puberty and menopause	Puberty is due to the activation of the previously quiescent HPG axis by the removal of a neural brake. Exactly how

	this works is not known. Puberty is initiated by a metabolic signal, possibly leptin, that encodes body mass. The menopause results from ovarian failure.
Prolactin (PRL)	Secreted by the anterior pituitary, prolactin stimulates breast tissue development during pregnancy and is responsible for reflex synthesis and secretion of milk by suckling in lactating women. PRL secretion, like that of GH, is subject to dual regulation by the hypothalamus. Dopamine acts on the anterior pituitary to inhibit PRL synthesis and release. Vasoactive intestinal peptide is the major PRL releasing factor, with estrogen and oxytocin having roles. Lactating women are relatively infertile because high concentrations of PRL suppress LH secretion.
Related topics	(D4) Dopamine (L2) Posterior pituitary function (L1) Anatomy and connections of the hypothalamus (L3) Neuroendocrine control of metabolism and growth

The hypothalamic–pituitary–gonadal (HPG) axis

The hypothalamic–pituitary–gonadal axis controls reproduction. In primates, neurons scattered throughout the hypothalamus synthesize a decapeptide, **gonadotrophin-releasing hormone** (**GnRH**, also referred to as **luteinizing hormone releasing hormone, LHRH**) from a large precursor. GnRH is secreted from axon terminals in the median eminence into the portal system. GnRH stimulates gonadotrophs of the anterior pituitary to secrete two gonadotrophins, **follicle stimulating hormone** (**FSH**) and **luteinizing hormone** (**LH**).

The gonadotrophins are large glycoproteins each consisting of two peptides, an α chain and a β chain. The α chains of FSH and LH are identical (and very similar to the α chain of TSH) but the β chains are distinct. Gonadotrophins stimulate the gonads to produce **sex steroids** and have effects on gamete development. Gonadotrophin secretion is cyclical in females but not in males. The secretion of gonadotrophins is pulsatile, as with other anterior lobe hormones, and is driven by bursts of GnRH from the hypothalamus. GnRH neurons have an intrinsic rhythmicity but this is modified by steroid hormones. In men, the pulses are regular, spaced about 3 hours apart, but in women the period varies between 1 and 12 hours depending on the phase of her reproductive cycle. Experimentally replacing pulsatile with continuous GnRH delivery in female rhesus monkeys abolishes gonadotrophin secretion, showing that *pulsatile* GnRH output is essential for proper HPG axis function.

Control in males

At the testis, LH stimulates **Leydig cells** to synthesize and secrete androgens, principally **testosterone**. FSH, together with testosterone, acts on **Sertoli cells** to organize the development of spermatozoa and secrete a glycoprotein, **inhibin**. Gonadotrophin secretion in males is subject to negative feedback control by both of these secretions from the testis. Testosterone acts both at the hypothalamus, decreasing the frequency of the episodic GnRH bursts, and at the anterior pituitary, making it less responsive to GnRH. Inhibin specifically suppresses only FSH secretion and acts only at the anterior lobe. Humans have a circadian rhythm in testosterone secretion and women have modest cyclical changes in blood testosterone concentration, higher around ovulation.

Control in females

In females, the situation is more complicated because the role of the HPG axis is to:

- Stimulate the growth of a group of ovarian follicles (one of which goes to maturity)

- Produce cyclical changes in sex steroid output, which prepares the reproductive tract for fertilization and implantation

- Trigger ovulation at the appropriate time

The first half of the cycle (in women this is typically 1–14 days) is the **follicular phase**, because it is dominated by the growth of the ovarian follicle which secretes **estradiol** and **inhibin**. The second half of the cycle is the **luteal phase** (15–28 days), because after ovulation the follicle becomes a **corpus luteum** which secretes **progesterone**.

Feedback depends on the phase of the cycle. For most of the follicular phase low or moderate levels of estradiol, and inhibin, exert negative feedback effect on gonadotrophin secretion (Figures 1 and 2).

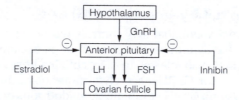

Figure 1. Negative feedback inhibition during the follicular phase of the ovarian cycle. Negative feedback in males is very similar except that luteinizing hormone (LH) is regulated by testosterone from the testis. FSH, follicle stimulating hormone; GnRH, gonadotrophin releasing hormone.

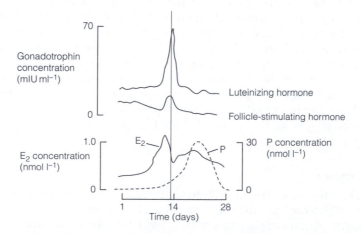

Figure 2. Pattern of hormone secretion during the human ovarian cycle. Ovulation, triggered by the estradiol (E_2) evoked surge in luteinizing hormone occurs around day 14. P, progesterone; mIU, milli-international units.

However, by about day 14, levels of estradiol become high enough to flip the HPG axis into a positive feedback mode. Now the estrogen stimulates a *rise* in LH and FSH secretion, which triggers ovulation, and switches the steroid metabolism of the post-ovulatory follicle to produce progesterone (Figure 2). The rise in progesterone secretion at the start of the luteal phase terminates the positive feedback LH surge and the system reverts to negative feedback mode.

How steroid feedback works

Feedback by steroids works at two levels (Figure 3). One is by changing the sensitivity of the anterior pituitary to GnRH. The other is by altering the size of the GnRH signal from the hypothalamus. Because GnRH neurons do not have steroid receptors, steroids must exert their effects on these cells via neuronal afferents. In general, estradiol reduces

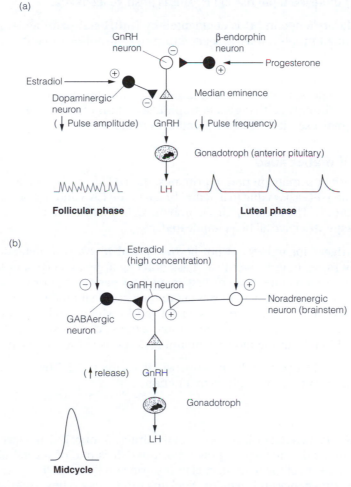

Figure 3. Model for steroid feedback regulation of gonadotrophin releasing hormone (GnRH) producing neurons: (a) negative feedback by estradiol (left) and progesterone (right); (b) positive feedback by estradiol works by inhibiting γ-aminobutyrate (GABA) inhibition and stimulating noradrenergic excitation of GnRH release. Excitatory neurons (o) inhibitory neurons (●). LH, luteinizing hormone.

GnRH pulse amplitude, while progesterone reduces pulse frequency. The frequency of the GnRH pulses influences the release of the gonadotrophins; low frequency favors FSH release, high frequency LH release.

At low estradiol concentrations the GnRH pulse frequency and amplitude are such that the numbers of *functional* GnRH receptors on the gonadotrophs are relatively low. Consequently the anterior pituitary is moderately insensitive to GnRH. This is negative feedback in the follicular phase, and the low GnRH pulse frequency favors FSH release. Estradiol may affect GnRH pulse amplitude by acting on dopaminergic neurons that make presynaptic synapses on GnRH terminals (Figure 3a).

Late in the follicular phase the high estradiol concentration, possibly acting via GAB-Aergic neurons in the mediobasal hypothalamus, causes GnRH release to occur in high-frequency low-amplitude pulses. This pattern allows upregulation of GnRH receptors on the anterior pituitary gonadotrophs which become exquisitely sensitive to the GnRH. Consequently there is a rapid rise in LH. This is positive feedback.

This pre-ovulatory surge in LH is augmented by **GnRH self-priming**. This refers to an increase in gonadotrophin release seen with repeated pulses of GnRH. It depends on early induction of (unknown) genes by GnRH and/or progesterone.

Subsequently, rising progesterone concentrations switch the secretion of GnRH into a low-frequency high-amplitude pattern of release. This downregulates GnRH receptors in the anterior pituitary so that gonadotrophin secretion plummets in the luteal phase. Progesterone may exert its effects via β-endorphin neurons.

Puberty and menopause

Except for a brief postnatal period in primates, the HPG axis is quiescent until puberty. Inactivity of the HPG axis is due to a neural brake in the CNS and not because of the lack of gonadal steroids. The nature of this neural brake is unclear, but GABA and **neuropeptide Y** and **kisspeptin** have all been implicated.

The precise trigger for puberty is not known, but it involves a metabolic signal that encodes body mass. In girls, a critical mass of 30 kg appears necessary for puberty to commence, with menarche (time of first menstrual period) occurring at about 47 kg. Female dancers, athletes, and anorexics fail to menstruate if their body mass falls below this. One candidate for the metabolic signal is **leptin** which is released by fat cells. The blood concentration of leptin is thought to signal the size of fat stores to the hypothalamus. Leptin inhibits neuropeptide-Y-containing neurons in the arcuate nucleus.

The end of reproductive life, the menopause, is characterized by ovarian failure; the hypothalamus and pituitary continue to function.

Prolactin

Prolactin (**PRL**) is secreted by lactotrophs of the anterior lobe. PRL is a glycoprotein with a similar amino acid sequence to growth hormone. It is one of several hormones that stimulates the growth of the mammary glands during pregnancy. PRL concentrations are highest during pregnancy and lactation. Suckling produces a reflex secretion of prolactin which stimulates the synthesis and secretion of milk. Like growth hormone, exercise and stress stimulate prolactin release.

PRL synthesis and secretion are inhibited by dopamine released into the portal system by dopaminergic neurons in the arcuate nucleus. Lactotrophs of the anterior pituitary express D2 dopamine receptors. These are GPCRs at which dopamine produces a fall in cAMP concentration, thereby reducing transcription of the prolactin gene.

PRL release is stimulated physiologically by:

- Vasoactive intestinal peptide (VIP) secreted by hypothalamic neurons into the portal system. This is the main short-term PRL releasing factor.

- Estrogens (producing a mid-cycle peak in PRL secretion); important in late pregnancy.

- Oxytocin from the posterior pituitary, which gains access to the anterior lobe via tiny blood vessels called the **short portal vessels**. This is important in lactating women.

With frequent suckling (every 2–3 hours) the amount of prolactin released is sufficient to block ovulation by suppressing LH and GnRH secretion, so lactating women are relatively infertile.

L5 Autonomic nervous system (ANS) function

Key Notes

Overview of the autonomic nervous system (ANS)	The autonomic nervous system (ANS) acts on smooth muscle, cardiac muscle, and glands to keep key physiological variables at a level appropriate to an animals' activity and its environment. Much ANS regulation is by negative feedback in that it operates to defend a set point in some variable (e.g., mean arterial blood pressure). In other circumstances, autonomic mechanisms are homeostatic in the sense that variables change to meet altered demands. The sympathetic division is activated in stress to produce "fright, fight, or flight" responses, whereas the parasympathetic division is more active in "rest and digest" situations. The two divisions often exert opposite effects on an organ (e.g., the heart). In a few instances ANS mechanisms are positive feedback in that they drive a physiological system away from its normal stable state (e.g., sexual responses).
ANS physiology	Acetylcholine (ACh) is the transmitter of autonomic ganglia and acts on both nicotinic (nAChR) and muscarinic (mAChR) receptors. The nAChR mediate fast ACh transmission. Activation of the mAChR greatly prolongs the time for which the postganglionic cell fires in response to the nicotinic receptor stimulation. Postganglionic sympathetic axons are usually noradrenergic and postganglionic parasympathetic terminals are invariably cholinergic. The adrenal medulla chromaffin cells are effectively postganglionic sympathetic cells and secrete mostly adrenaline (epinephrine). Postganglionic terminals co-release a variety of peptide transmitters which modulate, or have additional actions to, the primary transmitters.
Related topics	(A3) Organization of the peripheral nervous system (L6) Control of vital functions

Overview of the autonomic nervous system (ANS)

The autonomic nervous system is the visceral motor system. It adjusts the contraction of smooth muscle and heart muscle and controls glandular secretion so that key physiological variables (e.g., core temperature, cardiac output, blood pressure, blood glucose) are maintained at levels appropriate to the environment or ongoing activities. The term autonomic (self-governing) is apt since the ANS usually operates without conscious awareness and has no cognitive component. Although by definition there are no afferents in the ANS its reflex activities are controlled by somatic and visceral afferent input.

The ANS is concerned with feedback regulation of physiological variables. For example, mean arterial blood pressure is kept fairly constant in the face of changes in posture. On suddenly standing from a lying position the gravitational pooling of blood in the legs is offset by autonomic reflexes which elicit vasoconstriction of leg blood vessels. This maintains blood pressure by negative feedback.

Many autonomic adjustments, though not negative feedback, are homeostatic in that they change physiological variables so as to cope with altered demands. In response to a wide variety of stressors, activation of the **sympathetic nervous system** (**SNS**) to targets such as the heart, blood vessels, airways, and liver results in increased cardiac output, regional alterations in blood flow, raised airflow through the lungs, and elevations

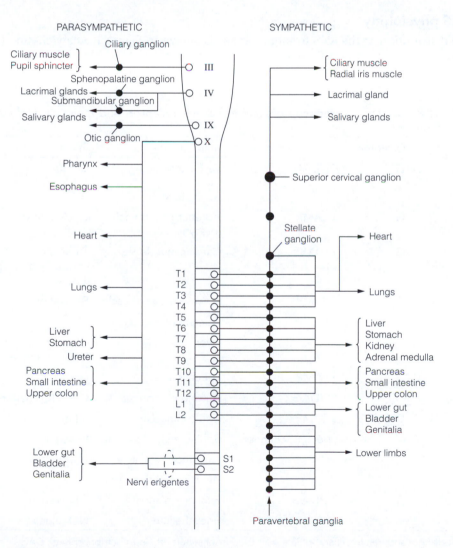

Figure 1. Distribution of the ANS to target organs.

in blood glucose concentrations, all adaptations which improve the chances of surviving the stress unscathed. In general the SNS mediates the response of "fright, fight, and flight." In contrast, **parasympathetic nervous system** (**PNS**) activation is seen when the body is in "rest and digest" mode; the PNS generally stimulates exocrine gland secretion and promotes anabolic processes.

The sympathetic and parasympathetic divisions of the ANS often have opposing effects on a system, for example on pupil diameter or heart rate, and it is the balance of activities in the two divisions that achieves the appropriate outcome.

In a few situations the ANS works by positive feedback. Sexual responses in humans require autonomic reflexes (both sympathetic and parasympathetic) in which the motor response (enlargement of the penis or clitoris by vasocongestion) increases the firing of the same visceral afferents which drive the reflex response. This is positive feedback because it carries the system away from its usual stable state.

ANS physiology

The distribution of the ANS to target organs is shown in Figure 1. Acetylcholine (ACh) is the major neurotransmitter at all autonomic ganglia. ACh released from preganglionic neurons acts on nicotinic cholinergic receptors to produce a fast excitatory

Table 1. Properties of adrenoceptors and muscarinic receptors and their principal actions at ANS targets

Receptor	G protein	Second messenger	Major tissues	Effect
α_1	G_q	IP_3/DAG	Vascular smooth muscle Sphincters[a]	Contraction
α_2	G_i	$\downarrow$ cAMP	Adrenergic terminals (presynaptic)	$\downarrow$ NA release
β_1	G_s	$\uparrow$ cAMP	Cardiac muscle	$\uparrow$ Force of contraction $\uparrow$ Rate
β_2	G_s	$\uparrow$ cAMP	Airway smooth muscle	Relaxation
			Gut smooth muscle[b]	
			Liver	Gluconeogenesis Glycogenolysis
β_3	G_s	$\uparrow$ cAMP	Fat cells	Lipolysis
M_1	G_q	IP_3/DAG	Autonomic ganglia	Close K_m channels
M_2	G_i and G_o	$\downarrow$ cAMP	Cardiac muscle	$\downarrow$ Rate
		opens K^+ channels	Sphincters[a]	Relaxation
			Gut smooth muscle[b]	Contraction
			Airway smooth muscle	
M_3	G_q	IP_3/DAG	Exocrine glands	$\uparrow$ Secretion
			Endothelium	NO release

[a] Includes gut and genito-urinary sphincters. [b] Except sphincters. IP_3, inositol triphosphate; DAG, diacylglycerol; NA, norepinephrine; cAMP, cyclic adenosine monophosphate; NO, nitric oxide.

postsynaptic potential (epsp) which, if sufficiently large, makes the postganglionic cell fire. In addition ACh acts on M1 muscarinic receptors producing a slow epsp—prolonging the firing of the postganglionic cell by many seconds—by closing a population of potassium (K_M) channels. In this way long-lasting autonomic responses are generated by brief stimuli.

Almost all terminals of the sympathetic postganglionic axons secrete noradrenaline (norepinephrine), the only important exception in humans being the *cholinergic* sympathetic supply to sweat glands. The adrenal medulla chromaffin cells are regarded as a postganglionic component of the SNS and secrete adrenaline (epinephrine; and noradrenaline, norepinephrine) directly into the blood as a result of activity in the preganglionic sympathetic fibers which supply it. There are four major types of adrenoreceptors that mediate the various effects of sympathetic stimulation. They are all GPCRs and their properties are summarized in Table 1.

All parasympathetic postganglionic axons release ACh, the effects of which are brought about by muscarinic cholinergic (mAChR) receptors. There are several subtypes of mAChR, all GPCRs, which account for the diverse effects of parasympathetic activity (Table 1).

Autonomic nerve terminals, in addition to secreting noradrenaline (norepinephrine) or ACh, also release ATP and peptides as co-transmitters. ATP, for example, acts on the smooth muscle of blood vessels to produce fast excitatory postsynaptic potentials and rapid contraction. This is followed by a slower response due to noradrenaline (norepinephrine). Peptide co-transmitters include neuropeptide Y (NPY) and **vasoactive intestinal peptide** (**VIP**). They prolong and modulate the effects of the primary transmitter. For example, NPY in sympathetic terminals enhances the vasoconstrictor response to noradrenaline (norepinephrine). VIP from parasympathetic terminals on salivary glands causes vasodilation which enables ACh to produce a greater salivary secretion.

L6 Control of vital functions

Key Notes

Thermoregulation

Both behavior and physiological mechanisms allow core temperature to be held at about 37°C. Physiological thermoregulation is achieved through negative feedback controlled by the hypothalamus. The pre-optic area (POA) monitors core temperature, and gets input from skin warm receptors. It regulates sympathetic output to vascular smooth muscle and hence adjusts the degree of vasoconstriction of skin blood vessels. The paraventricular and dorsomedial hypothalamic nuclei get input from skin cold receptors and stimulate brown fat and shivering. These nuclei are inhibited by the POA when heat loss is required. The set point for the negative feedback (the "thermostat") is provided by temperature-insensitive hypothalamic interneurons and is modified by a variety of factors.

Cardiovascular regulation

Short-term negative feedback control of mean arterial blood pressure (MAP) is achieved by baroreceptor reflexes. A rise in MAP excites arterial baroreceptors, reflexly activating parasympathetic, but inhibiting sympathetic, neurons via neural circuits in the brainstem. Adjustments to the tonic output of sympathetic and parasympathetic supply to the heart alters its rate and force; increasing them when the sympathetic dominates. Tonic sympathetic discharge to vascular smooth muscle controls blood vessel diameter. Increased activity causes vasoconstriction which elevates MAP.

Control of breathing

Voluntary control of breathing is possible for short periods via the pyramidal motor system. Normally this is overridden by a central pattern generator (CPG) in the medulla and pons which drive spinal cord inspiratory and expiratory neurons that supply the diaphragm and other respiratory muscles. Pacemaker cells in the medulla may be the origin of the basic respiratory rhythm. Visceral inputs to the CPG (many via the vagus nerve) allow breathing to be modified in swallowing, speech, exercise, and sleep. Airway receptors and baroreceptor stimulation inhibit inspiration. Peripheral chemoreceptors that respond to reduced blood oxygen concentration, and central chemoreceptors responsive to a rise in CO_2 concentration or fall in pH, stimulate breathing.

Related topics

(F1) Sensory receptors   (M1) Emotion
(L5) Autonomic nervous
system (ANS) function

Thermoregulation

A core temperature of around 37°C is maintained by behavioral and physiological mechanisms. The core is defined as the inside of the head, trunk, and limbs, from deep to the subcutaneous fat layer. Behavior, for example seeking sun or shade, curling up when cold (to minimize the surface area for radiation), wearing clothes, and building shelter, is needed to defend against all but modest changes in environmental temperatures.

Physiological heat loss or heat gain mechanisms are activated whenever the ambient temperature moves outside the **thermoneutral zone**, a window about 1°C wide in which an individual feels comfortable. The thermoneutral zone depends on humidity, wind velocity, and clothing. For naked humans in still air at 50% relative humidity it is 28°C.

The first response to ambient temperature moving outside the thermoneutral zone is adjustment in firing of sympathetic nerves to smooth muscles of skin arterioles. In heat stress, firing is reduced and the fall in noradrenaline (norepinephrine) evoked vascular smooth muscle contraction causes **cutaneous vasodilation**. This warms the skin, increasing heat loss by radiation. In the cold, increased sympathetic activity causes **cutaneous vasoconstriction**.

Larger excursions from the thermoneutral zone evoke either sweating or shivering. Sweat glands are innervated by sympathetic neurons that are atypical in secreting acetylcholine rather than noradrenaline (norepinephrine). ACh acts on muscarinic receptors to trigger sweat production which causes skin cooling by evaporation. Shivering is the almost simultaneous contraction of agonist–antagonist muscle pairs. It starts in jaw muscles and spreads to the trunk and proximal limb muscles. Shivering is brought about by activation of brainstem reticular neurons that synapse with γ-fusimotor neurons. The contraction of intrafusal fibers excites stretch reflexes. So, shivering is mediated peripherally by the somatic *not* the autonomic nervous system. Muscle contraction, both in shivering or exercise, generates heat.

A further heat gain mechanism, **non-shivering thermogenesis**, is particularly important in human babies. It is caused by increased sympathetic activity to **brown adipose tissue** (**BAT**), mostly located in the neck and between the shoulder blades. Released noradrenaline (norepinephrine) acts on β_3 adrenoceptors to stimulate a rise in cAMP. This activates lipolysis, liberating free fatty acids which are metabolized by β-oxidation in BAT mitochondria, and at the same time uncoupling oxidative phosphorylation in the mitochondria, generating heat.

The **pre-optic area** (**POA**) of the hypothalamus has **internal warm thermoreceptors** that monitor core temperature, and gets input from skin warm receptors (Figure 1). Hence the POA has numerous warm-sensitive neurons. It projects to the periaqueductal gray which then relays via the raphe nucleus to preganglionic sympathetic neurons in the intermediolateral column of the spinal cord. This is the route by which vasomotor tone is adjusted. In addition the POA stimulates the supraoptic nucleus to secrete antidiuretic hormone from the posterior pituitary. This helps conserve water. The paraventricular (PVN) and dorsomedial hypothalamic (DMH) nuclei get input from skin cold receptors and stimulate brown fat and shivering.

Warm sensitive neurons in the POA modulate both heat loss, via the vasomotor pathway, and the heat gain mechanisms by *inhibiting* the PVN and DMH. The integration of thermoregulatory responses by the hypothalamus is exemplified by the fact that thresholds for activating either sweating or shivering can change. Hence during exercise sweating, triggered by internal warm thermoreceptors in the POA as core temperature rises, is reduced in a linear fashion the colder the skin temperature.

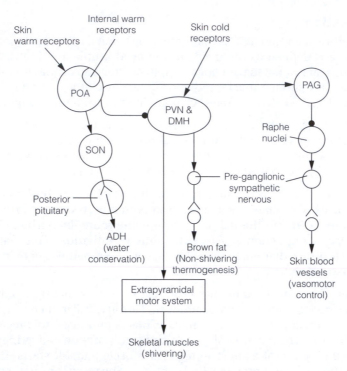

Figure 1. Neural control of thermoregulation. PAG, periaqueductal gray; POA, pre-optic area; PVN, paraventricular nucleus; DMH, dorsomedial hypothalamus; SON, supraoptic nucleus. → excitatory, —● inhibitory.

The core temperature maintained by thermoregulation is the set point. It is equivalent to a thermostat and is provided by the activity of interneurons in the hypothalamus that are *not* temperature sensitive. These are regulated by catecholaminergic neurons in the pontine reticular formation, and the set point is not constant. It shows a circadian rhythm, falling about 0.5°C during sleep, and is increased by progesterone during the luteal phase of the menstrual cycle by about the same amount. Chronic exposure to hot or cold environments cause gradual long-term shifts (adaptation) of the set point.

Cardiovascular regulation

Long-term regulation of blood pressure does not require the ANS and relies on control of blood volume and osmolality via vasopressin and the renin–angiotensin–aldosterone cascade. However, the ANS is crucial for the short-term regulation of **mean arterial blood pressure** (**MAP**). At rest the ANS operates to maintain a roughly constant MAP by negative feedback. Mean arterial pressure is the product of cardiac output (Q), the volume output of the left ventricle per minute, and the peripheral resistance (R), which is related to the radius of the arterioles.

Now, the cardiac output is in turn a product of stroke volume, the volume ejected from the left ventricle per beat, determined by the contractile force of the heart, and heart rate. Hence cardiac output can be raised (lowered) either by increasing (decreasing) stroke volume or heart rate or both. The ANS regulates cardiac output via both sympathetic and parasympathetic supply to the heart. Both are tonically active at rest and increases (decreases) in cardiac output are achieved by raising (lowering) sympathetic activity and reducing (elevating) parasympathetic activity. In humans sympathetic activity raises both force and rate, whereas parasympathetic activity lowers rate, but has little effect on force because few parasympathetic fibers innervate the ventricles.

Peripheral resistance is controlled solely by altering the tonic firing rates of sympathetic neurons going to vascular smooth muscle. Increased firing frequency causes vasoconstriction which raises peripheral resistance.

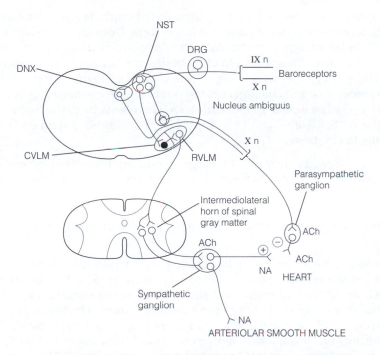

Figure 2. Brainstem circuits controlling mean arterial blood pressure. DRG, dorsal root ganglion; DNX, dorsal vagal nucleus; CVLM, caudal ventrolateral medulla; RVLM, rostral ventrolateral medulla; NST, nucleus of the solitary tract.

Circuitry for the negative feedback regulation of MAP resides in the medulla (Figure 2). Baroreceptors are stretch receptors located in the carotid sinus and the aortic arch that are sensitive to rapid alterations in MAP. Their afferents run in the glossopharyngeal (IX) and vagus (X) cranial nerves respectively and terminate in the nucleus of the solitary tract (NST), a structure involved in a wide variety of visceral reflexes (e.g., swallowing, chemoreceptor responses). The NST projects to the **dorsal vagal nucleus** (**DNX**) and to the **nucleus ambiguus** (**NA**), both of which give rise to preganglionic parasympathetic axons that run in the vagus nerve to the heart. The NTS controls sympathetic outflow to heart and blood vessels by input to the **caudal ventrolateral medulla** (**CVLM**). This contains GABAergic inhibitory neurons which synapse in the **rostral ventrolateral medulla**, axons of which run down the spinal cord, terminating on preganglionic sympathetic neurons. A rise in MAP increases the firing rate of baroreceptor afferents, and this directly activates the parasympathetic innervation to the heart, slowing its rate. However, the presence of inhibitory neurons in the CVLM means that baroreceptor discharge suppresses sympathetic outflow to the heart, reducing its rate and force of contraction, and arterioles, which reduces peripheral resistance. The net effect is a fall in blood pressure back to the set point. Responses occur in the opposite direction to an initial drop in MAP.

Cardiovascular variables are modified to match circumstances. During exercise the cerebellar and cerebral cortex act to modify hypothalamic autonomic regulation. Similarly cardiovascular responses seen in emotional states require limbic system components such as the amygdala and the cingulate cortex.

Control of breathing

The diaphragm and muscles of the chest wall used in breathing are skeletal muscles supplied by motor neurons of the somatic nervous system. Despite this we only have voluntary control over breathing (via pyramidal tract control of spinal motor neurons) for short periods of time, since this is overridden by brainstem circuits that drive and regulate breathing. These circuits get sensory input from visceral afferents—for example pharynx, airways baroreceptors, and chemoreceptors—and are interconnected with CNS circuits controlling the cardiovascular system and sleep. This allows breathing to be modified in swallowing, speech, exercise, and sleep. Inputs from the limbic system evoke changes in breathing due to emotions.

Breathing results from the rhythmic discharge of spinal motor neurons which supply ventilatory muscles. Axons of motor neurons in spinal segments C3–C5 run in the phrenic nerves to the **diaphragm**, contraction of which increases chest volume during **inspiration**. Motor neurons in C4–L3 supply neck muscles and external intercostal muscles that aid inspiration, and internal intercostal muscles and abdominal muscles responsible for **expiration**.

The spinal motor neurons are driven by a central pattern generator that consists of heavily interconnected neurons in the medulla and pons (Figure 3). The **dorsal respiratory group** (**DRG**) are clusters of neurons in the **nucleus of the solitary tract** (**NST**) of the medulla. One cluster are inspiratory upper motor neurons which synapse with spinal motor neurons supplying the diaphragm and external intercostal muscles. These cells are active during quiet inspiration. They receive inputs from another cluster of DRG cells that integrate the visceral inputs. The **ventral respiratory group** (**VRG**), in the ventrolateral medulla, are silent in quiet breathing but active in heavy breathing. The VRG houses both inspiratory and expiratory neurons which are reciprocally innervated by inhibitory connections. Hence, when the inspiratory neurons are firing the expiratory neurons are silent and *vice versa*. The **pontine respiratory group** (**PRG**) includes the **Kölliker–Fuse nucleus** (**KFN**), which drives the medullary respiratory neurons so as to prolong inspiration (**apneusis**), and cells in the **parabrachial nucleus** (**PBN**) of the rostral pons which overrides apneusis, thereby terminating inspiration.

The basic respiratory rhythm could originate from the **pre-Botzinger complex** (**pBOT**), a nucleus just rostral to the VRG which makes extensive connections with the VRG, DRG, and pons. Pre-Botzinger complex cells have intrinsic pacemaker properties.

Integrating neurons in the DRG receive sensory input, largely via the vagus nerve, from a variety of sources which modify the basic rhythm of breathing. These include:

- Nociceptors (lung irritant receptors) which trigger the cough reflex.
- Pulmonary stretch receptors which inhibit inspiration.
- Baroreceptors, which inhibit inspiration: if blood pressure falls (as a result of hemorrhage, for example) depth of inspiration increases.
- Peripheral chemoreceptors in the carotid body and aortic arch which are activated by large reductions in partial pressure of O_2.

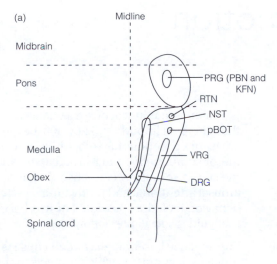

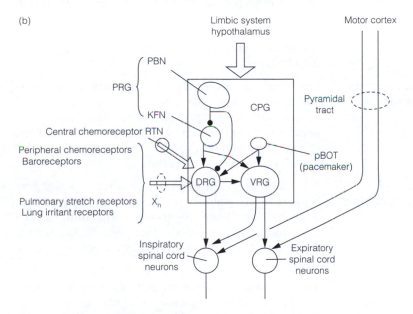

Figure 3. Neural control of breathing. (a) Dorsal view of the right side of the brainstem showing implicated structures. (b) A highly simplified model. CPG, central pattern generator; DRG, dorsal respiratory group; KFN, Kölliker–Fuse nucleus; NST, nucleus of the solitary tract; PBN, parabrachial nucleus; pBOT, pre Botzinger complex; PRG, pontine respiratory group; VRG, ventral respiratory group; RTN, retrotrapezoid nucleus; X_n, vagus nerve. ⟶ excitatory connections; ⟶• inhibitory connections; ⇒ inputs.

- Central chemoreceptors which are excited by a rise in brain extracellular fluid concentrations of CO_2 and fall in pH and drive deeper breathing in response. Central chemoreceptors have been identified in the **retrotrapezoid nucleus** which is adjacent to the nucleus of the facial (VII) nerve.

M1 Emotion

Key Notes

Functions of emotion

Emotions occur in response to changes in our environment thought to be important. They may be hard wired, but many are learnt. Emotions have cognitive, visceral sensory, and motor components, and learned associations arise between the cognitive and visceral sensory aspects. By activating arousal attentional and motivational systems they alter behavior, including providing nonverbal communication that guides social interactions.

Affective basal ganglia circuit

The affective basal ganglia circuit consists of ventral striatum (nucleus accumbens) which projects to the ventral pallidum of the globus pallidus that relays via the mediodorsal thalamus to limbic cortex. The cortex connects back to the nucleus accumbens. Nucleus accumbens output to the substantia nigra goes to the motor and cognitive basal ganglia loops.

Affective motor pathways

Output of the amygdala goes to the hypothalamus and a variety of brainstem structures that organize motor, arousal, and visceral aspects of emotions. Facial expressions engendered by emotions are brought about by extrapyramidal motor pathways not the corticobulbar fibers used for intentional movement.

Affect and explicit learning

The limbic cortex projects via the entorhinal cortex to the hippocampus, which in turn sends output to the hypothalamus. Pathways from hypothalamus to anterior thalamus and then limbic cortex close the Papez circuit, thought to be concerned with explicit learning associated with emotions.

Amygdala and fear learning

The amygdala is in the temporal lobe. Olfactory input enters the corticomedial nucleus. Other input (sensory, arousal, and cognitive information) goes to the basolateral nuclei. Output from the central nucleus goes to the hypothalamus, septum, several brainstem nuclei, and the nucleus accumbens. The amygdala is concerned with innate and learned fear responses and organizes the appropriate avoidance behavior. It is necessary for recognizing facial expressions of emotion.

Neocortex and emotion

A number of cortical areas are implicated in how cognition influences emotion and *vice versa*, including the orbital prefrontal cortex and anterior cingulate cortex.

Related topics

(K6) Anatomy of the basal ganglia

(N1) Types of learning

Functions of emotion

Emotions arise in response to changes in our surroundings that could have important consequences. These consequences may be immediate (e.g., being confronted by a mugger) or delayed (e.g., anticipating an examination). Some emotions are short-lived (surprise, humor) and these often arise when there is a discrepancy between what is expected and what actually happens, others are long-lived (e.g., jealousy, hatred). Some emotional responses are hard wired—executed by neural circuits that are genetically specified during development—such as the universal aversive reaction of infants to bitter-tasting (potentially toxic) foods, but most (e.g., love) are probably learnt. Apparently hard-wired responses need not be forever fixed; most adults come to like the bitter foods their culture teaches them is safe (e.g., coffee).

Emotional states have three components:

- A conscious cognitive component

- Visceral sensations arising from autonomic and endocrine events (e.g., feeling a rise in heart rate)

- Motor actions (e.g., characteristic facial expressions)

There are learned associations between visceral sensations and cognitive aspects of the emotional state which make them self-reinforcing. Realization of just how bad or good a situation is drives visceral changes, while conscious efforts to stem visceral sensation (e.g., controlled breathing) lessens emotional intensity.

Emotions presumably enhance survival, for several reasons:

1. They are arousing and direct attention to important aspects of a situation so that it can be assessed as threatening or beneficial.

2. Emotions are goads to useful action. We usually avoid snakes.

3. The motor component (e.g., laughing or crying) communicates our emotional state to others, altering *their* behavior. Before it acquires language an infant can only communicate its needs and desires by expressing its emotions. But also in adults emotions act as a powerful form of nonverbal communication. This is crucial for social interactions.

The subjective experience of emotion, especially that leading to action, is termed **affect**.

Affective basal ganglia circuit

The core of the limbic system consists of the **affective striato-thalamo-cortical circuit** and its connections with the amygdala (Figure 1). This is laid out in much the same pattern as the motor loop, except that the striatal component is the **ventral striatum** (**nucleus accumbens, nAc**) which projects to the **ventral pallidum** (ventral part of the globus pallidus). The ventral pallidum relays via the **mediodorsal thalamus** to the **anterior cingulate cortex** and **medial orbital prefrontal cortex**. The loop is closed by connections from the cortex back to the nucleus accumbens. The affective loop has reciprocal connections with the amygdala, which is responsible for fear learning, and is modulated by the dopaminergic mesolimbic system which is concerned with reward learning.

Output of the nucleus accumbens goes to the compact part of the substantia nigra. This allows the activity of the motor and cognitive striato-thalamo-cortical circuits to be modified, providing for some of the motor and the cognitive aspects of emotional states.

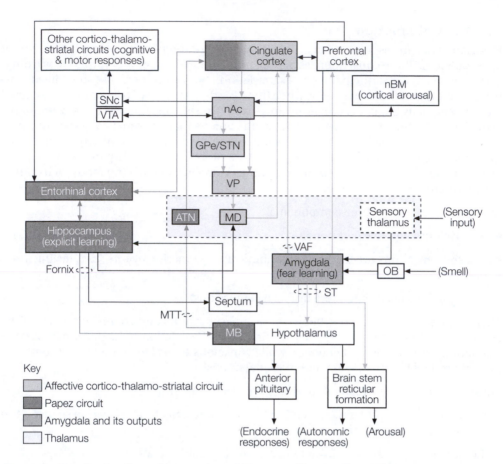

Figure 1. Circuitry implicated in emotions. MB, mammillary bodies; ATN, anterior thalamic nucleus; MD, mediodorsal nucleus; nAc, nucleus accumbens; GPe/STN, globus pallidus (external part)/subthalamic nucleus; VP, ventral pallidum; SNc, substantia nigra (compact part); VTA, ventral tegmental area; nBM, basal nucleus of Meynert; VAF, ventral amygdalofugal pathway; ST, stria terminalis; MTT, mammillothalamic tract; OB, olfactory bulb.

Affective motor pathways

Output of the amygdala goes to the hypothalamus and a variety of brainstem structures that organize motor, arousal, and visceral aspects of emotions. Facial expressions engendered by emotions—smiling, crying, and so forth—are brought about by extrapyramidal motor pathways that run in the brainstem reticular formation. Patients with unilateral damage to corticobulbar fibers descending from the motor cortex have voluntary motor paresis on the opposite side. When asked to smile on demand their smile is lopsided. However, when genuinely amused their smile is natural and bilateral (**Duchenne smile**) because different emotion-driven motor pathways are engaged.

Affect and explicit learning

The affective loop is wired into a second circuit. The anterior cingulate cortex projects to the entorhinal cortex that acts as a gateway for all neocortical input to the hippocampus. Efferents leave the hippocampus by way of the fornix for the hypothalamus. Output from the mammillary bodies (by way of the mammillothalamic tract) goes via the **anterior thalamic nuclei** back to the anterior cingulate cortex. The hypothalamus also has connections with the prefrontal cortex. This **Papez circuit** was originally thought to be *the* circuit for emotion, but the well-established function of the hippocampus in **explicit learning** (learning that can be consciously recalled) implies that the Papez circuit has the more restricted role of mediating explicit learning during emotional states (e.g., remembering the location of a hornet's nest so as to avoid being stung again).

Amygdala and fear learning

The **amygdala** is a cluster of nuclei in the white matter of the temporal lobe. Olfactory input runs from the olfactory bulb to the **corticomedial nucleus**. Sensory information from other modalities (vision, hearing, somatosensory) enters the **basolateral nuclei** from specific thalamic nuclei and their corresponding areas of sensory cortex. The basolateral nuclei also receive information about:

- The state of the viscera from the hypothalamus

- Arousal status from the locus coeruleus and nucleus basalis of Meynert

- Cognitive processing by the orbital prefrontal cortex

The output of the amygdala is from its **central nucleus** and follows two anatomical pathways. Efferents to the hypothalamus, septum, and several brainstem nuclei go via the **stria terminalis**, while the **ventral amygdalofugal pathway** conveys connections to the nucleus accumbens. The functions of these outputs in the expression of emotions is summarized in Table 1.

Table 1. Effects of amygdala central nucleus efferents

Target nucleus/pathway	Effect
Periaqueductal gray → raphe nuclei	Anti-nociception
→ medullary reticular nuclei	Freezing
Locus coeruleus	Arousal
Noradrenergic medullary neurons → preganglionic sympathetic neurons	↑ Heart rate, vasoconstriction
Hypothalamus → dorsal nucleus of vagus	↓ Heart rate (mediates vaso-vagal syncope)
Hypothalamus	Corticotrophin releasing hormone secretion → activation of HPA axis (stress response)
Parabrachial nucleus → respiratory CPG	Hyperventilation

The amygdala is implicated in innate fear responses. It has access to species-specific hard-wired neural representations of scary things (e.g., crocodiles, snakes, raptors) to drive automatic fear responses.

It is also essential for aversive learning. Fear conditioning occurs when a neutral stimulus (CS), such as a tone, is paired with a noxious stimulus (US), such as a brief electric foot shock. After several tone–shock pairings the tone becomes a negative reinforcer and it elicits conditioned fear responses (CR), including autonomic, endocrine, and behavioral signs of fear. The role of the amygdala in fear learning is well documented:

- The connections of the amygdala (Table 1 and Figure 1) supports its role in aversive learning because it activates the cholinergic attentional system, the sympathetic nervous system, and the release of stress hormones.

- Firing of amygdala (central nucleus) neurons correlates with the development of the fear responses.

- Lesions of the amygdala prevent acquisition of new conditioned fear responses or expression of preexisting ones, although they do not affect *autonomic* responses to aversive stimuli (e.g., the defense reaction) which are organized by the hypothalamus.

- Electrical stimulation of the amygdala in humans during surgery evokes feelings of apprehension and fear.

- Brain scans show increased activity in the amygdala in humans shown fearful faces. This response is impaired in people in whom the amygdala is calcified, even though they are still able to identify individual faces. Hence the neural system for emotional memory is distinct from that for explicit memory of faces.

Armed with representations of both innate and learned fear-inducing stimuli to compare with the ongoing data stream, the amygdala evaluates the significance of (i.e., the threat posed by) the current situation, and organizes the appropriate visceral and avoidance responses. Several characteristics of amygdala function are noteworthy:

- The evaluation of a stimulus by the amygdala begins *earlier* than any conscious cognitive appraisal of the situation.

- Amygdala fear learning is **implicit learning**, which means that the fear responses cannot be consciously generated. The amygdala develops more rapidly during infancy than the hippocampus (responsible for explicit memory). During this time fearful memories may be acquired which cannot later be consciously accounted for. This could underlie specific **phobias**.

- Long-term emotional memories are probably stored in the cerebral cortex rather than the amygdala: presentation of fear-evoking stimuli activates visual association cortex and orbital prefrontal cortex as well as the amygdala.

- At the cellular level fear learning by the amygdala involves NMDA receptor-dependent long-term potentiation similar to that underlying hippocampal learning.

Neocortex and emotion

Extensive interconnections between different parts of association neocortex allow cognition to influence emotional states (e.g., although feeling very anxious we do sit the examination, because we recognize that in the long term doing so will bring benefits), and *vice versa* (e.g., we know that the chance of our child being harmed on walking to school is negligible, but we feel sufficiently anxious to drive her anyway). Lesions of the orbital prefrontal cortex reduce emotional responses (e.g., aggression) in primates, while in humans lesions of the anterior cingulate cortex reduce the emotional distress of chronic intractable pain.

M2 Motivation and addiction

Key Notes

Motivated behavior	Motivated (goal-directed) behaviors are aimed at achieving a specific outcome, and driven by internal states and external cues. Some are homeostatic. Physiological deficits produce internal states (e.g., hunger) that motivate appetitive behavior (e.g., the search for food) and consummatory behavior (e.g., eating) leading to satiation. Most motivated behaviors are not homeostatic and their neurobiology is poorly understood. A stimulus that increases the chance of a motivated response is a positive reinforcer; if it decreases the probability it is a negative reinforcer.
Dopamine reward system	Dopaminergic neurons from the ventral tegmental area to the nucleus accumbens constitute a mesolimbic reward system. Mesolimbic neurons release dopamine most strongly in response to unexpected rewards or stimuli associated with rewards by learning. The nucleus accumbens has GABAergic medium spiny neurons which project to the GABAergic cells of the ventral pallidum (VP) which is part of the affective basal ganglia circuit. Phasic dopamine release by mesolimbic cells disinhibits this circuit. The cingulate and orbitofrontal cortices respond to rewards and learned associations and regulate, or initiate, goal-directed behaviors.
Drug addiction	Addictive drugs are positive reinforcers that take the place of natural reinforcers (e.g., food, sex) in driving the brain dopaminergic reward system. There are four aspects to addiction: tolerance, repeated doses of a drug become progressively less effective; dependence, normal functioning is only possible in the drugged state; withdrawal, unpleasant effects result if the drug is not taken; craving, a long-lasting intense desire for the drug, which seems to be due to learned associations between the pleasure produced by the drug and the context in which it is taken.
Neurobiology of addiction	Most addictive drugs stimulate the release of dopamine from the mesolimbic system, but more persistently than natural rewards, resulting in the dysregulation of mesolimbic and affective basal ganglia circuits that occurs in addiction. The effectiveness with which dopamine transmission occurs is initially compromised, causing tolerance, though once addiction is well established the mesolimbic system becomes sensitized to the drug. The withdrawal syndrome is partly due to the increased expression of cAMP response element binding protein in response to activation of D_1 receptors. Enduring glutamate neuroplasticity, akin to that which

<table>
<tr><td></td><td colspan="2">underpins learning (LTP and LTD) eventually occurs in the VTA, nAc, and prefrontal cortex, and this is responsible for the craving, compulsive drug-seeking behavior and vulnerability to relapse experienced by addicts even after long drug-free periods.</td></tr>
<tr><td>Related topics</td><td>(D4) Dopamine
(K6) Anatomy of the basal ganglia</td><td>(M3) Control of feeding
(N1) Types of learning</td></tr>
</table>

Motivated behavior

Behavior that is driven by internal states or external events and which is aimed at achieving a particular outcome is **motivated** or **goal-directed** behavior.

Some motivated behavior occurs in order to satisfy physiological needs. Chemical and neural signals give rise to an internal state, hunger. This drives **appetitive** or goal-seeking behavior, that is, foraging for food, and subsequently **consummatory** behavior, eating. Goal-directed behaviors like this are homeostatic and normally self-limiting as the internal state is switched off (sated) by consumption.

Much motivated behavior is not so straightforward because it occurs in the absence of any obvious physiological deficit. Sexual behavior leading to copulation, although not homeostatic, is driven by an internal state (libido) that can be sated. However, parenting behavior is neither homeostatic nor self-limiting and little is understood about the internal state that motivates it, although hormones (e.g., oxytocin) seem to be important. What motivates many activities, listening to music, exploring a forest path, engaging in a sport or academic study, is currently a mystery.

Any stimulus that increases the probability of a motivated response occurring is a **positive reinforcer**. An animal will work to get access to a positive reinforcer. By contrast a reinforcer is said to be negative if the animal works to *avoid* the stimulus, in which case it is displaying **aversive behavior**. The reinforcing quality of a stimulus depends on context. For example, food is a powerful reward to a hungry person but its positive reinforcing quality diminishes with satiety. However, a particular food may still be a positive reinforcer if it is novel and sufficiently delicious even if the person is not hungry. Hence the motivation to eat is a complex interplay of internal state, external cues, and memory. Similar qualifications apply to other goal-directed behaviors.

Dopamine reward system

Dopaminergic neurons ascend from the **ventral tegmental area** (**VTA**) to the nucleus accumbens (nAc, ventral striatum) as the **mesolimbic system** and to the frontal cortex (including the cingulate and orbitofrontal cortex) as the **mesocortical system** (Figure 1 in Section D4). The mesolimbic system is frequently described as a reward system because:

- Firing of mesolimbic neurons increases in the presence of natural reinforcers such as food.

- Conscious, behaving rats will forego food or sex in order to stimulate their own mesolimbic neurons by pressing a lever to deliver a small current through electrodes chronically implanted into the medial forebrain bundle; a technique termed **intracranial self-stimulation** (**ICSS**).

- Natural rewards, ICSS, and addictive drugs all increase the release of dopamine from mesolimbic terminals in the nucleus accumbens, and the reinforcing properties of all three are blocked by dopamine D1 receptor antagonists.

VTA neurons fire in response to a natural reward and the dopamine release facilitates learned associations with the reward, in animal studies a conditional stimulus—for example, light or tone—with which the reward has been paired. The learning occurs via the nucleus accumbens and the affective basal ganglia circuit (Figure 1). Subsequent firing of the VTA neurons depends on the *predictability* of the reward. Unexpected or novel rewards elicit a strong response, although this declines with repeated presentation—perhaps explaining why we eat more of a meal that consists of six courses of gourmet foods than a single bowl of rice. Predicted rewards have little effect, though conditioned stimuli associated with them continue to elicit dopamine release. Omission of a predicted reward *reduces* mesolimbic activity. The immediate response to omission of an expected reward is to persevere with the activity that usually provides it. So, low activity of mesolimbic neurons when a predicted reward is missed promotes reward-seeking behavior.

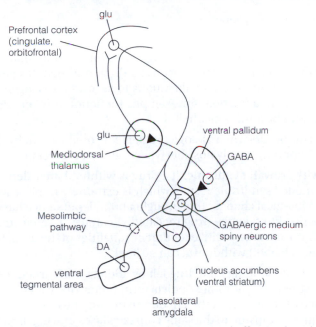

Figure 1. Simplified model of Mesolimbic reward system and affective basal ganglia circuit. DA, dopamine; GABA, γ-aminobutyric acid; glu, glutamate. —< excitatory, —◀ inhibitory.

The nucleus accumbens is the major target of the mesolimbic system. Like its dorsal striatum counterpart it has GABAergic **medium spiny neurons** (**MSNs**). These project to the GABAergic cells of the ventral pallidum (VP). Phasic dopamine release from terminals of VTA cells excites the MSNs, particularly if this coincides with excitatory input from the cortex or amygdala. Activity of the MSNs inhibits the GABAergic VP cells, hence *disinhibiting* the affective basal ganglia circuit. Distinct ensembles of MSNs respond differentially to cues associated with specific rewards.

The cingulate and orbitofrontal cortices respond to rewards and associated stimuli in a manner that depends on their predictability and seems to be concerned with evaluating their overall salience (importance) and determining the intensity of the behavioral response. In some cases at least the prefrontal cortex *initiates* goal-directed behavior via glutamatergic connections to the nucleus accumbens.

Drug addiction

All addictive drugs have positively reinforcing properties and these are primarily responsible for drug-seeking behavior in addicts. Animal operant learning studies have proved useful in investigating both behavioral and physiological aspects of addiction. The positive reinforcing properties of a drug can be assessed by **self-stimulation studies** which measure the extent to which animals (usually rats or monkeys) will work to get a dose of the drug. **Conditioned place preference studies** reveal that the context in which drugs are taken is important. Animals are first exposed to one environment when drugged and to a different environment when non-drugged. Next, the animals are given a choice between the two environments (they can now move freely between them) and the time spent in each is recorded. With positively reinforcing drugs, animals spend more time in the environment they experienced in the drugged state. This is **context-dependent learning** and shows that learning is important in addictive behavior.

Drug addiction has four components:

- **Tolerance**. On repeated administration a drug becomes progressively less effective, so the dose has to be increased if the original action is to be maintained. The precise mechanism for tolerance depends on the drug, but includes enzyme induction, changes in receptor numbers, and alterations to second messengers. Tolerance does not necessarily lead to addiction; for example, moderate drinkers develop tolerance for ethanol without becoming alcoholics.

- **Dependence**. This occurs when biological changes brought about by the drug are such that normal functioning is only possible when the drug is present.

- **Abstinence** (**withdrawal**) **syndrome**. If a drug is withheld after dependence is established an abstinence syndrome results which is extremely unpleasant and lasts until the long-term biological changes that brought about dependence have abated. Hence, **addiction** (the need to take the drug repeatedly) can be driven as much by the aversion to withdrawal as by the positive reinforcing qualities of the drug. In rare cases with some drugs (e.g., alcohol) withdrawal can be fatal.

- **Craving**. The intense longing for a drug felt by addicts is a learned response that long outlasts the abstinence syndrome. Addicts form memories which associate the pleasure produced by the drug with the environment and cues that accompany the drug taking. Subsequent exposure to the same context causes craving. Brain imaging shows that cocaine-addicted subjects have increased activity in the cingulate and orbitofrontal cortices in response to stimuli associated with cocaine availability but decreased activity when presented with stimuli associated with natural rewards, compared to non-addicts. This corresponds with the much more intense motivation of addicts to seek the drug than natural rewards. Craving is the major barrier to the successful permanent rehabilitation of addicts.

Neurobiology of addiction

Almost all addictive drugs act on the brain reward system. However, unlike natural rewards, which when familiar cease to elicit dopamine release, addictive drugs continue to cause dopamine release with repeated exposure, resulting in dysregulation of the reward and affective basal ganglia circuits, and eventually long-term neuroplasticity that underpins craving and compulsive drug-seeking behavior.

Cocaine addiction can be induced and studied in animals. Cocaine produces tolerance by blocking the **dopamine transporter** (**DAT**) in the presynaptic terminals of mesolimbic neurons, limiting dopamine reuptake, so the concentration of transmitter in the synaptic cleft is raised. This causes downregulation of postsynaptic dopamine receptors. In addition dopamine binding to presynaptic receptors reduces dopamine synthesis and release. The overall effect is that higher amounts of transmitter, and hence drug, are needed to achieve the same level of dopamine transmission. When drug use stops, dopamine transmission in the mesolimbic system drops below normal. This loss in the effectiveness of the brain reward system is probably responsible for the lack of pleasure given by natural rewards seen with drug withdrawal.

Activation of D_1 receptors, which are positively coupled to cAMP, switches on transcription of the *creb* gene. **CREB** (**cAMP response element binding protein**) is itself a transcription factor that influences the expression of genes with the cAMP response elements (*cre*) in their regulatory domains; that is, genes switched on by increases in cAMP. When the drug is withheld, the CREB changes reverse with a time course that matches the abstinence syndrome; hence *creb* overexpression contributes to this unpleasant phenomenon.

With continued drug usage eventually, and paradoxically, the mesolimbic system becomes *more* sensitive to the effects of the drug, overwriting the preexisting tolerance. This appears to be due to a gradual increase in the expression of a Fos family transcription factor, ΔFosB, in the nucleus accumbens. This molecule increases the expression of $GluR_2$ AMPA receptors, cell signaling molecules, and **brain-derived neurotrophic factor** (**BDNF**), which can stimulate dendritic growth. ΔFosB expression may be the switch for transition from acute drug responses to chronic plastic changes in addiction.

In the early stages of drug use craving arises from the release of dopamine in the nucleus accumbens. As addiction takes hold and dopamine transmission in the mesolimbic system reduces, increased metabolic activity in the orbitofrontal cortex contributes to craving. This transition is due to glutamate-mediated plastic changes in ventral tegmental area, nucleus accumbens, and cortex akin to that which underpins learning in the hippocampus and elsewhere, namely long-term potentiation (LTP) and long-term depression (LTD).

Recent history of drug experience alters the direction of plasticity at nAc excitatory synapses. Thus LTP occurs in the medium spiny neurons during a drug-free period *after* cocaine addiction has been established. This is accompanied by increased numbers of AMPA receptors and dendritic spines on MSN cells, making them more sensitive to excitatory inputs from the orbitofrontal and cingulate cortex, and amygdala, some of which encode drug-contextual learning. Remarkably, just a single dose of cocaine reverses this to bring about LTD, a reduced responsiveness to glutamate, most probably by NMDA receptor-controlled endocytosis of AMPA receptors. There is increasing evidence that inhibition of nucleus accumbens MSNs (GABAergic projection neurons), for example by LTD, promotes reward-seeking behavior, perhaps by disinhibiting downstream regions such as the ventral pallidum.

Vulnerability to relapse in long-term addicts lasts for years, implying that the neuroplasticity is extremely enduring or even irreversible by this stage. Moreover, reinstatement of drug-seeking now depends on dopamine release not in the nucleus accumbens but in the amygdala and prefrontal cortex, probably due to plastic changes there.

M3 Control of feeding

Key Notes

Control of food intake

For a constant body weight, food intake and energy expenditure must balance in the long term. The hypothalamus and nucleus of the solitary tract (NST) regulate feeding and metabolic rate in response to hunger and satiety signals. But control of eating is complicated by affect, learning, and social and cultural factors.

Hunger signals

Ghrelin and neuropeptide Y are both gut hormones that promote feeding. Ghrelin, secreted by the empty stomach, acts at the NST either via the circulation or by stimulating vagus nerve terminals in the gut.

Short-term satiety signals

Cholecystokinin (CCK), somatostatin, and peptide YY are gut hormones, secreted in response to food in the gut, which inhibit feeding. CCK acts via the circulation or by stimulating gut vagal afferents to the NST. Stretch receptor afferents monitoring gastric distension run in the vagus nerve to the NST and inhibit feeding.

Long-term satiety signals

Leptin and insulin are long-term satiety signals. Leptin is secreted by white adipose cells in a manner that reflects the size of the fat store and balances food intake and energy expenditure by inhibiting feeding and increasing basal metabolic rate. Hence it regulates fat stores by negative feedback.

Central regulation of feeding and satiety

Food intake is controlled by two brain pathways which originate in the arcuate nucleus of the hypothalamus. An orexigenic pathway that involves the lateral hypothalamus promotes feeding and reduces energy expenditure, whilst an anorexigenic pathway via the paraventricular nucleus reduces feeding and increases energy expenditure. Cells in the orexigenic pathway which use neuropeptide Y as a transmitter are activated by ghrelin. The anorexigenic pathway, or melanocortin system, uses several peptide transmitters including melanocortin, oxytocin, and corticotrophin releasing hormone. Both pathways produce their effect via the NST. Leptin reduces food intake by inhibiting the orexigenic pathway and stimulating the anorexigenic pathway. Stress acts on CRH neurons to suppress feeding via the melanocortin system. This may have a role in anorexia nervosa.

Obesity

Genetic obesity can result from lack of functional leptin or leptin receptors (leptin resistance) or excessive ghrelin concentrations. However, generally obesity is produced by a long-term mismatch between food intake and energy expenditure. Obese individuals probably respond more to external cues than internal (physiological) cues for feeding. Obese animals and humans have elevated plasma leptin that reflects the size of their fat stores but animals have leptin resistance so the set point of body fat they defend is higher. In mice leptin resistance is triggered by a high fat diet.

Related topics	(L1) Anatomy and connections of the hypothalamus	(L3) Neuroendocrine control of metabolism and growth
	(L2) Posterior pituitary function	(M2) Motivation and addiction

Control of food intake

Over the long-term food intake and energy expenditure must be in balance if body weight is to remain steady. Eating is a goal-directed behavior controlled in the shorter term by hunger signals, which motivate feeding, and satiety signals, which reduce it. An agent, neuron or pathway which stimulates appetite is said to be **orexigenic**, those which suppress appetite are **anorexigenic**. Food intake and energy expenditure are additionally regulated by hormones released from white fat cells (**adipokines**). The neural networks responsible for central regulation of feeding and energy expenditure lie in the hypothalamus and in the nucleus of the solitary tract, which lies in the medulla.

However, the control of eating is a complex phenomenon that involves the reward system, affect, learning, and memory, all of which is colored by social and cultural contexts. Wealthy humans have *ad libitum* access to high fat, high carbohydrate foods rarely encountered in the past, eat when they are not hungry, and have far less need or opportunities for exercise. Obesity is the result for many.

Hunger signals

The peptide **ghrelin** is a hunger signal that acts in both the hypothalamus and the reward system to increase food intake. In general blood concentrations of ghrelin are lower in obese and higher in anorexic individuals compared to those with normal body mass implying that ghrelin release is inversely related to energy intake. There is also an inverse relation between hours of sleep per night and blood ghrelin concentrations. The fewer hours slept the higher the ghrelin, and individuals who sleep less are more likely to be obese.

Ghrelin cannot cross the blood–brain barrier. It exerts its effects by three routes:

- As a peripheral signal secreted from the epithelium of the empty stomach it enters the circulation (i.e., it is a gut hormone) and acts on ghrelin receptors on the area postrema.

- Locally in the gut it acts on terminals of vagus nerve axons that run to the NST.

- It is a transmitter of neurons in the hypothalamus. It acts presynaptically to excite orexigenic (NPY/AgRP) neurons, and on corticotrophin releasing hormone (CRH) neurons in the hypothalamus.

Neuropeptide Y (**NPY**) is a gut hormone and the most potent orexigenic agent known. It is also a transmitter in hypothalamic orexigenic pathways.

The **area postrema** is a circumventricular organ located adjacent to the nucleus of the solitary tract (NST). It has chemosensory cells that respond to a variety of peptides. In response to toxins in the blood it triggers vomiting.

Short-term satiety signals

Satiety signals may be humoral or neural (Figure 1) **Cholecystokinin** (**CCK**) is secreted by the gut in response to food. As a circulating gut hormone it acts at the area postrema. As a local hormone it acts on gut vagal afferents that project to the NST. CCK also inhibits feeding by acting as a transmitter in hypothalamic anorexigenic pathways. **Somatostatin** and **peptide YY** are also gut hormone satiety signals.

Afferent input from taste buds and pharynx, stomach distension, and neural signals related to energy metabolism from the liver all inhibit eating. Stretch receptor afferents conveying gastric distension signals from the gut run in the vagus nerve to the NST. The inhibition of food intake by CCK and gastric distention is mediated by the paraventricular nucleus (PVN) of the hypothalamus which receives input from the NST.

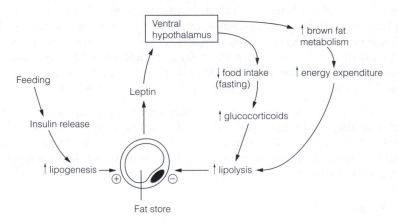

Figure 1. Negative feedback control of fat stores by leptin. An increase in the size of the fat store causes a rise in leptin release. This activates processes which result in increased fat breakdown.

Long-term satiety signals

Despite day-to-day imbalances between food intake and energy expenditure, people are able to match energy input and output closely over years. This is done largely by defending the size of the body store of white adipose tissue (Figure 1).

Leptin is secreted by fat cells and crosses the blood–brain barrier to act at the hypothalamus where it inhibits feeding (i.e., it is a satiety factor) and increases energy expenditure. Leptin acts as an **adipostat**. It has a plasma concentration that correlates well with body fat content and so is a molecule that reflects the size of the fat store and regulates it homeostatically. Leptin secretion is enhanced by the insulin-stimulated lipogenesis that occurs on feeding and is suppressed by the glucocorticoid-stimulated lipolysis that accompanies fasting, so leptin regulates fat mass by means of a negative feedback loop.

Leptin raises energy expenditure by increasing the expression of uncoupling proteins in mitochondria of fat and skeletal muscle, and by acting centrally to increase sympathetic activity to brown adipose tissue. Both effects raise basal metabolic rate.

Insulin is also a long-term satiety signal.

Central regulation of feeding and satiety

Food intake is regulated in the CNS by two parallel brain pathways which originate in the **arcuate nucleus** of the hypothalamus. An **orexigenic pathway** promotes feeding and reduces energy expenditure whilst an **anorexigenic pathway** reduces feeding and increases energy expenditure (Figure 2). Both exert their effects by modulating the nucleus of the solitary tract (NST). This structure organizes feeding reflexes and regulates energy expenditure by altering sympathetic nervous system output, basal metabolic rate, and locomotor activity.

Arcuate nucleus **NPY/AgRP neurons**, so called because they use neuropeptide Y and agouti-related protein as co-transmitters, are the first-order neurons in the orexigenic pathway. NPY secretion from the NPY/AgRP neurons is stimulated by ghrelin. NPY/AgRP neurons project to the **lateral hypothalamus**, synapsing with second-order neurons that use peptides termed **orexins** as neurotransmitters. These relay to neurons in the nucleus of the solitary tract (NST). NPY/AgRP cells also produce GABA and inhibit the anorexigenic pathway.

First-order cells in the anorexigenic pathway in the arcuate nucleus contain **pro-opiomelanocortin** (**POMC**), the precursor protein for several biologically active peptides, one of which, **melanocortin**, is a neurotransmitter of these neurons. Hence

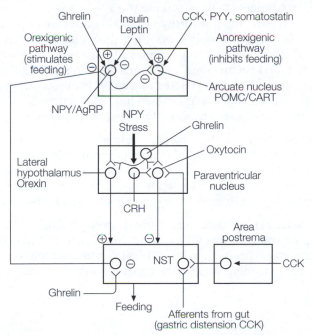

Figure 2. Central pathways controlling feeding. CRH, corticotrophin releasing hormone; POMC, proopiomelanocortin neuron (releases melanocortin); NPY, neuropeptide Y; AgRP, agouti-related protein; CART, cocaine- and amphetamine-related transcript; NST, nucleus of the solitary tract; CCK, cholecystokinin.

the anorexigenic pathway is often described as the **melanocortin system**. Other first-order cells in the anorexigenic pathway express **cocaine- and amphetamine-related transcript** (**CART**). POMC/CART cell axons run to the **paraventricular nucleus** (**PVN**) of the hypothalamus which contains neurons that use oxytocin, thyrotrophin releasing hormone, or corticotrophin releasing hormone as transmitters. The melanocortin system gets a wealth of inputs relaying short- and long-term satiety signals, and also possibly from a **glucostat** in the arcuate nucleus. Activation of the melanocortin system inhibits feeding and increases basal metabolic rate via outputs from the NST.

Leptin and insulin receptors exist on first-order neurons of both orexigenic and anorexigenic pathways. Leptin and insulin inhibit the orexigenic pathway and stimulate the anorexigenic pathway, thereby reducing feeding.

Stress and neuropeptide Y stimulate CRH neurons in the PVN which synapse with melanocortin system neurons to inhibit feeding. Stress is a key component of **anorexia nervosa** in which patients self-starve. Brain imaging shows that anorexics produce amygdala fear responses to their own body image. Once established, starvation naturally maintains high activity in the hypothalamic pituitary stress axis.

Endocannabinoid pathways are activated in fasting and stimulate feeding by modulating the action of leptin on the melanocortin system, and by stimulating the mesolimbic reward system.

Obesity

Over half of USA and UK citizens are overweight or obese. In a few instances obesity is genetic; animals and humans that lack either functional leptin or leptin receptors (and hence have **leptin resistance**) are obese, as are those with excessive ghrelin concentrations. However, the overwhelming majority of people are overweight because over the long term their energy expenditure is lower than their energy intake. Experiments in which obese individuals must work to get food suggest they respond more to external cues (how appetizing food seems) than to internal cues (hunger and satiety) than lean people. Basal metabolic rate (BMR) is related to lean body mass and because obese people have much the same lean body mass as lean individuals their BMRs do not differ on average. However, they have poorer capacity for energy expenditure.

Most humans with obesity have elevated plasma leptin concentrations—reflecting the size of their fat stores—but fail to respond to it. This leptin resistance means that despite having leptin concentrations that reflect their weight, the set point of body fat they defend by energy homeostasis is raised. In mice leptin resistance is triggered by a high fat diet, which acts by suppressing leptin receptor signal transduction in the arcuate nucleus.

Potential pharmacological treatments for acquired obesity include NPY, orexin, or endocannabinoid receptor antagonists, and melanocortin, CART, or CRH receptor agonists. Some of these interventions are likely to be problematic; for example, orexins are involved in wakefulness as well as feeding, and CRH receptor agonists are anxiogenic.

M4 Brain biological clocks

Key Notes

Intrinsic rhythms	Many physiological parameters show circadian (about a day) variation. In the complete absence of external time cues, for example, light, circadian rhythms free run initially with a period of 25 rather than 24 hours. After a couple of weeks, sleep/waking cycles decouple from other functions to free run with a period of 30 hours. Hence, circadian clocks must exist that are entrained by external events.
Suprachiasmatic nucleus	An intrinsically active neural oscillator in the suprachiasmatic nucleus (SCN) fires with a frequency that varies sinusoidally, peaking in the daytime. The SCN projects to other parts of the hypothalamus to regulate sleep/waking cycles, autonomic, and endocrine functions. Light signals from the retina arrive at the SCN to entrain it to a 24-hour cycle.
Pineal gland	The pineal gland secretes melatonin during the hours of darkness. The duration of the melatonin pulse is a direct measure of the length of the night, and so signals the time of year for animals living at latitudes away from the equator. For seasonal breeders, melatonin secretion acting at the SCN controls reproductive cycles. Light inhibits melatonin synthesis by means of a pathway from the SCN via the sympathetic nervous system to the pineal.
Related topics	(L1) Anatomy and connections of the hypothalamus (L3) Neuroendocrine control of metabolism and growth	(L4) Neuroendocrine control of reproduction (M5) Sleep (N4) Long-term potentiation

---

Intrinsic rhythms

Many body functions vary cyclically with a period of about a day; they have a **circadian rhythm**. Functions regulated in this way include sleep/wakefulness, core temperature, and the secretion of anterior pituitary hormones. Humans isolated from all external time cues show intrinsic circadian rhythms with a period of about 25 hours initially. This decoupling of circadian rhythms from the normal 24-hour period is called **free running** and shows that there exist intrinsic **circadian clocks** that are usually entrained by environmental cues called **zeitgebers** (German "time-giver"). Zeitgebers include light, exercise, social interactions, and work schedules. Light is the strongest. A powerful light pulse given during subjective night, produces shifts in the circadian rhythm. In humans with normal sleep patterns the nadir in core temperature occurs at about 5 a.m. A light pulse given during the night before this time causes circadian rhythms to be delayed (**phase delay**) whereas a light pulse after this time causes **phase advance**.

After 1–2 weeks free running, physiological variables often desynchronize from each other. For example, typically, fluctuations in core temperature, secretion of ACTH and glucocorticoids, and rapid eye movement (REM) sleep continue with a period of about 25 hours, but the cycles of sleep–wakefulness and secretion of growth hormone (GH) lengthen to over 30 hours. This suggests that there are two circadian clocks. Both are normally entrained by light–dark cycles of day and night.

Suprachiasmatic nucleus

The circadian clock that regulates sleep–waking cycles resides in the **suprachiasmatic nucleus** (**SCN**) of the anterior hypothalamus. The clock function of the SCN is due to pacemaker neurons which fire with a frequency that varies in a circadian fashion even when isolated from the rest of the nervous system. The firing frequency of SCN neurons varies sinusoidally with a period of 24 hours, peaking during the day, and dropping to its lowest rate during the night. The SCN projects largely to other hypothalamic structures to regulate sleep–wake cycles, autonomic, and endocrine functions, but also sends output to the thalamus and basal forebrain (e.g., septal nucleus) which probably accounts for circadian variation in memory and cognitive functions. Most SCN neurons are GABAergic and co-release peptides, and are assumed to be inhibitory on their targets.

Light signals encoding total luminance are relayed to the SCN by the **retinohypothalamic tract** (**RHT**). This pathway consists of the axons of a population of small retinal ganglion cells driven by cone photoreceptors over a wide area, which synapse directly with neurons in the core of the SCN. The RHT uses glutamate as a transmitter.

Pineal gland

The pineal gland is a circumventricular organ which secretes **melatonin** into the blood during the hours of darkness; the duration of the pineal melatonin pulse is a direct measure of the length of night, and hence also of day length, the **photoperiod**. Melatonin secreted into the blood is transported across the blood–brain barrier to act on the

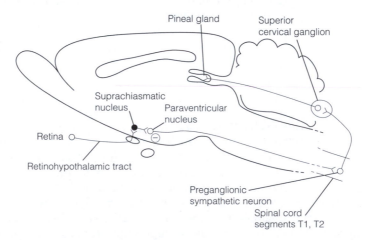

Figure 1. The brain circuitry by which light inhibits the secretion of melatonin from the pineal gland.

SCN. For animals living at latitudes other than the equator, day length varies during the year, so melatonin secretion acts as a signal which codes for the time of the year. For seasonal breeders, the length of the melatonin pulse regulates the hypothalamic–pituitary–gonadal (HPG) axis of both males and females via its action on the SCN. For example, in sheep the longer melatonin signals produced in the shorter photoperiods of November (in the Northern hemisphere) activates estrous cycles in ewes, and testicular growth—with a consequent rise in testosterone secretion and spermatogenesis—in rams. Although photoperiodic control of melatonin secretion is not important for reproduction in humans, it does exert feedback effects on the function of the SCN circadian clock and so affects sleep–wake cycles.

The pathway by which light inhibits melatonin synthesis is circuitous (Figure 1). Neurons in the SCN which get retinal input from the RHT inhibit central autonomic neurons in the paraventricular nucleus (PVN). The PVN sends axons through the brainstem to synapse with preganglionic sympathetic neurons in the intermediolateral horn of spinal cord segments T1 and T2. These project to the superior cervical ganglion (SCG), the postganglionic cells of which innervate the pineal gland. At night the activity of SCG neurons is increased, and the secretion of noradrenaline (norepinephrine) from sympathetic terminals acts on β adrenoceptors of **pinealocytes** to synthesize melatonin. The biosynthetic pathway is illustrated in Figure 2.

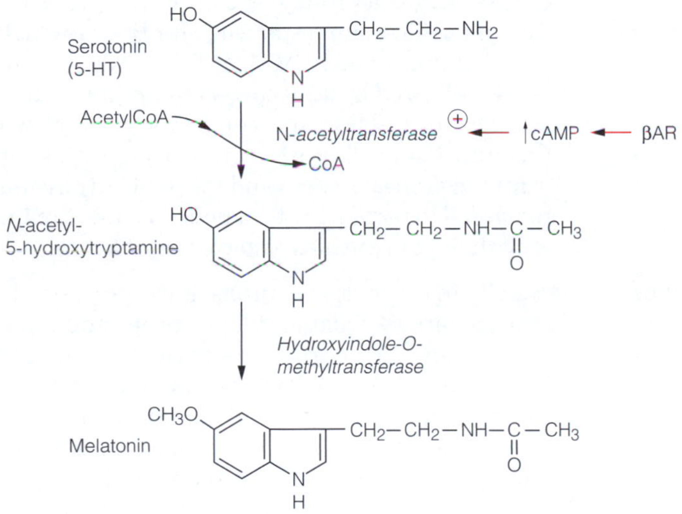

Figure 2 Synthesis of melatonin in the pineal. N-acetyl transferase is activated by norepinephrine, secreted from sympathetic terminals, acting at β adrenoceptors (βAR). cAMP, cyclic adenosine monophosphate.

Melatonin can entrain the circadian clock in the SCN, reset sleep–wake cycles in animals and humans, and reduce the symptoms of **jet-lag**, the sleep disturbance that arises when light–dark cycles and circadian rhythms are suddenly desynchronized by air travel over several time zones.

M5 Sleep

Key Notes

States of sleep

There are two states of sleep. Non-rapid eye movement (NREM) sleep has low-frequency, high-voltage, synchronized EEG, low cerebral blood flow and brain glucose utilization, but muscle tone is retained. There are four stages of NREM through which an individual passes on going to sleep; stages 3 and 4 constitute deep, slow-wave sleep. In rapid eye movement (REM) sleep the brain has a desynchronized EEG and metabolic activity similar to the awake state, but muscle tone is largely absent. Dreaming occurs mostly in REM sleep.

Sleep pathways

The ascending reticular formation is an arousal system. It has two branches. One, containing monoaminergic neurons, goes to the cerebral cortex from brainstem, hypothalamus, and basal forebrain. These are responsible for maintaining arousal and go silent in REM sleep. The other branch consists of cholinergic neurons in the pons that project to the thalamus; these are responsible for EEG desynchronization in wakefulness and REM sleep. Thalamic relay neurons are depolarized by arousal system input and this allows thalamocortical transmission. During NREM sleep the thalamic relay cells are hyperpolarized by loss of input from the arousal system and this blocks transmission from thalamus to cortex. Cortical neurons now fire in isolation with their own intrinsic (slow wave) rhythm.

Physiology of NREM sleep

Activity in GABAergic neurons in the pre-optic hypothalamus hyperpolarizes thalamic relay neurons and inhibits the monoaminergic neurons that project to the cortex. This induces NREM sleep. The triggers for NREM sleep, which act through the pre-optic hypothalamus, include endogenous sleep-promoting substances such as adenosine, the brain biological clock, and elevated core temperature.

Physiology of REM sleep

REM sleep may be brought about by GABAergic inhibition from the periaqueductal gray matter. This shuts down the noradrenergic and serotonergic neurons, but *not* the pontine cholinergic cells. The cholinergic neurons depolarize the thalamic relay neurons allowing EEG desynchronization, and inhibit sensory input and motor output, disconnecting the brain from the external world. What organizes the flip between NREM and REM sleep is not known but orexin neurons are implicated in stabilizing the current state.

Functions of sleep

The metabolic hypothesis is that sleep is a homeostatic mechanism to correct a metabolic energy deficit that accrues during waking hours. It is based on the fact that sleep-

	deprived rats die of failure of their thermoregulatory and immune systems.	
Related topics	(D5) Noradrenaline (norepinephrine)	(D6) Serotonin (D7) Acetylcholine

States of sleep

Two sleep states, **non-rapid eye movement** (**NREM**) sleep and **rapid eye movement** (**REM**) sleep, can be distinguished on the basis of several physiological measures. Large numbers of cerebral cortical cells fire in synchrony and consequently their summed activity produces potentials large enough that they can be recorded with an array of scalp electrodes as **electroencephalography** (**EEG**). The EEG waveform varies in frequency and the frequency ranges are conventionally grouped: alpha (8–13 Hz), beta (13–30 Hz), delta (0.5–2 Hz), and theta (4–7 Hz). When awake, the EEG waveforms are of low amplitude and high (alpha) frequency (Figure 1) and are described as **desynchronized**. Non-REM (NREM) sleep, has high-amplitude, low-frequency (synchronized) EEG waveforms.

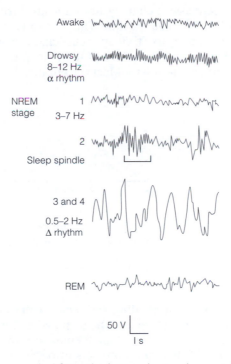

Figure 1. EEG waveforms recorded from the human brain when awake and asleep.

During NREM sleep, muscle tone is retained and postural adjustments (turning over) are occasionally made. Respiration rate, heart rate, and mean arterial blood pressure all fall, though gastrointestinal motility increases. Most growth hormone secretion occurs during NREM sleep.

On falling asleep the EEG frequency progressively decreases, passing through four stages of NREM sleep (stages 1–4). Stage 2 is interspersed by higher frequency bursts called **sleep spindles** and large spikes called **K complexes**. Sleep spindles may have a role in motor learning. K complexes are brief synchronized episodes of widespread cortical inactivity (cortical down time). Stages 3 and 4 are collectively referred to as **slow-wave sleep** because it is characterized by delta waves. In passing from stage 1 through to 4 it becomes increasingly difficult to arouse the sleeper. Brain scan shows that cerebral blood flow and glucose utilization fall by as much as 40% in NREM sleep.

In REM (**paradoxical**) sleep the EEG resembles that of the awake state. Muscle tone is absent, except for transient contractions of extraocular eye muscles (hence rapid eye movement sleep). Respiration rate, heart rate, mean arterial blood pressure, and core temperature become irregular. People aroused from REM sleep usually report that they were dreaming. At this time dream content is in short-term memory and is rapidly forgotten unless rehearsed. Dreaming can also occur in NREM sleep. In REM sleep the brain is as metabolically active as it is when awake.

During a typical night's sleep (Figure 2) adults drop rapidly into deep (stage 4) NREM sleep and then REM and NREM sleep alternate about every 90 min with increasingly longer periods of REM sleep as the night progresses. After sleep deprivation an individual spends more time in NREM.

The proportion of time spent asleep changes dramatically during development. Human fetuses sleep (mostly REM) almost all the time. This falls to 17–18 hours sleep (50% REM) for babies born at term. The amount of time spent in stage 4 sleep falls exponentially with age. Between 10 and 70 years the proportion of REM sleep is constant at about 25% of total sleep time and declines in the elderly. Most vertebrates sleep but only homeotherms have REM sleep.

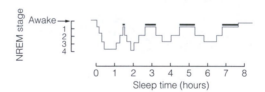

Figure 2. Distribution of sleep stages during a typical night's sleep. Dark bars are rapid eye movement sleep periods. NREM, Non-rapid eye movement sleep.

Sleep pathways

Monoaminergic neurons in the **ascending reticular formation**, required to generate the awake state and to increase arousal and the responsiveness of the cortex to sensory input, constitute an ascending arousal system. It splits into two branches at the level of the diencephalon (Figure 3). One branch projects through the lateral hypothalamus to the cerebral cortex from a variety of sources:

- Noradrenergic neurons of the locus coeruleus

- Serotonergic cells in the raphe nuclei

- Histaminergic neurons in the tuberomammillary nucleus (TMN) of the hypothalamus

This branch is augmented by cholinergic neurons in the basal forebrain.

The noradrenergic and serotonergic neurons fire at the highest rate in alert animals, have low firing rates during NREM sleep and go silent during REM sleep, hence are called **wake-on/REM-off cells**.

The second branch projects to the thalamus and contains:

- Histaminergic neurons in the TMN which make excitatory synapses with thalamic relay neurons

- Cholinergic neurons in the pons (**pedunculopontine nucleus, PPN**, and **lateral dorsal tegmental nucleus, LdT**) which inhibit GABAergic thalamic reticular neurons and hence excite thalamic relay cells

The *pontine* cholinergic cells are active during wakefulness and REM sleep—and are responsible for the desynchronization of the EEG in these states—but become quiescent during NREM sleep, and hence are described as **wake-on/REM-on cells**.

Neurons which use *orexins*, located in the tuberal region of the hypothalamus excite all the groups of neurons listed above and are active during the awake state and REM sleep. Orexigenic neurons probably stabilize whichever state (awake or REM) the brain is in.

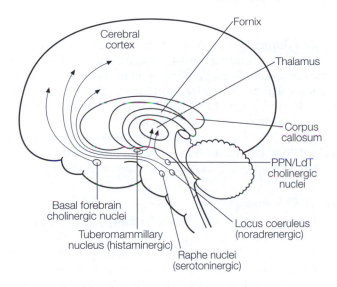

Figure 3. Ascending arousal system.

Only 5–10% of synapses in the thalamus are made by afferent terminals. Most of the thalamus is concerned with controlling which sensory information is sent to the cortex (attention). **Thalamic relay cells** project to the cortex and get reciprocal connections from the cortex (Figure 4). Both relay cells and cortical projection neurons use glutamate and are excitatory. Specific sensory input (e.g., from the retina) directly excites the relay cells. The relay cells are subject to inhibition from GABAergic interneurons and from the **thalamic reticular nucleus** (**TRN**), a sheet of GABAergic neurons covering the thalamus.

Thalamic relay neurons have two modes of firing:

- Tonic firing of single action potentials occurs in the awake state when the relay cells are depolarized by input from the ascending arousal system. In this mode, transmission of sensory input from thalamus to cortex takes place.

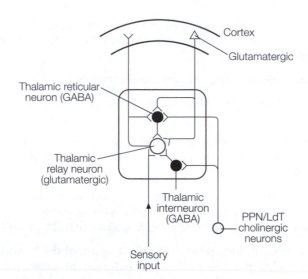

Figure 4. Neural circuitry of the thalamus.

- Burst firing is seen during NREM sleep when the relay cells are hyperpolarized by a loss of input from the ascending arousal system. Burst firing occurs because relay cells have T-type Ca^{2+} channels that are activated by hyperpolarization. Opening these channels causes a calcium depolarization that triggers a burst of 4–5 conventional action potentials. Burst firing of the relay cells drives synchronized bursting of cortical cells. The conventional argument is that this limits information transmission between thalamus and cortex. With deepest NREM sleep the thalamocortical neurons become so hyperpolarized they go silent and cortical neurons, now completely decoupled from the thalamus, fire with their own intrinsic (delta) rhythm.

Physiology of NREM sleep

The start of NREM sleep is organized by hypothalamic nuclei and involves suppressing both branches of the ascending reticular formation (Figure 5). GABAergic neurons in the **ventrolateral pre-optic area** (**VLPO**) inhibit histaminergic neurons in the tuberomammillary nucleus. The loss of excitation on thalamic relay neurons causes them to hyperpolarize so they go into burst firing, NREM sleep, mode. VLPO cells suppress the cortical branch of the ascending reticular formation by inhibiting all the monoaminergic and the orexigenic neurons. Inhibition of the orexin neurons:

- Shuts off the cholinergic basal forebrain cortical arousal system
- Reduces activity of the noradrenergic and serotonergic (wake-on/REM-off) cells and of the pontine cholinergic (wake-on/REM-on) cells

Although the loss of activity in the wake-on/REM-off cells lifts their inhibition on the pontine cholinergic cells, the lack of tonic excitation from the orexin neurons keeps them quiet.

The triggers for NREM sleep, which activate the GABAergic VLPO cells, are not well understood but include the biological clock in the suprachiasmatic nucleus, and elevated core temperature detected by warm receptors in the hypothalamus. A number of molecules have also been proposed as candidate endogenous sleep-producing substances including adenosine, melatonin, and interleukin-1. **Adenosine**, derived from ATP by neural activity is secreted by astrocytes into the brain extracellular space; an example of gliotransmission. Adenosine concentration rises during wakefulness and it decreases the activity of the cholinergic cells of the basal forebrain cortical arousal system by acting at adenosine A_1 receptors. During sleep, adenosine concentrations decrease as it is degraded by adenosine deaminase.

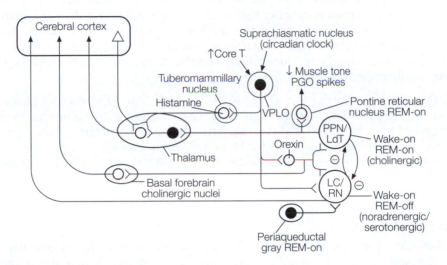

Figure 5. A model for the neural circuitry regulating sleep and wakefulness. LC/RN, locus coeruleus/Raphe nuclei; PPT/LdT, pedunculopontine nucleus/lateral dorsal tegmental nucleus; VLPO, ventral lateral pre-optic hypothalamus. Open circles, excitatory neurons; filled circles, GABAergic inhibitory neurons.

Physiology of REM sleep

During REM sleep (Figure 5) GABAergic inhibition comes from a different source, REM-on cells, possibly located in the periaqueductal gray matter. These shut down the noradrenergic and serotonergic (wake-on/REM-off) neurons, but *not* the orexin cells. Now, pontine cholinergic wake-on/REM-on cells are activated by the combination of continued excitation by orexinergic cells plus disinhibition from the aminergic neurons. In consequence:

● High levels of activity in these cholinergic neurons causes depolarization of the thalamic relay neurons which go into tonic firing mode.

● The continual activity of the orexin neurons keeps the basal forebrain cholinergic cells responsive.

Both of the above effects conspire to desynchronize the EEG.

The pontine cholinergic neurons also organize—via relays through pontine and medullary reticular nuclei—two major features of REM sleep. The first is the powerful suppression of sensory input and motor output. Presynaptic GABAergic inhibition on afferent terminals reduces sensory input. Glycinergic postsynaptic inhibition of motor neurons is the route by which muscle **atonia** (loss of muscle tone) is brought about. Thus, during REM sleep the brain is uncoupled from the external world; it is "off-line." Lesions of the pons which prevent the muscle atonia produces animals which express stereotyped behaviors during REM sleep. This suggests that during normal REM sleep motor patterns are generated but not executed; we cannot act out our dreams!

A second major feature of REM sleep are periodic **pontine geniculate-occipital** (**PGO**) **spikes** in the EEG. These originate from cholinergic (PGO-on) cells in the pons. They drive vestibular and reticular neurons to excite oculomotor neurons (causing the rapid eye movements) and other cells to produce the phasic alterations in respiration, heart rate, blood flow, and muscle twitches seen in REM sleep. PGO-on cells also initiate the spread of activity to the lateral geniculate nucleus and visual cortex recorded as PGO spikes. During wakefulness PGO-on cells are usually inhibited by serotonergic cells, however, PGO spikes can be produced in awake subjects by sudden stimuli, so they may underlie startle responses.

It is not known what causes the brain to switch from NREM to REM sleep and back several times during the night, and there are several candidates for the GABAergic neurons that trigger REM sleep. However, it is thought that orexigenic neurons stabilize the waking or REM state once the brain has made the transition. This idea is supported by the discovery of deficits in orexin neurotransmission in animals and humans with **narcolepsy**. In this condition patients experience frequent and undesired flips into REM sleep during the day.

Functions of sleep

Remarkably, the function of sleep is not known. Among the numerous ideas that have been proposed two have emerged as front runners, the metabolic hypothesis and the memory hypothesis (Section N4).

The **metabolic hypothesis** postulates that NREM sleep provides a period of low metabolic demand needed to replenish neural energy resources depleted during waking. One possibility is that wakefulness decreases the ATP:AMP ratio, resulting in accumulation of extracellular adenosine which, acting through adenosine A_1 receptors, inhibits neuronal activity and lowers brain metabolic demand. However, the need for NREM sleep seems to extend way beyond the nervous system. Sleep-deprived rats suffer anorexia (even though their food intake increases), lose the ability to thermoregulate, and die as a consequence of immune system failure after about 4 weeks. Hence, sleep may have global anabolic functions that conserve energy stores and core temperature. Circumstantial evidence for this includes:

- Smaller mammals (with the highest metabolic rates) sleep the most.

- Heat stress or experimental warming of the pre-optic hypothalamus (which is involved in both thermoregulation and triggering NREM sleep) can trigger or prolong NREM sleep.

N1 Types of learning

Key Notes

Definition of learning

Learning is the acquisition of altered behavior as a result of experience and occurs by rewiring of neural pathways (plasticity). Storing the changes over time is memory. Prior learning is tested by recall, elicited by the appropriate stimuli.

Declarative and procedural memory

Declarative memory is memory for facts, is fast, and is consciously recalled. It includes episodic memory in which associations that relate to a single event are learned, and semantic memory in which facts are learnt. Procedural memory is memory for motor skills. It is slow and not recalled consciously. Many learning situations include elements of declarative and procedural learning.

Working memory and long-term memory

Declarative memory has at least two temporal phases. Working memory (WM) is brief, of limited capacity and requires continual rehearsal. Long-term memory (LTM) is long-lasting and apparently of unlimited capacity.

Consolidation and reconsolidation

Consolidation is the process that makes memories increasingly resistant to disruption. For declarative learning, serial models postulate that consolidation transfers selected material from WM to LTM. In some instances recalling a consolidated memory returns it to a fragile state and it must be reconsolidated for the memory to persist. Both consolidation and reconsolidation require protein synthesis.

Nonassociative and associative learning

Only a single type of stimulus is needed for nonassociative learning. In habituation, repetitive delivery of a weak stimulus causes the loss of a motor response. Sensitization is the enhancement of a response to innocuous stimuli seen after an unpleasant stimulus. Associative learning requires pairing of two events within a short time. In classical conditioning, animals learn an association between one stimulus (the conditioned stimulus) and the appearance of a second that may be rewarding or unpleasant (unconditioned stimulus). The conditioned stimulus must always be presented immediately before the unconditioned stimulus. In operant conditioning, animals learn an association between some action they perform and the arrival of a stimulus which may be either rewarding or aversive.

Memory modulation

The degree of arousal determines the probability that specific memories will be consolidated. Arousal is signaled by the brain noradrenergic arousal system, and hypothalamic–pituitary–adrenal stress axis, both of which modify the amygdala and hippocampus. Optimal learning occurs with moderate catecholamine or glucocorticoid concentration. High levels of these hormones are detrimental.

Related topics	(N2) Working memory	(N4) Long-term potentiation
	(N3) Hippocampus and episodic learning	

Definition of learning

Some neural pathways establish connections during development that subsequently remain unaltered. These pathways are often said to be **hard wired** and the generic term for those processes that ensure the pathway is properly connected is **specificity**. However, pathways subject to continual re-wiring, either during development or as a result of experience, are referred to as **plastic**, and the re-wiring processes described as **plasticity**. **Developmental plasticity** is wiring that is conditional on early sensory experience and shapes subsequent perception. **Learning** is also plasticity and is the *acquisition* of reproducible alterations in behavior as a result of particular experiences. The storage of the altered behavior over time is **memory**. In animals, learning and memory can only be tested operationally by **recall**, in which the previously learned behavior is elicited by the appropriate stimuli.

Declarative and procedural memory

Learning occurs in a variety of distinct situations, differing in time course, stimulus requirements, and outcomes, so is classified in several ways to reflect these features (Figure 1). A major distinction is between declarative and procedural memory.

Declarative (explicit) memory is memory for facts. Declarative learning is fast, it requires few trials, requires conscious recall, and may be readily forgotten. It has two components that are dissociable in patients with cortical damage. **Episodic (recollection) memory**

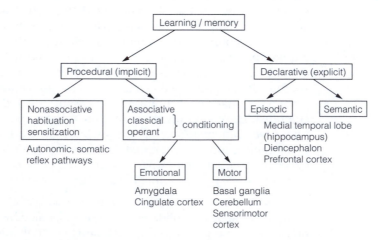

Figure 1. Types of memory. Structures responsible for a particular category are shown beneath. Episodic learning is also associative.

is memory for specific events, in which associations are established at a specific time and place (e.g., going to a gamelan concert whilst on holiday on Bali). The ability of rats to navigate through a maze in which they must learn to associate their positions in the maze with cues in their surroundings, **spatial navigation learning**, is an extensively studied example of episodic learning. The different components of an episodic memory are distributed across disparate regions of the brain and must be bound together for recall of the episode. The extensive connections of the medial temporal lobe allow it to act as a convergence zone for the different information streams.

The second component is **semantic (familiarity) memory** which is memory of facts unrelated to events; that Bali is an Indonesian island can be recalled without ever having been there, so semantic memory is about "knowing that." Studies of brain-damaged patients show that semantic memories are sorted into categories (sets of related objects), which appear to be stored in different areas of brain. Recall of specific items seems to need activation of multiple brain sites, each of which codes for a given attribute (e.g., color, function, name) of the item.

Procedural (motor) memory is memory for skills, such as learning to walk, swim, ride a bike, or play a musical instrument. It is "knowing how" memory. Procedural memory is slow, it needs many trials—in other words, a lot of rehearsal—and it is incremental in that improvement occurs gradually over time. Performance of procedural tasks does not involve conscious recall. For this reason procedural memory (like emotional memory) is described as **implicit memory**. Once established, procedural memories are not forgotten even after many years without rehearsal.

Many tasks have both factual and skill memory components. Playing the flute requires declarative memory for the musical notation of the score and procedural memory for the sequence of finger movements and the breathing pattern needed to create the sounds.

Working memory and long-term memory

Declarative memory has at least two phases categorized by their time course. **Working memory** (short-term memory) is temporary, limited in capacity and requires continuous rehearsal to keep it.

Long-term (remote) memory (LTM) is, if not permanent, at least long-lasting, has no obvious upper limit to its capacity and does not require continual rehearsal. **Amnesias** (loss of memory) due to brain damage can affect working memory and LTM independently. Amnesias of LTM are of two types, depending on whether memories are lost for events and facts acquired before, **retrograde amnesia**, or after, **anterograde amnesia**, the brain damage.

Consolidation and reconsolidation

Consolidation is the process that makes both declarative and procedural memories increasingly resistant to disruption or interference from similar learning over time. Serial models postulate that elements may be selected from working memory for consolidation into LTM by attention and arousal mechanisms. Experiments in which animals are injected shortly after training on a novel task with antibiotics which inhibit protein synthesis (e.g., anisomycin, cyclohexamide) show that consolidation requires protein synthesis.

In at least some cases recalling a consolidated memory returns it to a labile state in which it becomes sensitive to **interference**. This is where a previously established memory can no longer be retrieved because of competition from new learning. To prevent this, the recalled memory must now be **reconsolidated**. For example, if established auditory fear memories are reactivated in rats, protein synthesis is required for the memory to be retained. The biochemistry of consolidation and reconsolidation appear to be different (Figure 2).

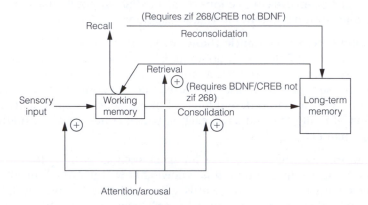

Figure 2. Possible functional relationships between working memory and long-term memory. Consolidation and reconsolidation are biochemically dissociable. BDNF, brain-derived neurotrophic factor; CREB, cyclic AMP response element binding protein; zif 268 (Egrl), zinc finger transcription factor 268 (Early growth response protein I).

Nonassociative and associative learning

Procedural memory is learning to produce a motor response to a particular input. It is divided into two types; nonassociative and associative. **Non-associative learning** occurs in response to only a single kind of stimulus. Two examples are **habituation**, in which repeated exposure to a weak stimulus results in a reduction or a loss of the response normally seen with occasional presentation of the stimulus, and **sensitization**, which is an exaggerated response to innocuous stimuli following a strong noxious (unpleasant) stimulus.

Associative learning needs the pairing of two different types of stimulus within a short time and in the correct order. It enables animals to behave as if they can predict relationships of the kind: if A then B. **Classical conditioning** was first investigated in dogs that learned to associate a sound with a subsequent food reward. Hungry dogs salivate at the sight or smell of food. The food is the **unconditioned stimulus** (**US**) and the salivation an **unconditioned response** (**UR**), so called because the nervous system is hard wired in such a way that salivation occurs as an autonomic reflex response to food. If, in a series of training trials, a sound, the **conditioned stimulus** (**CS**), is presented shortly before the arrival of the food, then in a subsequent test, presentation of the sound *alone* will elicit salivation. The salivation response is now a **conditioned response** (**CR**) because the animals have learnt to salivate when the CS (sound) is presented. Classical conditioning is characterized by **temporal contiguity**, the requirement that the CS must be presented *before* the US, and **contingency**, that animals learn that a predictive relationship exists between the CS and the US. **Extinction** of the conditioned response occurs if the CS is repeatedly presented without the US or if the temporal pairing of the CS and US is disrupted (i.e., if they are presented randomly). Extinction is *not* the same as forgetting. If after extinction the pairing of CS and US is restored the CR returns much more rapidly than it does in naive animals. Classical conditioning in which the US is noxious and which results in fear responses to normally neutral stimuli is **aversive conditioning**.

In **operant** (**instrumental**) **conditioning** an animal learns an association between a motor activity it performs (e.g., pressing a lever) and the arrival of a stimulus, termed the **reinforcer** (e.g., a food pellet). Reinforcers may be positive, in which case they increase the probability that an animal will act to obtain it, or negative (an aversive stimulus, such as an electric foot shock) in which case the animal will work to avoid it. Operant conditioning is used to investigate motivated behaviors.

Memory modulation

The arousal levels associated with an event modulate the likelihood of specific memories being consolidated. The arousal signals to which the brain memory circuits respond are adrenal hormones (both catecholamines and steroids), and several CNS peptide neurotransmitters released in response to stress.

Evidence for the involvement of catecholamines includes:

- Enhanced recall of emotionally neutral learning tasks by noradrenaline (norepinephrine) or adrenaline (epinephrine) given within a short time of the learning trials.

- No better recall of an emotionally charged version of a story compared with the neutral version after administration of the β-adrenoceptor antagonist propranolol.

People with higher levels of sympathetic activity are more likely to suffer from post-traumatic stress disorder after a traumatic experience.

The catecholamine dose–response curve has an inverted U shape; moderate concentrations are more effective enhancers of memory than either high or low levels. As neither of these hormones crosses the blood–brain barrier their actions on the CNS must be exerted peripherally. The catecholamines act at β-adrenoceptors on visceral afferents that run in the vagus (X) nerve to the nucleus of the solitary tract. This results in activation of noradrenergic neurons of the locus coeruleus that are part of a brain arousal system. This system projects to the amygdala and hippocampus to modulate learning. Electrically stimulating the vagus nerve immediately after training improves recall in an inverted U relationship with firing frequency. Cutting the vagus nerves or lesioning the nucleus of the solitary tract blocks the effects of systematically administered catecholamines on memory.

Glucocorticoids released by activation of the hypothalamic–pituitary adrenal axis in stress also have effects on learning and memory. These hormones readily cross the blood–brain barrier to act on steroid receptors that are located in high density in the amygdala and hippocampus. Low doses of glucocorticoids enhance, while high doses (or chronic exposure in long-term stress), impairs memory. Low concentrations occupy the high-affinity mineralocorticoid receptors (MR) and this facilitates strengthening of synapses thought to be crucial for learning. In contrast, high glucocorticoid concentrations fully saturate the low-affinity glucocorticoid receptors (GR) and this blocks the synaptic strengthening necessary for learning.

The anterior pituitary corticotrophs manufacture, from a single precursor, adrenocorticotrophic hormone (ACTH) and the opioid peptide, β endorphin, both of which impair learning by direct action on the CNS. Enkephalins, also opioid peptides, are co-released from the adrenal medulla along with catecholamines and impair memory by a peripheral action. Naloxone, an antagonist of opioid receptors facilitates memory.

Cholinergic enhancement of memory is well documented. Muscarinic receptor antagonists impair memory, while inhibitors of acetylcholinesterase improve it. Acetylcholine modulation of memory is mediated by the septohippocampal pathway and the cholinergic nuclei of the basal forebrain. The amygdala may enhance consolidation by activating the cholinergic attentional system in the basal forebrain.

N2 Working memory

Key Notes

Working memory

Working (short-term) memory is an on-line memory. Its capacity is limited by a fixed number of slots rather than the complexity of the information it can contain. It contains two slave systems; a phonological loop which holds verbal information and requires the left cerebral hemisphere, and a visuospatial sketch pad that holds visual, spatial, and kinesthetic information and requires the right hemisphere. An episodic buffer produces consciously recalled episodes from the disparate information in the slave systems and long-term memory which it accesses via a central executive.

Prefrontal cortex

The connections of the prefrontal cortex with temporal lobe and diencephalic structures involved in learning and the effects of lesioning it, both in monkeys and humans, implies that it is concerned with tasks requiring working memory.

Related topics

(G7) Oculomotor control and visual attention

(K6) Anatomy of the basal ganglia
(N1) Types of learning

Working memory

Working memory (**WM**) is an on-line memory system. It is used, for example, to hold a novel phone number while it is dialed. On average humans can hold up to four items *simultaneously* in WM, though there are individual differences. EEG, brain imaging, and behavioral studies all suggest that the number of items that can be held is unaffected by the complexity of the items. Hence, the capacity of WM seems to be defined by the number of items rather than the total quantity of information. This supports the **discrete resource model** of WM which argues that any item represented in WM must be assigned to one of a limited number of slots. It contrasts with the **flexible resource model** in which each item is assigned a share of WM resources and that performance is limited for large numbers of items because each item gets only a small share of WM resources. **Interference** can occur between representations held in WM. If items to be remembered are similar to those being processed, accuracy of recall is reduced. Delaying recall has little effect on recall accuracy so a simple decay model, in which item storage in WM fades over time, cannot be the case.

WM capacity is positively correlated with several complex cognitive skills: for example, reading comprehension, problem-solving ability, maintaining attention in the face of distractions.

Brain imaging and lesion studies imply the existence of several independent subsystems for working memory (Figure 1). The **phonological loop** allows speech sounds to be held and rehearsed for long enough to give continuity to spoken language, so that phrases and sentences can be comprehended. It requires the left cerebral hemisphere.

The **visuospatial sketch pad** is a temporary store for visual, spatial, and kinesthetic input that brain imaging indicates involves several regions in the right hemisphere. The phonological loop and visuospatial sketchpad are together referred to as **slave systems**. An **episodic buffer** is proposed as a temporary interface between the slave systems and LTM via the **central executive**, which binds information from a variety of sources into coherent episodes. The contents of the episodic buffer are assumed to be consciously retrievable through activity of the central executive. The executive is also postulated to mediate directed attention, which transfers information into the episodic buffer from the slave systems and LTM where it can used to create new cognitive representations and for problem solving. The central executive may plan how to execute complex cognitive activities.

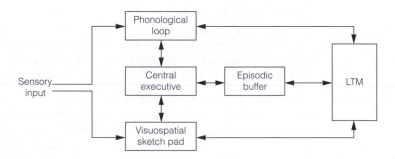

Figure 1. Model of working memory.

Prefrontal cortex

The **prefrontal cortex** (**PFC**) is involved in complex problem solving and planning future actions and there are good reasons for supposing that these executive tasks require working memory. The connectivity of the prefrontal cortex argues for its role in working memory. Firstly, there are reciprocal connections between the PFC and other cortical areas, so the PFC receives visual, auditory, and somatosensory information. Secondly, the PFC is interconnected with the medial temporal lobe and dorsomedial thalamus that have a well-documented role in learning and memory. Thirdly, the PFC is a component of the executive thalamocortical-basal ganglia circuit which allows it to modulated by reward and salience.

In **spatial delayed response tasks**, monkeys see a food reward placed in one of several covered locations. After a delay, which can be varied over trials, the animals are tested to see if they remember the location of the food. Monkeys with lesions of the prefrontal cortex have deficits in these tasks, and performance degrades progressively as the delay is lengthened. Recording from the PFC in alert behaving monkeys reveals cells that fire in predictable ways during delayed-response tasks. For example, many cells fire throughout the delay period, others fire when the food is placed in the location and when the animal is allowed to choose the location. Particular regions of the PFC seem to be modality-specific; evidence for specific subsystems of working memory.

Humans with prefrontal lesions also show deficits on working memory tasks in which they are required to use recent data to make correct decisions. Such individuals have great difficulty in tracing a path through a drawing of a maze. They will make the same errors repeatedly, and start right from the beginning of the maze after making a mistake, rather than from the position in the maze just before they made the error.

N3 Hippocampus and episodic learning

<table>
<tr><td colspan="2">Key Notes</td></tr>
<tr>
<td>Anatomy of learning</td>
<td>The medial temporal lobe and its connections with the hypothalamus and thalamus are required for declarative learning. Hippocampal lesions cause severe anterograde amnesia, an inability to form new long-term memories.</td>
</tr>
<tr>
<td>Spatial navigation learning</td>
<td>The hippocampus is used for spatial navigation learning, a type of episodic learning in which rats are thought to acquire a cognitive map of their surroundings while they explore. Over a period of a few weeks the hippocampal map is transferred to the neocortex.</td>
</tr>
<tr>
<td>Place cells</td>
<td>Pyramidal cells in the hippocampus fire when a rat is in a particular location. These are called place cells and the location which causes the cell to fire is its place field (by analogy with a sensory receptive field). Place cell firing rates increase with an animal's speed. The synchronized neural activity of all cells in the hippocampus produces a theta rhythm, seen while an animal is exploring. How a place cell fires in relation to the theta rhythm encodes distance and sensory aspects of the environment.</td>
</tr>
<tr>
<td>Related topics</td>
<td>(E4) Temporal coding   (N1) Types of learning
(M5) Sleep   (N4) Long-term potentiation</td>
</tr>
</table>

Anatomy of learning

Considerable evidence implicates the medial temporal lobe (Figure 1) and its connections as critical for declarative learning. Bilateral lesions of the hippocampus, even when confined to the selective loss of pyramidal cells from just one region, result in an anterograde amnesia so extreme that patients are no longer able to form new long-term memories. Working memory and procedural memory are unharmed, as are remote long-term memories, although some retrograde amnesia is seen.

Bilateral medial temporal lobe lesions in macaque monkeys provide an animal model for human amnesias. Typically, the animals are trained on a **delayed nonmatching to sample task**. In this, the monkey is trained to select an object to get a food reward. Following a variable delay during which the animal cannot see any manipulations, it is given a choice between the same object and a novel object and is required to select the novel object to get the reward, that is, it needs to remember which object it saw first. Lesioned animals show an anterograde amnesia and selective lesioning shows that the most severe deficit occurs with damage to the perirhinal and parahippocampal cortex of the temporal lobe.

Three diencephalon structures are extensively connected with the temporal lobe and play a role in memory. A major output of the hippocampus is the **fornix** which projects

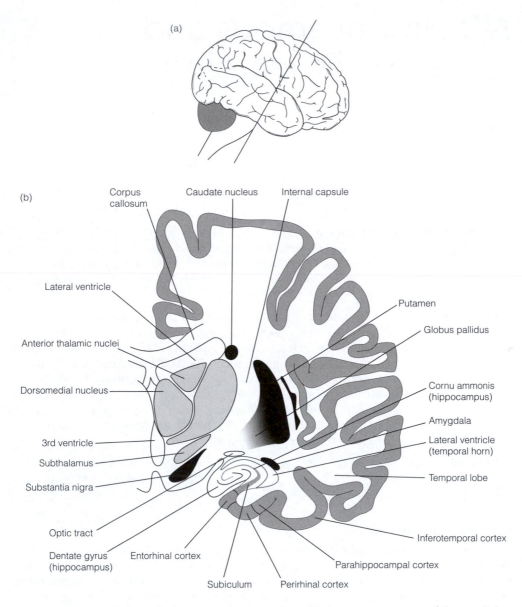

Figure 1. Section (a) through the human brain to show (b) the gross anatomy of the medial temporal lobe.

largely to the **mammillary bodies** of the hypothalamus, output from which goes to the **anterior thalamus**. Furthermore, areas of the temporal cortex and amygdala make connections with the **dorsomedial nucleus** of the thalamus. Bilateral lesions confined to just one of these diencephalonic structures in monkeys modestly impairs performance in the delayed nonmatching to samples tasks, but larger lesions affecting all three produce very severe deficits. The proposed circuitry for declarative memory is summarized in Figure 2.

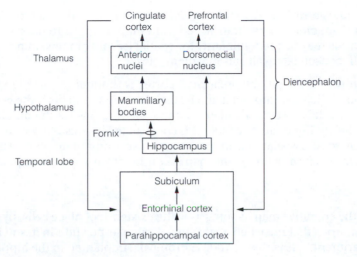

Figure 2. Mammalian forebrain memory circuitry.

Spatial navigation learning

The hippocampus of rats is particularly important for **spatial navigation (place) learning**, by which animals acquire a memory for their location. One widely used way of investigating this is the **Morris water maze**. This is a circular pool filled with opaque warm water. Hidden just below the surface is a small platform. During learning trials rats swim in the pool, discover the platform and learn its position in the pool relative to cues in the laboratory. The motivation is that the platform provides an escape from the water. Learning takes several trials and can be tested by measuring the time taken for a rat to reach the platform or the length of the path it swims to reach it as recorded on a video camera. Cued control experiments in which the platform is raised just above the level of the water ensure that any differences in behavior are not attributable to motivational, perceptual, or locomotor factors.

Rats with hippocampal, but not selective neocortical, lesions are seriously compromised in the learning, but not the cued versions of this task. Rats given a microinjection of colchicine to destroy a specific population of cells (dentate granule cells) in their hippocampus either 1, 4, 8, or 12 weeks after learning the location of a submerged platform, and tested 2 weeks later (Figure 3), reveal that the hippocampus is not a permanent site for the spatial memory. The 12-week group remembered the location as well as control (uninjected) animals, but performance got progressively worse for 8-, 4-, and 1-week groups. This study shows that the hippocampus is needed for consolidation of spatial learning, but that over successive weeks the site of the memory store is transferred elsewhere, probably the neocortex.

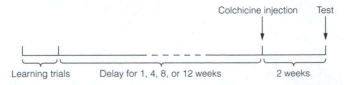

Figure 3. Protocol for investigating the time course of place learning by rats in the Morris water maze.

Retrograde amnesia in humans with medial temporal lobe damage probably results from the loss of memory not yet transferred from hippocampus to neocortex. Although the hippocampus may be predominantly for spatial learning in rats, in primates it has a broader role in consolidating all episodic memories.

The hippocampus of the rat and associated cortex is thought to provide the rat with a representation of the space around it and its location within it. This is the **cognitive map hypothesis** and it has several postulates. Firstly, the map allows the animal to find its way through the environment. Secondly, it is constructed by episodic learning as specific locations come to be associated with particular sensory and motor cues. Thirdly, it does not require reinforcers; and finally, the map is continually updated by exploration.

Place cells

Evidence for the cognitive map hypothesis is the existence of **place cells**, pyramidal cells in the hippocampus that fire when a rat is in a particular position in its environment. In a typical experiment, rats with electrodes chronically implanted in the hippocampus for extracellular recording, are allowed to explore a plus-shaped maze. The animals learn the spatial relationships between the maze and visual cues in the surrounding laboratory so as to find a food reward located in one of the four arms of the maze. The location in the maze which causes the place cell to fire is the cell's **place field** (analogous to the sensory field of sensory neurons) and the entire maze is encoded by an array of place cells.

Place cells fire at a higher frequency the faster the animal moves, so are encoding loco-motor cues, although place fields are influenced by sensory cues. The higher the firing frequency of a place cell, the smaller its place field. This sets an upper limit on distance coding of about 50 cm; that is, locations and objects > 50 cm ahead will be indistinguish-able in a rat's cognitive map, but will be resolved as the animal moves forward. A place cell may have several place fields, each for a different environment. New place fields arise as an animal explores a novel environment and altering familiar surroundings disrupts preexisting place fields.

When rats explore a maze, the EEG shows a theta (θ) rhythm with a frequency of 4–10 Hz which reflects periodic firing of all hippocampal neurons (including place cells) orga-nized by GABAergic interneurons. During θ discharge, place cells encoding new infor-mation about the environment are entrained by the θ rhythm while all other pyramidal cells are silenced by inhibition from GABAergic interneurons. Thus, θ activity ensures that only cells involved in learning a particular environment are active.

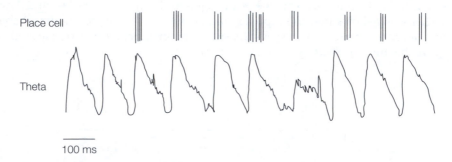

100 ms

Figure 4. Hippocampal theta and place cell firing in CAI as a rat walks through a familiar environment. Note the theta phase precession.

The firing of hippocampal place cells shows a systematic phase relationship to the theta rhythm. Individual place cells fire whenever the animal enters a particular region in its environment but this firing is modulated at a slightly higher frequency than the theta rhythm. Hence, as the animal moves through the place field the place cell fires at a progressively earlier phase of the theta rhythm. This is called **theta phase precession**. It is an example of temporal coding and encodes the distance traveled. Theta phase precession is also modulated by sensory cues in the maze (Figure 4).

N4 Long-term potentiation

Key Notes

Hippocampal circuitry

Excitatory input from the entorhinal cortex goes directly, or indirectly via granule cells in the dentate gyrus, to pyramidal cells in CA3. CA3 cell axon branches make recurrent connections back on CA3 dendrites and go to the contralateral hippocampus, the hypothalamus, and CA1 pyramidal cells. Axons of CA1 cells go to the entorhinal cortex. All these principal neurons use glutamate and are excitatory. In addition, the hippocampus contains GABAergic inhibitory interneurons.

Long-term potentiation (LTP)

LTP may be associative (Hebbian) or nonassociative. Associative LTP has been studied at synapses between CA3 and CA1 cells in the hippocampus, where it is produced by applying a high-frequency (tetanic) stimulus to the CA3 axons. Subsequent single stimuli elicit a larger excitatory postsynaptic potential in the CA1 cell than before the tetanic stimulus. Early and late LTP can be distinguished by the fact that the latter requires gene expression and protein synthesis.

Associative LTP biochemistry

Induction of LTP in CA1 cells requires activation of NMDA glutamate receptors. This needs coincident firing of CA3 cells and depolarization of the CA1 cell which is produced by a tetanic stimulus. Ca^{2+} entry through NMDA triggers phosphorylation of AMPA receptors increasing their sensitivity to glutamate. This, plus increased glutamate release triggered by a retrograde transmitter, enhances the postsynaptic response. Maintenance of early LTP occurs by activation of kinases such as CaMKII. LTP beyond 2 hours requires activation of ERK, transcription and translation, and synaptic tagging so that only activated synapses are modified.

CREB, LTP, and long-term memory

The activated catalytic subunit of protein kinase A translocates to the nucleus where it phosphorylates a transcription factor that binds to cAMP response elements (*cres*) in genes regulated by cAMP. The transcription factor is termed cAMP response element binding protein (CREB). When phosphorylated, CREB binds to *cres*, initiating transcription. CREB is involved in long-term memory in several learning models including spatial navigation and fear learning.

Hippocampal function and brain oscillations

Modulations of gamma and theta oscillations and phase relations between them synchronize neurons, thereby generating LTP that underlies hippocampal learning, and controlling the flow of information through the hippocampus that determines encoding or retrieval of information. Brain oscillations in synchrony with those in the hippocampus are seen in several other brain areas and seem to be related to memory functions.

Memory and sleep   Some memory consolidation appears to happen during sleep. Slow-wave sleep seems to be important for declarative memory of events and facts, whilst REM sleep may be required for procedural learning and emotional learning. Episodic information initially stored in the hippocampus is transferred to neocortical long-term memory during awake immobility and slow-wave sleep. During these times sequential activation of place cells that occurred during spatial exploration are replayed at a faster rate. This is mediated by sharp wave/ripple complexes that hugely increase the excitability of hippocampal cells and their cortical targets. Enhanced recall of important emotional memories is seen after REM-rich sleep in humans; and REM sleep, when increased activity is seen in limbic cortex and amygdala, is increased in stress. All this implies a role for REM sleep in emotional memory.

Related topics   (D1) Ligand-gated ion channel receptors   (N3) Hippocampus and episodic learning

(E4) Temporal coding   (N5) Motor learning in the cerebellum: LTD

Hippocampal circuitry

The hippocampal formation is folded archaecortex (ancient cortex) consisting of the **dentate gyrus** and the **cornu ammonis** (**CA**)—collectively termed the **hippocampus**—plus the subiculum. The cortex of the dentate gyrus and CA has three layers, while the subiculum is transitional cortex between the hippocampus proper and the six-layered neocortex of the **entorhinal area**. A major input to the hippocampus from the entorhinal cortex comes via the **perforant pathway**, axons of which synapse with **granule cells** of the dentate gyrus or pyramidal cells in the CA3 region of the CA (Figure 1). Axons of the granule cells (mossy fibers) also synapse with CA3 pyramidal cells.

The CA3 pyramidal cell axons branch, forming:

- **Commissural fibers** which pass to the opposite hippocampus

- Efferents which leave the hippocampus via the **fornix** to terminate largely in the hypothalamus or thalamus

- Collaterals which turn back to form synapses on the same and neighboring CA3 cells (**recurrent collaterals**), or which synapse with cells in the CA1 region of the CA (**Schaffer collaterals**)

CA1 cell axons go to the subiculum and entorhinal cortex. The perforant pathway, granule cells, and pyramidal cells are glutamatergic and excitatory. The hippocampus also

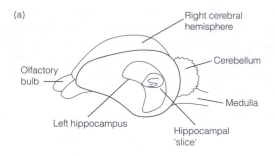

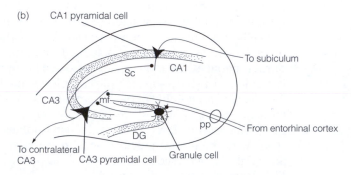

Figure 1. Hippocampus. (a) Location of the left hippocampus in rat brain; a hippocampal slice is at right angles to the long axis of the hippocampus. (b) Structure of a hippocampal slice showing the principal excitatory neurons. From Revest P & Longstaff A (1998) *Molecular Neuroscience*. BIOS Scientific Publishers. DG, dentate gyrus; pp, perforant pathway; Sc, Schaffer collateral; mf, mossy fibers.

harbors inhibitory interneurons that are GABAergic. These help generate intrinsic oscillations (theta and gamma rhythms) that are essential for learning and memory. Other inputs to the hippocampus include a cholinergic pathway from the septum (required for theta rhythm) and modulatory noradrenergic and serotinergic axons from the brainstem reticular system.

Theoretical modeling of the hippocampal circuitry suggests how it might work. The dentate gyrus may act to keep sensory representations coming from the entorhinal cortex segregated. Intriguingly, recent evidence suggests that this requires new neurons to be born (**neurogenesis**) in spatial learning. The recurrent connections of CA3 (Figure 1c in Section E3) allow associations between entorhinal inputs (i.e., episodic memories) to be rapidly established and then incorporated into neocortical long-term memory (i.e., consolidation) over a period of a few weeks. CA1 acts as a novelty detector by comparing stored memories with ongoing sensory information. Encoding in CA1 corresponds to input from entorhinal cortex while retrieval from CA1 corresponds to CA3 input. At each of the synapses in the hippocampus spike timing-dependent plasticity is seen; long-term potentiation in the hippocampus has been extensively researched.

Long-term potentiation (LTP)

LTP can be either associative (Hebbian) or nonassociative. LTP at the synapses between CA3 Schaffer collaterals (Scs) and CA1 cells in the hippocampus is Hebbian. It can be studied in hippocampal brain slices by intracellular or extracellular recording from CA1

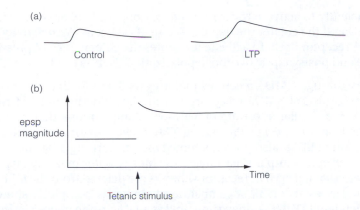

Figure 2. LTP in a hippocampal slice. (a) Excitatory postsynaptic potentials (epsps) recorded from CA1 pyramidal cells before (control) and after tetanic stimulus (LTP); (b) epsp magnitude remains elevated over several hours.

neurons whilst electrically stimulating a bundle of Scs (Figure 2). In response to brief, low frequency stimulation of the Scs, the CA1 cells show a brief epsp due to glutamate release. If a brief tetanic burst of high-frequency stimulation is given (typically 100 Hz for 0.5 s), *subsequent* low frequency pulses now elicit a larger epsp. This is LTP. It can also be studied using chronically implanted electrodes in awake, behaving animals. LTP has several properties:

- Input specificity. Delivery of low frequency stimuli to the CA1 cell via a different untetanized bundle of Scs does not elicit the enhanced epsp.

- Cooperativity. The probability of producing LTP increases with the number of *afferent* fibers (Scs) tetanically stimulated. While weak (i.e., low current) high-frequency stimuli often fail to generate LTP, because they excite only a few afferents, strong tetanic stimuli are successful because they recruit many afferents.

- Associativity. A given CA1 cell receives Scs from CA3 cells on the same side and commissural axons that come from CA3 cells in the contralateral hippocampus. A weak tetanic stimulus to either pathway that fails to generate LTP will do so if it is paired with a strong tetanic stimulus in the other pathway.

- Persistence. It lasts for many minutes (in brain slices) to months when induced *in vivo*. This marks it out from other forms of synaptic plasticity.

Two phases of LTP are recognized, early (**E-LTP**) which does not require protein synthesis, and late (**L-LTP**) which does. Biochemical processes associated with the induction, maintenance, and expression of each phase have been defined.

Associative LTP biochemistry

CA1 cells have AMPA and NMDA glutamate receptors. In order to be activated, NMDA receptors must bind glutamate *and* experience a depolarization big enough to remove Mg^{2+} ions from the channel (voltage-dependent blockade). This condition is not provided by low-frequency Sc stimulation. The amount of glutamate released is low, few AMPA receptors are activated, and the resulting epsp is too small to open NMDA receptors. However, high-frequency stimulation opens numerous AMPA receptors and so depolarizes the cell sufficiently to activate NMDA receptors. Cooperativity arises because the more afferents activated the greater the depolarization of CA1 cells. Associativity is a property of the NMDA receptor itself: to activate it requires the coincidence of presynaptic glutamate release and postsynaptic neuron depolarization (Hebb's rule).

It is Ca^{2+} entry through NMDA receptors that triggers the induction of E-LTP at CA3-CA1 synapses. Antagonists of NMDA receptors prevent the induction of LTP. Crucially for the idea that LTP is a cellular substrate of learning, manipulations that impair LTP cause learning deficits and *vice versa*. Preventing NMDA receptors from functioning not only blocks induction of LTP, it also prevents some types of learning. Pharmacological blockade of NMDA receptors impairs learning of the Morris water maze by rats. Mice genetically engineered so that NMDA receptor subunits are deleted from the CA1 region in the third postnatal week (after hippocampal development is complete) show no NMDA-receptor dependent LTP in CA1, have impaired spatial learning in the Morris water maze, and show a loss of the correlated firing of place cells that is seen in normal animals.

NMDA receptors are required for induction of LTP in most locations in the CNS, but at some synapses metabotropic glutamate receptors or kainate receptors are required.

Both presynaptic and postsynaptic sites are implicated in the expression of LTP, but the specifics depend very much on the type of LTP and the synapses involved. At the CA3–CA1 synapses postsynaptic changes include:

- An increase in responsiveness of the postsynaptic membrane as AMPA receptors become more sensitive to glutamate. The biochemistry of this has been worked out. **Calcium–calmodulin dependent protein kinase II** (**CaMKII**), a major protein of the postsynaptic density, is thought to be activated by Ca^{2+} entry through NMDA receptors, and it then phosphorylates AMPA receptors ($GluR_1$ subunits), enhancing their response to glutamate.

- Activation of **silent synapses**. These are synapses that harbor only NMDA receptors. They are normally silent because low-frequency stimulation does not activate NMDA receptors. Calcium entry during LTP induction causes them to acquire AMPA receptors and so become responsive to low-frequency input.

A presynaptic component to CA3–CA1 synapse LTP has been proposed in which there is increased glutamate release. This requires that the NMDA Ca^{2+} signal generates a **retrograde messenger** in the postsynaptic cell that travels the "wrong" way across the synaptic cleft. One molecule proposed for this role is **nitric oxide** (**NO**). In neurons NO is synthesized by a Ca^{2+}-dependent **nitric oxide synthase** (**NOS**). As a small, freely diffusible molecule, NO rapidly diffuses out of the postsynaptic cell, across the cleft and into the presynaptic cell where it stimulates guanylyl cyclase, thereby enhancing the probability of glutamate release.

With brief tetanic stimulation, LTP lasts for about 2 hours. Several mechanisms probably have a role in this maintenance of E-LTP. They must explain how synaptic alterations are retained, even while individual molecules are being turned over, after the original signals that brought about the alterations have gone. Two well-documented processes are as follows:

- CaMKII consists of four subunits. Ca^{2+} activation phosphorylates them and once the Ca^{2+} concentration has fallen to resting levels they remain phosphorylated. This is because if a subunit becomes dephosphorylated it will immediately become autonomously phosphorylated by one of the other subunits. In this way CaMKII remains persistently active.

- Ca^{2+} stimulates an isoform of adenylyl cyclase and consequently cAMP concentrations increase in LTP. Protein kinase A becomes persistently activated and has effects on gene expression.

A number of kinases are activated in E-LTP, but **extracellular signal-related kinase (ERK)** is thought to be important in the transition to L-LTP that occurs if several tetanic stimuli are delivered. This long-lasting L-LTP is blocked by drugs that inhibit mRNA or protein synthesis which shows that it depends on transcription and translation. In L-LTP morphological as well as functional changes are seen. This includes the changes to the geometry of dendritic spines and the formation of new synapses by the splitting of old ones into two.

Late phase synaptic plasticity depends on synthesis of proteins that must function only in activated synapses; **synaptic tagging**. The idea is that a specific protein or perhaps a process (altered mRNA translation or cytoskeletal assembly) marks out an activated synapse so that plasticity-related proteins can then assemble to produce the functional and morphological changes that result in persistent synapse strengthening. Synaptic tagging was proposed as a theoretical means to ensure input specificity as long ago as 1997 but only recently has direct evidence for it come to light. NMDA receptor activation has been shown to trigger input-specific spine entry of specific proteins which could serve as synaptic tags.

Key events in LTP are summarized diagrammatically in Figure 3. Somewhat different mechanisms are responsible for nonassociative LTP that occurs at some other synapses.

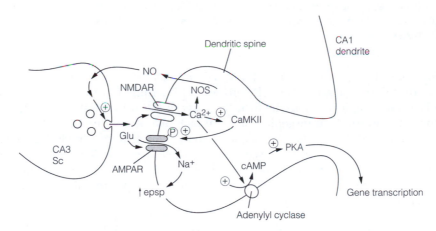

Figure 3. Key events in LTP. Sc, Schaffer collateral; Glu, glutamate; AMPAR, AMPA receptor; NMDAR, NMDA receptor; NO, nitric oxide; NOS, nitric oxide synthase; PKA, protein kinase A; CaMKII, Ca^{2+}–calmodulin-dependent kinase II.

CREB, LTP, and long-term memory

One of the biochemical processes involved in the maintenance of L-LTP which is also implicated in several learning models is the modification of gene expression by the cyclic AMP second messenger system. The activated catalytic subunit of protein kinase A translocates to the nucleus where it phosphorylates a transcription factor that binds to cAMP response elements (CREs) in upstream regions of genes regulated by cAMP. The transcription factor is termed **cAMP response element binding protein (CREB)**. When it is phosphorylated CREB binds to *cre* and this engages the other components needed for transcription (Figure 4). CREB can also be phosphorylated by calcium–calmodulin-dependent kinases (e.g., CaMKIV) which will be activated by Ca^{2+} entry via NMDA receptors, or any other mechanism that increases cytoplasmic calcium concentration. CREB-mediated gene transcription is controlled in at least two ways:

- By protein phosphatases that are activated by the calcium-dependent protein calcineurin

- By repressor proteins termed **cAMP response element modulators** (**CREMs**) that bind to *cres* and so prevent CREBs from doing so

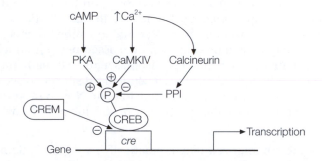

Figure 4. Cyclic AMP response element binding protein (CREB) modulation of gene transcription. *cre*, cyclic AMP response element; CREM, cAMP response element modulator; CaMKIV, calcium–calmodulin-dependent kinase IV; PKA, protein kinase A; PPI, protein phosphatase I.

Considerable evidence suggests that CREB is implicated in long-term memory in a number of different learning models: procedural learning in a marine snail (*Aplysia*), odor discrimination learning in the fruit-fly *Drosophila*, and several types of learning in mammals. For example, transgenic mice that lack two of the three CREB isoforms are impaired in three different tasks that depend on the hippocampus (including the Morris water maze) and fear conditioning that requires the amygdala. LTP is impaired by protein kinase A inhibitors and in CREB-deficient mice. Selectively deleting neurons overexpressing CREB in the lateral amygdala, but not a similar proportion of randomly selected neurons, abolished fear memory in mice.

Hippocampal function and brain oscillations

The importance of brain oscillations for hippocampal learning was first suggested by the fact that tetanic stimulus protocols effective in producing LTP *in vitro* are similar to hippocampal theta rhythms *in vivo*. It is now clear that modulations in gamma frequency (30–100 Hz) and theta frequency (4–8 Hz) bands and in phase relations between them can provide the synchrony needed to generate LTP in the hippocampus and control information flow that brings about encoding or retrieval of information.

Gamma oscillations of hippocampal pyramidal cell membrane potential arise from rhythmic inhibition imposed by activity in local GABAergic hub interneurons. The effectiveness of incoming excitatory inputs to the pyramidal cell is greatest when it occurs out of phase with the inhibition and this synchronizes the cells subjected to the gamma inhibition to fire within 10 ms of each other. This has several effects:

- It can generate STDP (10–20 ms is the necessary time window).

- If the synchronized cells converge onto downstream neurons their near simultaneous activity will result in significant temporal summation, increasing the probability that the downstream neuron will fire.

- Because stronger excitatory inputs overwhelm inhibition more easily than weak ones, more active neurons spike earlier during a gamma band oscillation than weakly activated neurons. Hence the gamma oscillations sort inputs so that the stronger arrive downstream earlier than the weak ones.

Theta oscillations can be produced by hippocampal neural networks in isolation but normally it requires cholinergic input from the septum, which makes pyramidal cells more excitable.

Correlations between gamma and theta oscillations are important. In rats learning associations between items and places (i.e., when we assume that place cells are acquiring their place fields) the amplitude of gamma oscillations is frequency modulated by the theta oscillations (i.e., the size of the gamma spikes rises and falls in phase with the theta rhythm). These correlations can be related to models of how the hippocampus works.

Gamma activity falls into two frequency bands, low and high. Low-frequency gamma in CA1 coincides with the falling phase of the theta band oscillations and is synchronous with low-frequency gamma in CA3. High-frequency gamma in CA1 happens during the rising phase of the theta band and is synchronized with high-frequency gamma in the entorhinal cortex. Hence theta rhythms seem to play a role in controlling information flow through entorhinal cortex and CA3 to CA1: inputs to CA1 from memory-related recurrent processing in CA3 occurs at one phase of theta and is characterized by low gamma frequency coherence, while sensory-related input from the entorhinal cortex occurs at a different theta phase and is characterized by high gamma coherence. These two different patterns are thought to correspond to different phases of the theta rhythm being needed for encoding or retrieval in CA1, with encoding corresponding to input from the entorhinal cortex while retrieval corresponds to CA3 input.

Hippocampal oscillations also influence activity in other areas. During spatial learning tasks theta activity coherent with that in the hippocampus is seen in the prefrontal cortex and the ventral striatum. In somatosensory and medial prefrontal regions bursts of locally produced gamma oscillations are briefly coherent with hippocampal theta rhythms during exploration and REM sleep. Synchronous activity in the gamma band is seen in the medal temporal lobe of rodents and primates (including humans) during memory tasks.

Memory and sleep

Some memory consolidation appears to happen during sleep. Slow-wave sleep seems to be important for declarative memory of events and facts, whilst REM sleep may be required for procedural learning and emotional learning.

Episodic information is rapidly stored in the recurrent connections in CA3 and then incorporated into neocortical long-term memory. It has been proposed that the transfer is mediated by coherent spiking of hippocampal neurons during high-frequency **sharp wave/ripple complexes** (**SPW-Rs**) that occur during awake immobility and slow-wave sleep (Figure 5). During SPW-Rs, sequential activation of place cells that occurred during spatial exploration are replayed at a faster rate. SPW–Rs are generated within CA3, in response to LTP induction there, and spread to CA1 then entorhinal cortex. They increase the excitability of hippocampal cells and their neocortical targets dramatically. Most CA1 pyramidal cells are powerfully inhibited during SPW–Rs, but the small number of active cells, those encoding the new spatial learning, increase their firing frequency to be in synchrony with the ripple component of SPW-R spikes. Firing of neocortical neurons by SPW-Rs during the long-lasting depolarizations of slow-wave sleep activity induces long-term potentiation that allows the neocortical connections to be re-specified.

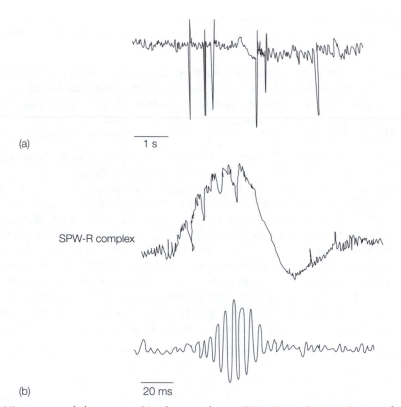

(a) 1 s

SPW-R complex

(b) 20 ms

Figure 5. Hippocampal sharp wave/ripple complexes (SPW-Rs) in CAI. (a) Series of SPW-Rs seen during awake immobility in a rat after a period of exploration, (b) Single SPW-R complex with (below) filtered recording to show 150–300 Hz ripples.

The enhanced recall of emotionally salient memories after particularly REM-rich sleep in humans suggests that REM sleep is also important for emotional memory:

- REM sleep is increased in stress generally and specifically in major depression, bereavement, and post-traumatic stress disorder.
- Brain imaging shows that during REM sleep while brain activity is reduced in the dorsolateral prefrontal cortex (part of the executive thalamocortical–basal ganglia circuit involved in working memory, planning, and problem solving), it is increased in the limbic cortex and the amygdala.

One problem with the REM sleep hypothesis of sleep is that monoamine oxidase inhibitors used for the treatment of depression severely reduce the proportion of REM sleep. However, there is no evidence that patients treated with these agents have any memory deficits even with long-term use.

N5 Motor learning in the cerebellum: LTD

Key Notes

The Marr–Albers–Ito model

Motor learning in the cerebellum comes about by a weakening of the strength of synapses between parallel fibers (pf) and Purkinje cells (PC) that are active at the same time as error signals arrive at the PC via climbing fibers. The synaptic weakening is long-term depression (LTD).

Classical conditioning of the eye blink reflex

A puff of air delivered to the eye normally causes a reflex eye blink. This can be classically conditioned by pairing a tone with the air puff. The tone activates pf–PC synapses just before the air puff signal arrives at the PC via the climbing fibers, and so the pf–PC synapses suffer LTD. Subsequent occurrence of the tone causes reduced PC excitation which translates into a larger cerebellar output to motor neurons driving the eye blink.

Long-term depression (LTD)

LTD is seen in the cortex and hippocampus as well as the cerebellum. In the cerebellar cortex, LTD requires simultaneous Ca^{2+} input into the PC (caused by climbing fibers) and activation of glutamate receptors at pf–PC synapses. The effect is to desensitize the AMPA receptors at the synapses.

Related topics

(K4) Cerebellar cortex circuitry

(K5) Cerebellar function
(N4) Long-term potentiation

The Marr–Albers–Ito model

Motor learning in the cerebellum involves alterations in the strengths of synapses between parallel fibers (pf) and Purkinje cells (PC). Those synapses that are active at exactly the same time that there is climbing fiber input to the Purkinje cell, experience a reduction in the synaptic strength, a type of spike timing dependent plasticity called **long-term depression (LTD)**.

In the **Marr–Albers–Ito model** of motor learning, the frontal cortex (via the corticopontine cerebellar tract) provides the mossy fiber–parallel fiber inputs and the climbing fibers from the inferior olive are thought to transmit error signals. All the pf–PC synapses that happen to be activated by a pattern of mossy fiber inputs at the same time as climbing fiber error signals arrive will show LTD. Synapses not concurrently active are unchanged. Subsequently, parallel fiber activity at the depressed synapses excites Purkinje cells less, thereby reducing their inhibitory output on deep cerebellar nuclei. The overall effect is that synapses at which LTD occurs enhance cerebellar output.

Classical conditioning of the eye blink reflex

Motor learning that occurs during the classical conditioning of the eye blink reflex has been extensively studied. In the eye blink reflex, a puff of air delivered to the eye (US) will produce an eye blink (UR). The eye blink reflex can be conditioned if the air puff is paired with a tone (CS). The circuitry involved in motor learning in this reflex is shown in Figure 1.

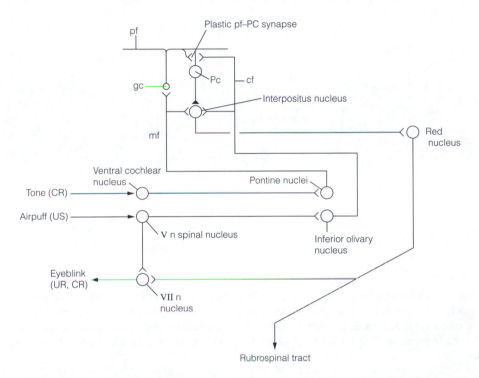

Figure 1. Motor learning in the cerebellum; circuitry implicated in the conditioned eye blink reflex. pf, parallel fiber; PC, Purkinje cell; CR, conditioned response; UR, unconditioned response; US, unconditioned stimulus; cf, climbing fiber; gc, granule cell; mf, mossy fiber.

The air puff (US) is sensed by neurons in the spinal nucleus of the trigeminal (5th) cranial nerve. The eye blink reflex (UR) is executed by connections between these cells and motor neurons in the facial (7th cranial) nerve. Conditioning of the reflex requires the cerebellum. The US signal is transmitted via climbing fibers that arise from the inferior olivary nucleus. The tone (CS) signal goes by way of the ventral cochlear nucleus and pontine nucleus, arriving at the cerebellum in mossy fibers. Activation of the pf–PC synapse by the CS, 250 ms before the arrival of the US via the climbing fiber, results in LTD of the pf–PC synapse. The effect of the LTD is that any subsequent arrival of the CS produces a smaller excitation of the PC. Hence PC inhibition of the interpositus neurons is diminished, so these cerebellar nucleus cells drive the eye blink via their connections with the red nucleus.

Long-term depression (LTD)

Long-term depression (LTD) is seen in the hippocampus and cerebral cortex where it can occur alongside LTP, and in the cerebellum (in which LTP is never seen). Induction of LTD in the cerebellum at the pf–PC synapse requires coincident Ca^{2+} influx into the Purkinje cell and activation of AMPA and metabotropic glutamate receptors at the synapse. The Ca^{2+} influx is provided by the large depolarization due to climbing fiber activity which opens voltage-dependent Ca^{2+} channels. The receptors are activated by the release of glutamate from the parallel fibers. The final cause of the synaptic depression is desensitization of the AMPA receptors (Figure 2) brought about by their phosphorylation by protein kinase C, and nitric oxide-activated protein kinase G.

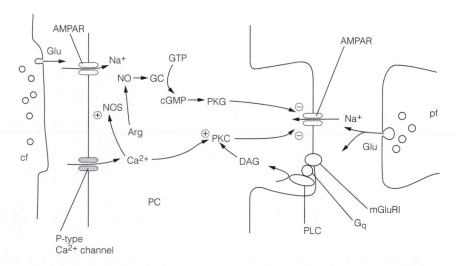

Figure 2. Cellular events in cerebellar LTD. Glu, Glutamate; AMPAR, AMPA receptor; mGluR1, type 1 metabotropic glutamate receptor; G_q, G protein; NO, nitric oxide; NOS, nitric oxide synthase; PKG, protein kinase G; PKC, protein kinase C; GC, guanylyl cyclase; PLC, phospholipase C; DAG, diacylglycerol; PC, Purkinje cell; pf, parallel fiber; cf, climbing fiber.

O1 Neuroimaging

Key Notes

Computer assisted tomography (CAT)
A CAT scan produces a series of X-ray images, each one a slice, of living brain. The technique can distinguish tissues that differ in their ability to transmit X-rays by as little as 1% and has a spatial resolution of 0.5 mm.

Positron emission tomography (PET)
PET scanning, by revealing the distribution in the brain of a positron emitting isotope (with a spatial resolution between 4 and 8 mm) can provide functional as well as anatomical information about the living brain. Using uptake of a positron-emitting glucose analog as a marker for neuron activity it can show which regions of the brain are involved in a variety of activities both in health and disease. Positron-emitting neurotransmitters and receptor ligands are used to study neurotransmitter pathways in the living brain.

Functional magnetic resonance imaging (fMRI)
Magnetic resonance imaging takes advantage of the fact that atomic nuclei with odd mass numbers aligned in a strong magnetic field resonate in response to a pulse of radio waves. When the radio pulses are switched off the nuclei relax in a way that depends on their chemical environment. fMRI has a spatial resolution < 1 mm. It can be used to measure changes in brain activity with a time resolution of a few seconds.

BOLD
A type of fMRI, blood oxygen level detection (BOLD) provides a sensitive measure of changes in cerebral cortical activity with a spatial resolution of a few mm^3 and time resolution of a few seconds. It depends on the ratio of oxygenated to deoxygenated hemoglobin and this varies with blood flow and metabolism. How the BOLD signal arises physiologically is complex and not fully understood.

Related topics
(A2) Glial cells
(G6) Parallel processing in the visual system
(G7) Oculomotor control and visual attention
(K1) Cortical control of voluntary movement
(N2) Working memory

Computer assisted tomography (CAT)

Computer assisted tomography (**CAT**) reveals the anatomy of the living brain. The head is placed between a source which emits a narrow beam of X-rays and an X-ray detector (Figure 1). A series of measurements is made of X-ray transmission. The source and detector are rotated as a pair through a small angle and a further series of measurements taken. This is repeated until the source and detector have rotated through 180°. The radiodensity of each region of the head is computed from the transmission data for all of the beams that have traversed that region, and the results visually displayed. This provides a view through a single slice of brain lying at a known orientation. By moving the head at right angles to the orientation plane for a short distance another section can be imaged. This is repeated until the whole brain has been scanned.

The key element here is the algorithm—and the computer software to implement it— that calculates the radiodensity for each point in the brain slice; this is **computerized tomography**. It is an example of an inverse problem which starts with a data set from which initial parameters, in this case source location, must be calculated. It contrasts with forward problems in which the source location is known and it is the data set which is calculated. The difficulty with inverse problems is they do not have unique solutions. Hence they have to be constrained by assumptions and prior modeling based on earlier results to find the most likely solution.

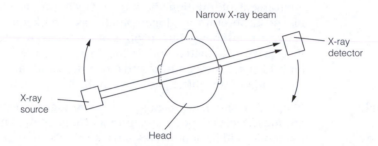

Figure 1. Computer assisted tomography (CAT). Arrows depict the rotation of the scanner.

CAT can distinguish tissues which differ in X-ray opacity by 1% (the lower the density the darker the image) with a spatial resolution of about 0.5 mm. Blood vessels can be seen by injection of radio-opaque dyes.

Positron emission tomography (PET)

Positron emission tomography (**PET**) provides insights into the *function* of the living brain as well as its anatomy. It uses the principles of computerized tomography in which γ-ray detectors are located around the head and the source is a positron-emitting compound, either injected or inhaled, which enters the brain (Figure 2). Compounds used include neurotransmitters, receptor ligands, and glucose analogs which are used for studying brain activity. Typically they are radiolabeled with $^{11}_{6}$C, $^{13}_{7}$N, $^{15}_{8}$O, or $^{18}_{9}$F (which substitutes for hydrogen). These isotopes have short half-lives, decaying to the element with atomic number one less; for example:

$$^{13}_{7}\text{N} \rightarrow {}^{13}_{6}\text{C} + \text{e}^+$$

The positron (e$^+$, the antiparticle of the electron) travels a short distance before colliding with an electron (e$^-$). The two particles annihilate with the production of two γ-ray photons that shoot off in exactly opposite directions. These are detected simultaneously by a pair of detectors 180° apart. This coincidence detection permits localization of the site of the γ-ray emission, which is between 2 and 8 mm from the positron source, depending on the isotope used.

The spatial resolution of PET is about 4–8 mm, not as good as CAT, but it can be used to follow brain events over time.

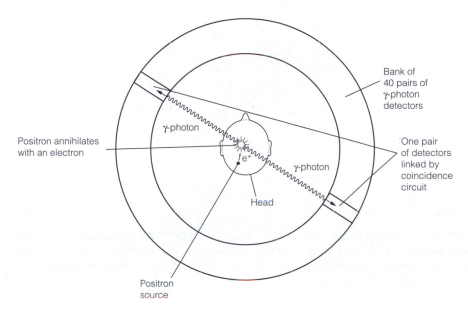

Figure 2. Positron emission tomography (PET).

The importance of PET in functional studies is illustrated by the use of the nonmetabo-lizable analog of glucose, **2-deoxyglucose** (**2-DG**). This molecule crosses the blood–brain barrier, is transported into neurons and phosphorylated to 2-DG-6-phosphate, so it remains in the cell. However, it cannot be metabolized further. This means it acts as a marker for local glucose uptake and therefore of neuron activity. Imaging the distribution of $[^{18}_{9}F]$2-DG while subjects engage in sensory, motor, or cognitive tasks reveals how these functions are localized in the brain. Related studies show that during transient increases in neuronal activity, the rise in local cerebral oxygen consumption (as measured by $^{15}_{8}O$ PET) does not match the increase in glucose utilization (as estimated from 2-DG PET). This implies that brief periods of brain activity can be supported by glycolysis.

Magnetic resonance imaging (MRI)

Like PET, magnetic resonance imaging provides information about brain function as well as anatomy. It combines computerized tomography with **nuclear magnetic resonance** (**NMR**). Nuclei with odd mass number, for example, $^{1}_{1}H$, generate a magnetic field along their spin axis. In the powerful magnetic field of an MRI scanner, hydrogen nuclei can adopt one of two orientations; with their magnetic fields either parallel or antiparallel to the external field. The parallel state has a slightly lower energy and normally a small excess of nuclei will be in this state (Figure 3). This gives rise to a net *longitudinal* magnetic field parallel to the scanner field.

A cylindrical coil placed around the head broadcasts a radio frequency (rf) pulse to a slice of head at right angles to the main scanner field. The rf pulse makes the nuclei wobble around their magnetic axis—rather like a spinning top as it slows down—with the rate of wobbling in resonance with the pulse frequency. The wobble generates an electric field that is received by the coil, producing a *transverse* magnetic field at right angles to the scanner field. When the rf pulse is turned off the nuclei return to their original state, and the longitudinal and transverse fields decay with **relaxation times** that are characteristic

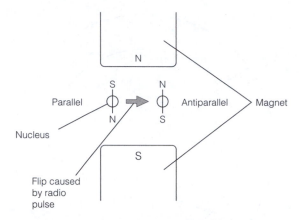

Figure 3. The principle of nuclear magnetic resonance (NMR). A radiofrequency pulse will excite atomic nuclei, flipping them from the parallel state into the higher energy antiparallel state. Relaxation of the nuclei back into the low energy state generates the magnetic resonance imaging (MRI) signal.

for the nucleus and its chemical environment (e.g., lipid or aqueous). Generating an MRI image actually requires a further three coils that produce magnetic field gradients in the x, y, and z directions. MRI has a resolution < 1 mm.

BOLD

An MRI method that records changes related to brain function in successive images is termed **functional MRI** (**fMRI**). The most important type of fMRI is **blood oxygen level detection** (**BOLD**) which provides a very sensitive measure of cerebral cortical activity with a **voxel** (volumetric pixel, the 3-D analog of a pixel in a 2-D image) size of 2 mm on each side, following changes in activity with a time resolution of a few seconds. It depends on the ratio of oxygenated to deoxygenated hemoglobin and this varies with blood flow and metabolism.

Precisely how brain physiology generates the BOLD signal is not completely understood. There is no simple relationship between the rise in cerebral blood flow that accompanies neural activity and the local metabolic demands, neither is it clear how any of these variables correlates with electrical activity in the brain. Despite that BOLD studies have shown that most of the energy expenditure of the brain is related to synaptic events rather than the generation and propagation of action potentials. Indeed it seems that action potentials are produced using only 30% more energy than the calculated theoretical minimum.

O2 Noninvasive electrophysiology

Key Notes

Electroencephalo-graphy (EEG)	Large numbers of cortical cells fire in synchrony and this activity can be recorded by scalp electrodes. The frequency bands of EEG signals correlate with behavioral state.
Magnetoence-phalography (MEG)	The synchronized flow of intracellular dendritic currents in large populations of similarly oriented cortical pyramidal cells, set up by synaptic activity, generates a weak magnetic field that can be detected by an array of very sensitive SQUID magnetometers. Magnetoencephalography has a temporal resolution of < 1 ms, comparable to intracranial electrodes, and much better than EEG.
Transcranial magnetic stimulation (TMS)	A magnetic field changing with time generated by an electromagnetic coil can be made to induce an electrical field down to a depth of about 3 cm in the brain. There are two modes of TMS, single or repetitive pulse. TMS can excite or inhibit brain functions depending on the stimulus characteristics. It has effects on visual perception, movement control, attention, memory, language, and decision making. Its effects on cognition are usually inhibitory, either increasing reaction time or increasing the number of errors in performance of cognitive tasks.
Related topics	(A4) Organization of the central nervous system (E4) Temporal coding (G7) Oculomotor control and visual attention	(K1) Cortical control of voluntary movement (M5) Sleep (N2) Working memory (N5) Motor learning in the cerebellum: LTD

Electroencephalography (EEG)

Recording the net electrical activity of the brain by means of surface electrodes attached to the scalp is termed **electroencephalography** (**EEG**). Large numbers of cerebral cortical cells fire in synchrony and consequently their summed activity produces **local field potentials** (**LFPs**) big enough that they can be recorded with scalp electrodes. By using an array of electrodes, activity of different brain areas can be examined simultaneously. The recording may be monopolar—each scalp electrode measures the potential with respect to a distant indifferent electrode—or bipolar, in which the potential is measured between a pair of scalp electrodes. The LFPs vary in frequency and the frequency ranges are conventionally grouped: alpha (8–13 Hz), beta (13–30 Hz), delta (1–4 Hz), theta (4–7 Hz). Activity in these frequency bands correlates with behavioral state, for example, sleep, arousal, or learning.

Sensory, perceptual or cognitive stimuli can generate brief fluctuations in the EEG termed **evoked potentials** (**EPs**) or **event-related potentials** (**ERPs**). These potentials are used to investigate the context, timing, and brain regions implicated in the process of interest.

Magnetoencephalography (MEG)

A current flowing in a wire sets up a magnetic field that flows around the current vector. Hence the synchronized flow of currents along dendrites of about 50 000 cortical pyramidal cells all oriented in the same direction is sufficient to set up a measurable, if weak, magnetic field (Figure 1). This is measured in **magnetoencephalography** (**MEG**) using an array of magnetometers called SQUIDS (superconducting quantum interference devices) which surround the head. The MEG signal is generated mostly by the flow of intracellular currents in dendrites generated by synaptic activity. Because cortical activity generates a field of order 10^{-13} tesla (T) compared with the Earth's magnetic field of 3.1×10^{-5} T. MEG must be done in a magnetically shielded environment. SQUIDs are sensitive to magnetic fields as small as 10^{-18} T.

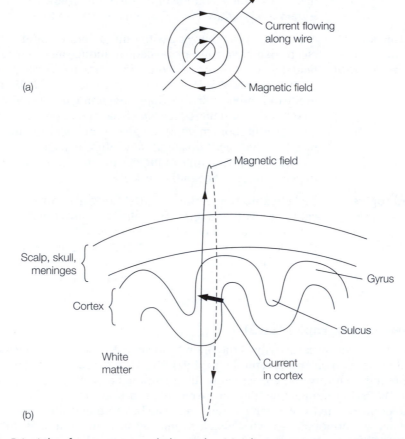

Figure 1. Principle of magnetoencephalography. (a) A linear current sets up a concentric cylindrical magnetic field with clockwise polarity for current entering the plane of the paper (corkscrew rule). (b) Magnetic field generated by current flow along radially oriented pyramidal cell dendrites due to synaptic input. The current vector (thick arrow) points in the direction of intracellular local circuit current flow.

A major advantage of MEG over EEG is its temporal resolution of better than 1 ms, which is comparable to intracranial electrodes. Also, magnetic fields are less distorted by skull anatomy than electric fields, which gives MEG a better spatial resolution.

Transcranial magnetic stimulation (TMS)

A magnetic field changing with time (in strength or direction or both) induces an electrical field. In **transcranial magnetic stimulation** (**TMS**) this is exploited by using electromagnetic coils to induce currents in the brain which influence firing of neurons directly beneath the coil. The currents attenuate with distance from the coil. TMS is typically effective to depths of 1.5–3 cm so the cerebral and cerebellar cortices are the usual targets, but more powerful coils can be used to affect subcortical structures. The shape of the induced electrical field is hard to model because of the nonuniform electrical properties of neural tissue but it has a high spatial resolution (< 1 cm) if a figure-of-eight coil is used which concentrates magnetic flux at the node of the coil, and a temporal resolution of tens of ms.

There are two modes of TMS, single pulse and repetitive pulse. Single pulse has the advantage that it can be time-locked to the delivery of a stimulus so can allow precise timing of any effect. However, single pulses may not always be effective and hence repetitive TMS (rTMS) can be used. This delivers exponentially rising and falling magnetic pulses (Figure 2) with a frequency up to 50 Hz for several seconds. TMS has effects on visual perception, movement control, attention, memory, language, and decision making. TMS can excite or inhibit depending on the stimulus characteristics. For example, 10 Hz rTMS to the motor cortex stimulates muscle contraction and improves performance in a motor learning task whereas I Hz rTMS impairs motor learning. When used to study cognition it is usually inhibitory so that it interferes with performance of cognitive tasks, either increasing reaction time or increasing the number of errors.

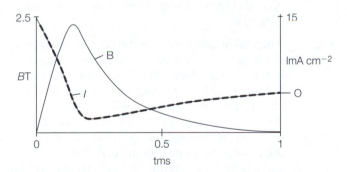

Figure 2. Transcranial magnetic stimulation (TMS). A rapidly changing magnetic field strength *B* (measured in tesla, T) produced by a single TMS pulse induces an electric current, *I*, in the brain.

O3 Classical electrophysiology

Key Notes

Extracellular recording	A technique for recording from single cells or groups of cells in a variety of situations both *in vitro* and *in vivo*, extracellular recording works by amplifying the potentials that arise between a focal electrode close to the neuron(s) and a distant, indifferent electrode.
Intracellular recording	This is a technique for measuring transmembrane potentials. It uses a fine, electrolyte-filled glass microelectrode to impale the cell. The microelectrode output goes to an amplifier and then to a computer for display, storage, and analysis.
Stimulating neurons	Neurons can be stimulated using a stimulator which delivers a current to the cell via a microelectrode. Inward currents cause neurons to depolarize (i.e., the membrane potential becomes smaller) whereas outward currents cause neurons to hyperpolarize.
Voltage clamping	Voltage clamping is a technique which allows the current that flows across nerve cell membranes to be measured. It does this by means of a negative feedback electronic circuit that holds the membrane potential constant. It provided the crucial evidence for the ionic currents that cause the action potential.
Patch clamping	Patch clamping measures the currents that flow through single ion channels. It depends upon electrically isolating the tiny patch of membrane that lies under the tip of a microelectrode. In the cell-attached mode the patch remains on the cell and single channel currents are recorded. Rupturing the patch gives the whole-cell mode, which allows recording of the macroscopic currents that flow through the plasma membrane of the cell. Alternatively the patch can be completely removed to give two other single channel configurations. In outside-out mode the effects of channel ligands applied to the bath can be studied. In inside-out mode agents can be added to the bath to investigate the role of second messengers in modulating channels.
Calcium imaging	Fluorescent dyes which alter the wavelength at which they absorb UV light on binding Ca^{2+} are used to visualize how Ca^{2+} signals spread in time and space through cells.
Related topics	(B1) Membrane potentials (C2) Neurotransmitter release (B2) Voltage-dependent (D1) Ligand-gated ion channel ion channels receptors

Extracellular recording

The firing patterns of either single neurons or clusters of neurons in living animals in response to physiological stimuli are obtained by extracellular recording. This technique uses two fine electrodes usually of tungsten or stainless steel. One, the **exploring (focal) electrode** is placed as close as possible to a neuron. The second, **indifferent electrode** is placed at a convenient distance. Neuron activity will cause currents to flow between the two electrodes. These currents are amplified and fed to a computer. By convention, if the exploring electrode is positive with respect to the indifferent electrode an upward deflection is recorded. The polarity, shape, amplitude, and timing of the recorded waveform generated by neural activity will depend on the position of the electrodes. The closer the exploring electrode is to the neuron, the larger the measured signal. Changing the distance between the two electrodes or altering their relative positions will modify all the above parameters. All of this can make extracellular recordings hard to interpret.

The technique can be used in brain slices or other *in vitro* preparations or *in vivo*, for example, in anesthetized animals. A particularly useful technique inserts electrodes into the brain that are attached to a connector cemented into the skull. This is done under anesthetic. The animal is allowed to recover. As required, the recording circuitry (amplifier to computer) is plugged into the connector. This allows electrophysiology in conscious, behaving animals, using sophisticated experimental protocols over long periods.

Intracellular recording

Being able to measure membrane potentials directly is crucial to understanding how excitable cells function. A standard technique for doing this in individual cells is **intracellular recording**.

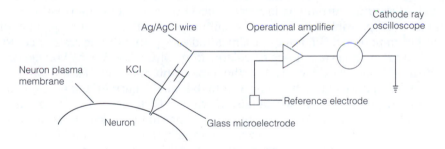

Figure 1. The circuitry used for intracellular recording.

To record the potential difference across a membrane it is necessary to have two electrodes, one inside the cell, the other outside, both connected to a voltmeter of some description (Figure 1). Because neurons are small the tip of the intracellular electrode impaling the cell needs to be very fine. To achieve this, glass micropipettes are manufactured to have a tip diameter of less than 1 μm. The micropipette is filled with an electrolyte (commonly KCl at a concentration of 0.15–3 M) to carry the current, so forming the **microelectrode**. Typically transmembrane potentials are less than 0.1 V and so must be amplified with an **operational amplifier**. This has inputs from both the intracellular microelectrode that impales the cell and the **reference** (**bath** or **indifferent**) electrode, which is placed in the solution bathing the cell. If no potential difference exists between the microelectrode and the reference electrode the amplifier output will be zero. If a potential difference exists between the electrodes, however, the amplifier generates a signal, the magnitude of which is proportional to the potential. The output of the amplifier goes to the analog-to-digital port of a computer running software which allows display, storage, and analysis of the data.

Stimulating neurons

In vivo, neurons are excited either by the cascade of synaptic inputs onto their dendrites and cell body from other neurons or by receptor potentials generated by sensory organs. Neurophysiologists often stimulate a neuron directly by injecting an electrical current into it via a stimulating microelectrode. The **stimulator** normally delivers a square wave current pulse. Three variables can be altered at will on most stimulators; the duration of the pulse, the amplitude of the injected current, and the frequency of the pulses. The direction of the current (which is defined as the flow of positive charge) determines the response of the neuron. If a small inward current is injected into a cell it will become a little more inside positive. This is a decrease in the membrane potential because V_m gets closer to zero and is called a depolarization. If, on the other hand, an outward current is injected (that is, if current is withdrawn from the cell) then the membrane potential increases; this is called hyperpolarization. The sizes and time courses of depolarizing and hyperpolarizing potentials seen in nerve cell injected with *small* currents are determined solely by the passive electrical properties of the neuron.

Voltage clamping

The ion fluxes that underlie neuron action potentials were discovered in the early 1950s using a technique called **voltage clamping** which remains a key technique in neurophysiology. It measures the currents that flow across an excitable cell membrane at a fixed potential. Measuring currents is important because it provides information on which ions might be responsible for changes in membrane potential. Current (I) cannot directly be found from potential (V); it requires, in addition, the membrane resistance (R). Only if *both* V and R are known can I be calculated from Ohm's law, $V = IR$. Voltage clamping circumvents this problem by measuring the transmembrane potential and having a feedback amplifier in the circuit which injects into the cell the current that is needed to keep the membrane potential constant; that is, the voltage is clamped. The current that must be injected by the circuit to keep the voltage fixed has to be the same size as the current flowing through the ion channels that would normally cause the potential to change. The voltage at which the membrane is clamped is called the **command voltage**. By examining the currents that flow across a membrane over a range of command voltages it is possible to determine which ions carry the currents.

The use of voltage clamping is illustrated in Figure 2, which shows an experiment in which the giant axon of a squid is initially clamped at its resting potential of –60 mV and the command voltage is then switched to zero. This first gives rise to a capacitance current as the change in voltage alters the amount of charge stored on the nerve cell membrane.

After the capacitance current comes an early inward current followed by a late outward current. These are the currents that normally flow during an action potential. When the squid axon is bathed in a solution that contains no sodium the early inward current is abolished. This shows that the early inward current is carried by Na$^+$. The same result is seen if an axon immersed in normal seawater is poisoned with the neurotoxin **tetrodotoxin** (**TTX**). By binding to the external mouth of the Na$_v$, TTX prevents Na$^+$ from permeating through the channel. TTX added to any nerve preparation abolishes action

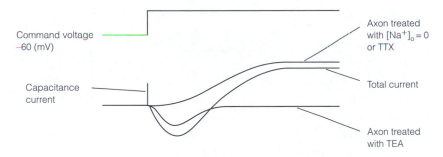

Figure 2. A voltage clamping experiment to show the currents that flow during an action potential. Outward currents are upward directions, inward currents downwards directions. Initially the neuron is clamped at −60 mV, and the step to 0 mV generates first a capacitance current, then the ionic currents. The experiment is done with the axon under three different conditions. See text for details. TTX, Tetrodotoxin; TEA, tetraethylammonium.

potentials. Similarly, **tetraethylammonium** (**TEA**) which blocks K_vs, when added to the bathing medium abolishes the late outward current, showing that it is carried by K^+.

Patch clamping

Patch clamping is an *in vitro* technique that makes it possible to study the electrophysiology of single ion channels. It works by forming a very high electrical resistance seal between a glass micropipette and the surface of a cell. Only currents flowing through the patch under the electrode will be recorded. This permits the extremely small currents that flow through single ion channels (about 1 pA) to be measured. The electronics allows the voltage of the patch to be clamped so voltage-clamping experiments can be performed (Figure 3a).

There are several configurations of patch clamping, each useful for particular types of experiment (Figure 3b):

- **Cell attached**. This is used for measuring single channel currents in intact cells. Second-messenger-induced modifications of the patched channels can be investigated in response to bathing the cell with specific agents, for example neurotransmitters.

- **Whole cell**. In this configuration the patch under the microelectrode is ruptured. The current flowing through the electrode represents the sum of all the currents flowing through individual ion channels on the cell surface. Hence whole cell patching measures **macroscopic currents**.

- **Outside-out**. This is one of two patch clamp modes used to study single-channel currents in which the patch is removed from the cell but remains sealed to the pipette tip. This configuration is used to study the effects of ligands such as neurotransmitters, hormones, or externally acting drugs on channels. These ligands are added to the bath because bath solutions can be changed much more easily and rapidly than the pipette solution. This has obvious advantages for performing complicated experiments such as investigating dose–response relationships.

- **Inside-out**. This is the second of the patch-only configurations, usually used for the detailed examination of second messengers because these can be applied directly to the inside face of the membrane via the bath solution.

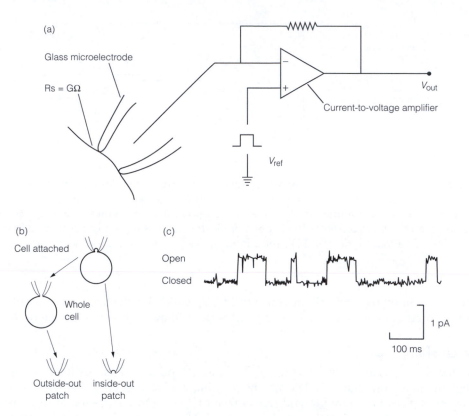

Figure 3. Patch clamping. (a) The circuitry, V_{ref} is used to clamp the potential of the neuron membrane, the method relies on forming a high resistance seal, Rs between the pipette tip and the cell membrane. (b) Patch clamp configurations. (c) Single channel currents through a GABA$_A$ receptor.

A representative patch clamp recording is shown in Figure 3c. Each of the square wave events is the opening of a single channel. The height of the wave is the **unitary channel current**, the duration of the wave is the time for which it is open. Statistical analysis of large numbers of opening events provides estimates for parameters such as **mean channel open time**, and allows models of channel kinetics to be tested. Such studies are very useful in fathoming out how neuroactive drugs act at the molecular level.

Calcium imaging

This makes visible how Ca^{2+} signals spread in time and space through cells. Fluorescent dyes are used that on binding Ca^{2+} absorb UV light at a different wavelength than they do in the unbound state. Neurons are preloaded with the dye and the emission of UV from the dye is observed in response to its excitation by the two distinct absorption wavelengths. This gives a quantitative measure of how the concentration of Ca^{2+} changes in the neuron in real time.

O4 Neuroanatomy imaging technologies

Key Notes

Confocal laser scanning microscopy

Conventional optical microscopy has a resolution limited approximately by half the wavelength of light. Confocal microscopy is fluorescence microscopy that allows living cells to be imaged at close to the diffraction limit for visible light. In confocal microscopy a small region of the sample is illuminated by focused laser light tuned to excite fluorescent dyes in the sample. The fluorescent light emitted by the specimen passes through a pinhole before entering a digital camera, eliminating all out of focus light, to achieve a resolution of 200–500 nm.

Stimulated-emission depletion microscopy

Several super-resolution imaging techniques have been developed which allow light microscopy to obtain information far below the diffraction limit of visible light. Stimulated-emission depletion microscopy circumvents the diffraction limit by targeted de-excitation of dye molecules, reducing the size of the excited region to a small central spot. It does this by using by using two lasers. An excitation laser makes the target fluoresce. An offset laser irradiates the excited dye molecules with light of similar wavelength to the fluorescence light, causing the molecule to return to the ground state by stimulated emission not fluorescence. This allows nanoscale resolution (< 10 nm) permitting detail to be visualized in living neural tissue that could hitherto only be approximated by three-dimensional reconstruction from ultrathin serial sections viewed under electron microscopy, a far more laborious technique.

Photoactivated localization microscopy

Photoactivated localization microscopy is one of several techniques that achieve better than diffraction-limited resolution by imaging individual fluorophores one at a time, making it possible to find the position of each molecule with high precision. The excitation light illuminates the entire sample but at low intensity so that only a few fluorophores are randomly excited at a given time, allowing different fluorophores to be excited and imaged individually at different times. Computer algorithms are used to construct an image of the sample from the locations of each fluorophore. The major disadvantage is the time it takes to acquire an image.

Brainbow technique

Individual neurons and glia in the living brain can be made to express specific fluorescent proteins, based on green fluorescent protein, using genetic engineering techniques. The cells can be imaged in brain sections with confocal microscopy, making it possible to map each one because it has a distinctive color. This technique allows detailed studies of neural connectivity and circuitry. It relies on the *cre/loxP* system.

Cre/loxP system

The *cre/loxP* system allows selective gene deletion or activation to be generated by recombination in specific cell types and at specific times in development. It requires two types of transgenic animal (usually mice) to be engineered which are subsequently crossed. *Cre* animals are created with a construct that has a gene for a recombinase enzyme, *cre*, downstream of a cell specific promoter. Floxed animals are created in which all cells contain the target gene flanked on both sides by a *loxP* sequence. The cre enzyme recognizes two *loxP* sequences if they are in the same orientation and causes recombination which excises the target gene between them. This happens in the offspring (F1 hybrids) of cre and floxed animal crosses. Because the *cre* construct is expressed only in the cell type targeted by the promoter the gene deletion is cell specific. The *cre/loxP* system can also activate a gene using a floxed construct in which *loxP* sequences flank a stop signal.

Related topics

(A4) Organization of the central nervous system	(K4) Cerebellar cortex circuitry
(E2) Coding of modality and location	(N3) Hippocampus and episodic learning
	(O5) Optogenetics

Confocal laser scanning microscopy

Conventional optical microscopy has a resolution limited approximately by half the wavelength of light. There are numerous types of light microscopy but the method that gets closest to the limits imposed by the physics of light diffraction is confocal microscopy. Moreover, because it uses fluorescence to image a sample and fluorescent dye can be taken up by living cells, unlike other types of light microscopy fluorescence microscopy can image living tissue.

A fluorescent dye molecule (**photoactivatable fluorophore**) is excited by light of a specific wavelength so that electrons are boosted from the ground state to high quantum energy states. After a few nanoseconds the molecule de-excites back to the ground state, emitting lower energy photons with longer wavelength because some of the energy of the exciting photon is dissipated internally within the molecule. In confocal microscopy the sample is illuminated by laser light tuned to excite the fluorescent dyes in the sample. However, rather than wide field illumination, as is the case with conventional light microscopy, light is brought to focus on a small region of the sample. In **confocal laser scanning microscopy** (Figure 1) scanning mirrors move the laser across the specimen so that each small region of it can be excited in turn. The fluorescent light emitted by the specimen passes through a pinhole before entering a digital camera. This eliminates all out of focus light; that is, only light from the focal plane reaches the camera.

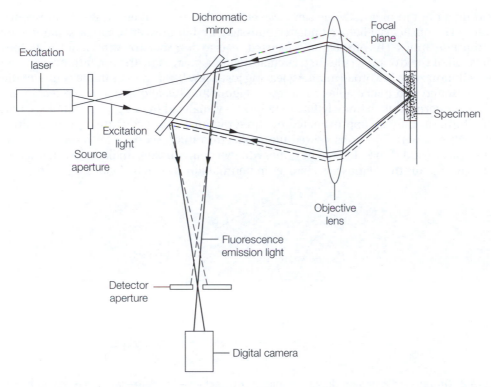

Figure 1. Principle of confocal microscopy. Excitation light is transmitted through a dichromatic mirror and brought to focus in the specimen. Emitted fluorescence light is reflected through a pinhole detector aperture into a digital camera. Fluorescence emitted from sources not on the focal plane (dashed lines) is cut off by the detector aperture.

The resolution of such a microscope is limited by the size to which the excited region can be focused and means that structures closer than 200 nm in the focal plane cannot be resolved. The depth resolution (along the optical axis) is worse, around 500 nm.

Stimulated-emission depletion microscopy

Several super-resolution imaging techniques have been developed which allow light microscopy to obtain information far below the $\lambda/2$ diffraction limit of visible light. One is **stimulated-emission depletion microscopy** (**STED**) which circumvents the diffraction limit by targeted de-excitation of dye molecules without making them fluoresce. Essentially this switches the dye molecules off, reducing the size of the excited region. STED works by using two lasers, an excitation laser and an offset laser. Fluorescent dyes which label specific sites of a sample are excited by light of a particular wavelength produced by the excitation laser. After a few nanoseconds the dye molecule spontaneously relaxes to the ground state by emitting a fluorescence photon of longer wavelength; that is, a redder light.

However, an excited molecule can also return to its ground state by stimulated emission instead of spontaneous relaxation and fluorescence. This can be triggered by irradiating the excited dye molecule with light of *similar* wavelength to the fluorescence light. This is done with the offset laser. In response, the dye molecule returns to the ground state, emitting a photon of exactly the same wavelength but without emitting a fluorescence photon. The offset laser light, stimulated emission, and fluorescence can be separated by appropriate optics. The offset laser is focused so as to de-excite (by stimulated emission) almost all of the dye molecules in a circular region activated by the excitation laser. The de-excitation zone is a broad annulus, leaving just a few fluorophores in the center of the region excited and hence able to fluoresce (Figure 2). This allows nanoscale resolution (< 10 nm) permitting extremely fine detail to be visualized in neuron morphology that could hitherto only be approximated by three-dimensional reconstruction from ultra-thin (~50 nm) serial sections viewed under electron microscopy, a far more laborious technique. In addition STED can be done with live neural tissue, unlike electron microscopy. This means that nanoscale changes in neuroanatomy can be followed over time.

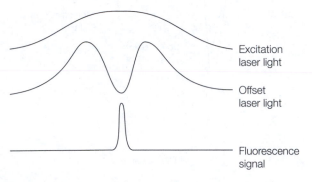

Excitation laser light

Offset laser light

Fluorescence signal

Figure 2. Stimulated emission depletion microscopy achieves super-resolution imaging by using an optical configuration that generates regions where fluorophores de-excite without fluorescing. Note the resemblance to surround inhibition!

Photoactivated localization microscopy

To spatially resolve closely spaced fluorescing dye molecules (fluorophores), a variety of techniques such as **photoactivated localization microscopy** (**PLM**), work by separating the fluorescence of each emitter in time. Instead of imaging all the fluorophores simultaneously, these techniques image each individual fluorophore one at a time, making it possible to find the position of each molecule with high precision. Once all of the positions have been found, they are plotted as points in space to construct an image. The spatial resolution is not limited by diffraction, but only by the precision with which each fluorophore can be localized.

To observe each protein individually the excitation light illuminates the entire sample but at low intensity so that only a few fluorophores are excited at a time, and this fluorophore excitation is stochastic; that is, whether a given dye molecule is excited at a given time is random. This enables different fluorophores to be excited at different times, so they can be imaged individually. Computer algorithms are used to construct an image of the sample from the locations of each fluorophore.

The precision of the position measurement is dependent on the contrast between the brightness of the fluorophore compared with the background; the greater the contrast, the higher the precision. An advantage of this stochastic fluorophore imaging is that each dye molecule only undergoes a few photoexcitation cycles and this avoids a problem called photobleaching. The major disadvantage is the time it takes to acquire an image. The greater the fluorophore density in the sample, the longer the imaging time. Because imaging time is determined by how rapidly each fluorophore turns on and off, acquisition time can be reduced by using higher excitation intensity (the fluorophore turns off within nanoseconds of it being excited), but this can limit resolution.

Brainbow technique

Individual neurons and glia in the living brain can be made to express specific fluorescent proteins using genetic engineering techniques. The cells can be imaged in brain sections with confocal microscopy, making it possible to map each one because it has a distinctive color. The importance of this **brainbow technique** is that it allows detailed studies of neural connectivity and circuitry.

Differential expression of five protein fluorophores—yellow, orange, red, green, and cyan derivatives of the original green fluorescent protein (GFP)—allows upwards of 90 distinct hues to color individual cells. The technique was invented in 2007 and originally allowed mapping of only a few neurons, but currently over 100 neurons can be mapped simultaneously.

The brainbow technique relies on the *cre/loxP* genetic engineering system which has been used extensively in brain research since 1998 to control gene expression in particular cell types.

Cre/loxP system

The ***cre/loxP*** system allows selective gene deletion or activation to be generated by recombination in specific cell types, and at particular times (e.g., stages in development). It requires two types of transgenic animal (usually mice) to be engineered which are subsequently crossed (Figure 3a).

1. *Cre* animals are created with a construct that has a gene for a viral recombinase enzyme, cre (causes recombination) downstream of a cell specific promoter. These can be specific for glia (glial fibrillary acidic protein promoter), neurons (e.g., synapsin 1 or neuron-specific enolase promoters), or even for specific neuron types (e.g., CaMKII promoter is selective for CA1 cells of the hippocampus).

2. Floxed animals are created in which all nucleated cells contain the target gene flanked on both sides by a 34 base pair *loxP* sequence, derived from bacteriophage P1. (Bacteriophages are viruses that infect bacteria.)

In both cases these animals are genetically engineered by having the construct inserted into a **vector** (often based on the double stranded circular DNA molecules, plasmids, that make up the genome of bacteria). The vector can be engineered with a variety of reporter genes so that its presence can be detected. It can then be injected into the pronucleus of fertilized oocytes or incorporated in embryonic stem cells that are then introduced into blastocysts (early embryos with 8–16 cells). The oocyte or blastocyst is subsequently implanted into a foster mother and the resulting offspring screened to ensure they have incorporated the constructs.

The principle of the *cre/loxP* system is that the cre enzyme recognizes two *loxP* sequences if they are in the same 5′–3′ orientation, bringing them together so that recombination effectively excises the target gene between them, leaving behind one *loxP* site. This happens in the offspring (F1 hybrids) of cre and floxed animal crosses. The floxed constructs are expressed in all nucleated cells in the resulting offspring, but the cre construct is expressed only in the cell type targeted by the promoter. Hence the *cre/loxP* recombination is cell specific.

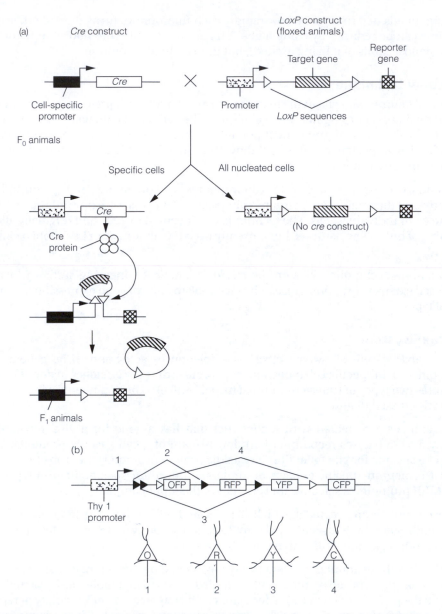

Figure 3. *Cre/loxP* system. (a) F0 transgenic animals containing *cre* or floxed gene constructs are crossed. The resulting F1 offspring will express the floxed gene in all of its nucleated cells. In the specific cell type targeted by the *cre* construct, the expression of cre causes deletion of floxed gene. (b) A floxed construct used for the brainbow technique houses genes for orange (OFP), red (RFP), yellow (YFP), and cyan (CFP) fluorescent proteins. Excisions 1–4 result in neurons fluorescing in orange, red, yellow, and cyan respectively. Other more elaborate constructs allow a huge diversity of color coding of individual cells.

The *cre/loxP* system can also be used to study the effect of *activating* a transgene. To achieve this, a floxed construct is produced in which *loxP* sequences flank a stop codon upstream of the target gene. Now recombination cuts out the stop codon in the cre-expressing cells and the transgene will be transcribed by DNA polymerase. The transgene will remain inactive in all other cells because these do not express cre. Hence the target gene is expressed in a cell-specific way.

In the brainbow technique constructs are created with several *loxP* sites flanking fluorescent protein genes. This allows a variety of excisions to take place resulting in a range of differently color-coded neurons (Figure 3b).

O5 Optogenetics

Key Notes

General principles

Optogenetics is a set of techniques which allow millisecond control and sensing of brain processes in targeted populations of neurons using light. Sensor proteins transduce physiological signals in cells to light emissions so as to make specific cell functions visible. Actuator proteins respond to light signals by altering a physiological process. In many cases the sensor or actuator proteins are produced by transgenic animals. Alternatively, sensors and actuators are delivered as plasmids or viruses. These are subsequently taken up by the cells for which these vectors are the targets.

Sensors

Protein-based optical sensors have been developed for membrane voltage, intracellular calcium concentration, neurotransmitter release, and for a number of second messengers, for example cAMP. Sensors are based on a green fluorescent protein (GFP) chromophore. Fusion proteins can be created between GFP and another protein which is able to report some biochemical change. For example, a GFP–Shaker potassium channel fusion protein is able to signal voltage-dependent conformational changes in the channel as alterations in GFP fluorescence. A potential problem arises if introducing the sensor protein changes the variable (e.g., membrane potential) being studied.

Actuators

Most actuators at present are ion channels, proteins which regulate ion channels, or ion pumps. Many are based on visual rhodopsin which is coupled to G protein or microbial rhodopsins that are proton or chloride pumps, and it is this which makes them light responsive. These allow control over membrane voltage, intracellular calcium concentration, and neurotransmitter secretion. Other actuators include G-protein-coupled receptors, second messengers, kinases, and transcription factors.

Optogenetic protocols

Channelrhodopsin-2 (ChR2) responds to blue light by opening a Na^+ channel and hence excites any neuron into which it has been incorporated. Fluorescently labeled ChR2 reveals light-stimulated axons and synapses in intact brain tissue. The gene archaerhodopsin-3 (Arch) codes for an outward proton pump. Illuminated with yellow light it silences mouse cortical neurons that express the protein. Another light-driven proton pump, Mac, silences neurons when illuminated by blue light. Hence using Arch and Mac together allows silencing of two neural populations independently with two different wavelengths of light.

The scope of optogenetic studies	Optogenetics has been used in a variety of animal models for three broad classes of investigation: tracing neuronal connections, studying neural network activity, and investigating neural mechanisms of sensation, reward, and cognition. It is being used in combination with other techniques, such as extracellular recording from awake behaving animals and with fMRI.	
Related topics	(A4) Organization of the central nervous system (B1) Membrane potentials (C2) Neurotransmitter release	(G3) Photoreceptors (O3) Classical electrophysiology (O4) Neuroanatomy imaging technologies

General principles

Optogenetics provides a set of techniques which allow millisecond control and sensing of brain processes in targeted populations of neurons using light. There are two types of optogenetic device. Sensors are proteins which transduce physiological signals in cells to light emissions so as to make specific cell functions visible. Actuators are proteins that respond to light signals by altering a physiological process. In many cases the sensor or actuator proteins are produced by animals genetically engineered to manufacture them by introducing the encoding DNA into their genome. Genetic engineering techniques can in some instances ensure that the DNA is restricted to particular cell types. However, the capacity for this is limited to date by our knowledge of gene expression in specific cell types. For example, whilst a requirement of dopaminergic cells is that the gene for the enzyme tyrosine hydroxylase (TH) must be switched on, this is also true of other catecholaminergic neurons, so animals engineered so that TH can be light activated or will emit light so as to indicate the enzyme's activity, cannot be interpreted as just reflecting dopaminergic neuron function.

However, not all optogenetic manipulations can be done by genetically engineering animals specifically. Some sensors and actuators require, in addition to genetically encoded proteins, small molecules that must be injected or ingested. Alternatively, the DNA is introduced in the form of plasmids or viruses. These are subsequently taken up by the cells for which these vectors are the targets. The disadvantage of this is that each animal has to be treated individually, making experiments time-consuming and laborious. However, it is sometimes needed to gain sufficiently high levels of expression of a protein.

Sensors

Protein-based optical sensors have been developed for membrane voltage, intracellular calcium concentration, neurotransmitter release, and for a number of second messengers, for example cAMP. Sensors are based on a green fluorescent protein (GFP) chromophore. In one guise fusion proteins are created between GFP and another protein which is able to report some biochemical change. For example, a GFP–Shaker potassium channel fusion protein is able to signal voltage-dependent conformational changes in the channel as alterations in GFP fluorescence. Light emission from GFP derivatives is modulated either by:

- Protonation/deprotonation which is in turn affected by the conformational state of the protein that controls access of the chromophore to the solvent; or

- Variations in the electrical properties of neighboring chromophores brought about by changes in their proximity or orientation to each other

To detect rapid events it is necessary to sample at a sufficiently high rate. Thus, to detect individual action potentials lasting one millisecond it is necessary to sample with a frequency greater than 1 kHz if all are to be captured. Short events produce fewer photons and so are harder to see. Increasing the expression of the sensor is often not the solution because the presence of sensor disturbs the variable under study. For example, the voltage sensors needed to detect action potentials change the capacitance of the cell membrane in which they are expressed, and this suppresses synaptic potentials thereby altering the behavior of the neurons under study.

Actuators

Most actuators at present are ion channels, proteins which regulate ion channels, or ion pumps. Many are based on visual rhodopsin which is coupled to G protein or microbial rhodopsins that are proton or chloride pumps, and it is this which makes them light responsive. These allow control over the same variables that the sensors respond to: membrane voltage, intracellular calcium concentration, and secretion of signaling molecules. There are also light-activated G-protein-coupled receptors, second messengers, kinases, and transcription factors. A **light-activated glutamate receptor** (**LiGluR**) is a calcium channel and hence allows control over calcium-dependent processes such as secretion.

Actuators are stimulated by LED or laser light delivered via optical fibers. These can be used *in vitro* (e.g., in brain slices) or *in vivo*, implanted into the brain ahead of time: this allows optogenetic experiments to be done in awake, behaving animals. Actuators available now allow control over processes with a timescale of a few milliseconds to seconds.

Optogenetic protocols

An extremely promising optogenetic device is **channelrhodopsin-2** (**ChR2**), an algal protein. It responds to blue light by opening a nonspecific cation conductance. The resulting inward Na^+ current excites any neuron that has incorporated ChR2 into its membrane. In addition to its use as an actuator fluorescently labeled ChR2 reveals light-stimulated axons and synapses in intact brain tissue. This has been used to study the molecular events in spike timing-dependent plasticity. In addition ChR2 has been used to map long-range cortico-cortical connections in the brain, and to map the spatial location of specific inputs on the dendritic tree of individual cortical pyramidal neurons.

The gene **archaerhodopsin-3** (**Arch**) codes for an outward proton pump. When virally expressed in mouse cortex and illuminated with yellow light it almost completely silences those neurons that express the protein. Arch spontaneously recovers from light-dependent inactivation, unlike light-driven chloride pumps that enter long-lasting inactive states in response to light. This means that Arch could mediate several cycles of optical silencing during an experiment. Expressed in specific cells this could be used to investigate their role in any particular neuron function by temporarily inhibiting the cells. Unlike gene knockout experiments, optogenetic silencing can be transient and can be repeated any number of times, allowing animals to be used as their own controls. Another actuator protein, **Mac**, a light-driven proton pump derived from a fungus, silences neurons when illuminated by blue light. Using Arch and Mac together allows silencing of two neural populations independently with two different wavelengths of light. A typical optogenetic protocol is shown in Figure 1.

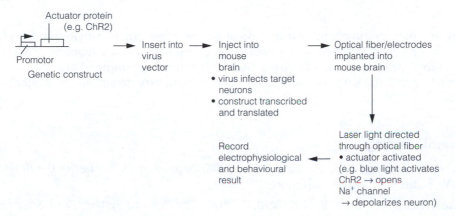

Figure 1. A typical optogenetic protocol.

The scope of optogenetic studies

Optogenetics is extremely young, the first experiment—triggering neuronal action potentials using light—was done in 2002. Since then it has been used in a variety of animal models including the nematode *Caenorhabditis elegans*, the fruit fly *Drosophila*, zebra fish, mice, and primates for three broad classes of investigation:

● Tracing of neuronal connections. This has shown, for example, that neocortical pyramidal cell dendritic trees are segregated into functional domains according to input with local, ascending, and descending axons going to specific domains.

● Neural network activity. This has revealed that brain circuits exist for behaviors an animal would not normally exhibit, such as male courtship display in a female.

● Neural mechanisms of sensation, reward, and cognition.

Optogenetics is being used in combination with other techniques, such as extracellular recording from awake behaving animals and with fMRI (in this case to establish precisely what cellular events were required to explain the fMRI signal).

Further reading

There are many comprehensive textbooks of neuroscience and no single volume is likely to satisfy all needs. Different readers will prefer different textbooks depending, for example, on their prior learning, so I do not think it helpful to recommend one over another. Rather I have listed some of the leading books which, experience shows, students find useful.

General reading

Bear, M.F., Connors, B.W., Paradiso, M.A. (2006) *Neuroscience: Exploring the Brain,* 3rd edn. Williams and Wilkins, Baltimore.

Michael-Titus, A., Revest, P., Shortland, P. (2010) *The Nervous System: Basic Science and Clinical conditions*, 2nd edn. Churchill Livingstone, Edinburgh.

Purves, D. *et al.* (eds) (2011) *Neuroscience*, 5th edn. Sinauer Associates, Sunderland, MA.

Squire, L.R., Bloom F.E., Spitzer N.C., Du Lac, S., Ghosh, A., Berg, D. (2008) *Fundamental Neuroscience,* 3rd edn. Academic Press, San Diego, CA.

More advanced reading

The following selected articles and books are recommended for those who wish to know more about specific subjects. Some of the papers discuss ideas reviewed in this book, others deal with material which could not be included for lack of space. Inevitably these sources vary in difficulty; some will be accessible to first year students, most will be more suitable for students later in their studies.

Section A

Fitzgerald, M.J.T., Gruener, G., Estomih, M. (2011) *Clinical Neuroanatomy and Neuroscience* , 6th edn. Saunders, Elsevier, Oxford.

Newman, E.A. (2003) New roles for astrocytes: regulation of synaptic transmission. *Trends Neurosci.* **26**, 536–542.

Pardridge, W.M. (2006) *Introduction to the Blood–Brain Barrier*, 2nd Cambridge University Press, Cambridge.

Parea, G., Navarrere, M., Argave, A. (2009) Tripartite synapses: astrocytes process and control synaptic formation. *Trends Neurosci.* **32**, 421–430.

Ransom, B., Behar, T., Nedergaard, M. (2003) New roles for astrocytes (stars at last). *Trends Neurosci.* **26**, 520–522.

Section B

Alle, H., Roth, A., Geiger, J.P.R. (2009) Energy efficient action potentials in hippocampal mossy fibres. *Science* **325**, 1405.

Hille, B. (2001) *Ionic Channels of Excitable Membranes*, 3rd edn. Sinauer Associates, Sunderland, MA.

Section C

Goldberg, J.H., Yuste, R. (2005) Space matters: local and global dendritic Ca^{2+} compartmentalization in cortical interneurons. *Trends Neurosci.* **28**, 158–167.

Larkam, M.E., Nevian, T., Sandler, M., Polsky, A., Schiller, J. (2009) Synaptic integration on tuft dendrites of layer 5 pyramidal cells: A new unifying principle. *Science* **325**, 756.

Ludwig, M., Pittman, Q.J. (2003) Talking back: dendritic neurotransmitter release. *Trends Neurosci.* **26**, 255–261.

Oláh, S., Füle, M., Komlósi, G., Varga, C., Báldi, R., Barzó, P., Tamás, G. (2009) Regulation of cortical microcircuits by unitary GABA-mediated volume transmission. *Nature* **461**, 1278.

Zhang, Q., Li, Y., Tsien, R.W. (2009) The dynamic control of kiss-and-run and vesicular release probed with single nanoparticles. *Science* **323**, 1448.

Section D

Baraban, S.C., Tallent, M.K. (2004) Interneuronal neuropeptides – endogenous regulators of neuronal excitability. *Trends Neurosci.* **27**, 135–142.

Barrera, N.P., Edwardson, J.M. (2008) The subunit arrangement and assembly of ionotropic receptors. *Trends Neurosci.* **31**, 569–576.

Ben-Ari, Y. (2005) The multiple facets of GABA. *Trends Neurosci.* **28**, 277 *et seq.* (Special issue.)

Björkland, A., Dunnett, S.B. (2007) Fifty years of dopamine. *Trends Neurosci.* **30**, 151 *et seq.* (Special issue.)

Cooper, J.R., Bloom, F.E., Roth, R.H. (2002) *The Biochemical Basis of Neuropharmacology*, 8th edn. Oxford University Press, Oxford.

Kuner, T., Seeburg, P.H., Guy, R.H. (2003) A common architecture for K^+ channels and ionotropic glutamate receptors? *Trends Neurosci.* **26**, 27–32.

Section E

Ben-Ari, Y. (2007) Physiological and pathological oscillations. *Trends Neurosci.* **30**, 307 *et seq.* (Special issue.)

Bonifazi, P., Goldin, M., Picardo, M.A., Jorquera, I., Cattani, A., Bianconi, G., Represa, A., Ben-Ari, Y., Cossart, R. (2009) GABAergic hub neurons orchestrate synchrony in developing hippocampal networks. *Science* **326**, 1419.

Grillner, S., Markam, H., De Schutter, E., Silderberg, G., Le Beau, F.E.N. (2005) Microcircuits in action – from CPGs to neocortex. *Trends Neurosci.* **28**, 525–533.

Huxter, J., Burgess, N., O'Keefe, J. (2003) Independent rate and temporal coding in hippocampal cells. *Nature* **425**, 828–832.

Singer, W. (2009) Distributed processing and temporal codes in neuronal networks. *Cogn. Neurodyn.* **3**, 189–196.

Thivierge, J-P., Marcus, G.F. (2007) The topographic brain: from neural connectivity to cognition. *Trends Neurosci.* **30**, 251–259.

Section F

Derbyshire, S.W.G. (2002) Measuring our natural painkiller *Trends Neurosci.* **25**, 67–69.

Kamchouchi, A. (2009) The neural basis of *Drosophila* gravity-sensing and hearing. *Nature* **458**, 165.

Maricich, S.M., *et al.* (2009) Merkel cells are essential for light touch responses. *Science* **324**, 1580.

McClesky, E.W. (2003) New player in pain. *Nature* **424**, 729–730.

Pernia-Andrade, A.J., Kato, A., Witschi, R., *et al.* (2009) Spinal endocannabinoids and CB_1 receptors mediate c fibre-induced heterosynaptic pain sensitization. *Science* **325**, 760.

Wall, P. (1999) *Pain: The Science of Suffering.* Weidenfeld and Nicholson, London.

Section G

Billock, V.A., Tsou, B.H. (2004) What do catastrophic visual binding failures look like? *Trends Neurosci.* **27**, 84–89.

Crick, F., Koch, C. (1995) Are we aware of neural activity in primary visual cortex? *Nature* **375**, 121–123.

Grieve, K.L., Acuna, C., Cudeiro, J. (2000) The primate pulvinar nuclei: vision and action. *Trends Neurosci.* **23**, 35–39.

Kara, K., Boyd J.D. (2009) A microarchitecture for binocular disparity and ocular dominance in visual cortex. *Nature* **558**, 627.

Knops, A., Thirion, B., Hubbard, E.M., Michel, V., Dehaene, S. (2009). Recruitment of an area involved in eye movements during mental arithmetic. *Science* **324**, 1583.

Zeki, S. (2003) Improbable areas in the visual brain. *Trends Neurosci.* **26**, 23–26.

Section H

McAlpine, D., Grothe, B. (2003) Sound localization and delay lines – do mammals fit the model? *Trends Neurosci.* **26**, 347–350.

Pollak, G.D., Burger, R.M., Klug, A. (2003) Dissecting the circuitry of the auditory system. *Trends Neurosci.* **26**, 33–39.

Scott, S.K., Johnsrude, I.S. (2003) The neuroanatomical and functional organization of speech perception. *Trends Neurosci.* **26**, 100–107.

Wiesz, C., Glowatsky, E., Fuchs, P. (2009) The postsynaptic function of type II cochlear afferents. *Nature* **461**, 1126.

Section I

Carleton, A., Accolla, R., Simon, S.A. (2010) Coding in the mammalian gustatory system. *Trends Neurosci.* **33**, 326–334.

Friedrich, F.R. (2002) Real time odor representations. *Trends Neurosci.* **25**, 487–489.

Schoppa, N.E., Urban, N.N. (2004) Dendritic processing within olfactory bulb circuits. *Trends Neurosci.* **27**, 501–506.

Section J

Clarac, F., Cattaert, D., Le ray, D. (2000) Central control components of a "simple" stretch reflex. *Trends Neurosci.* **23**, 199–208.

Pearson, K., Ekeberg, O., Büsches, A. (2006) Assessing sensory function in locomotor systems using neurochemical simulations. *Trends Neurosci.* **29**, 625–631.

Poppele, R., Bosco, G. (2003) Sophisticated spinal contributions to motor control. *Trends Neurosci.* **26**, 269–276.

Section K

Botvinick, M. (2004) Probing the neural basis of body ownership. *Science* **305**, 782–783.

Capaday, C. (2003) The special nature of human walking and its neural control. *Trends Neurosci.* **26**, 370–376.

Grillner, S., Hellgren, J., Ménard, A., Saitoh, K., Wikström, M.A. (2005) Mechanisms for selection of basic motor programs – roles for the striatum and pallidum. *Trends Neurosci.* **28**, 364–370.

Matsumoto, M., Hikosaka, O. (2009) Two types of dopamine neuron distinctly convey positive and negative motivational signals. *Nature* **459**, 837.

McHaffie, J.G., Stanford, T.R., Stein, B.E., Coizet, V., Redgrave, P. (2005) Subcortical loops through the basal ganglia. *Trends Neurosci.* **28**, 401–407.

Nakanishi, S. (2005) Synaptic mechanisms of the cerebellar cortical network. *Trends Neurosci.* **28**, 93–100.

Ohyama, T., Nores, W.J., Murphy, M., Mauk, M.D. (2003) What the cerebellum computes. *Trends Neurosci.* **26**, 222–227.

Tsai, H-C., Zhang, F., Adamantidis, A., Stuber, G.D., Bonci, A., de Lecea, L., Deisseroth, K. (2009) Phasic firing of dopaminergic neurons is sufficient for behavioural conditioning. *Science* **324**, 1080.

Zaghloul, K.A. Blanco, J.A., Weidemann, C.T., McGill, K., Jaggi, J.L., Baltuch, G.H., Kahana, M.J. (2009) Human substantia nigra neurons encode unexpected financial rewards. *Science* **323**, 1496.

Section L

Hadley, M.E., Levine, J. (2006) *Endocrinology*, 6th edn. Prentice-Hall International, New Jersey.

Kauffman, A.S., Clifton, D.K., Steiner, R.A. (2007) Emerging ideas about kisspeptin-GPR54 signaling in the neuroendocrine regulation of reproduction. *Trends Neurosci.* **30**, 504–511.

Johnson, M., Everitt, B. (2000) *Essential Reproduction*, 5th edn. Blackwell Scientific Publications, Oxford.

Section M

Antle, M.C., Silver, R. (2005) Orchestrating time: arrangements of the brain circadian clock. *Trends Neurosci.* **28**, 145–151.

Balaban, E. (2004) Why voles stick together. *Nature* **429**, 711–712.

Berridge, K.C., Robinson, T.E. (2004) Parsing reward. *Trends Neurosci.* **27**, 507–513.

Berson, D.M. (2003) Strange vision: ganglion cells as circadian photoreceptors. *Trends Neurosci.* **26**, 314–320.

Carlezon, W.A., Nestler, E.J. (2002) Elevated levels of GluR1 in the midbrain: a trigger for sensitization to drugs of abuse. *Trends Neurosci.* **25**, 610–615.

Cash, S.S., Halgren, E., Dehghani, N., *et al.* (2009) The human K complex represents an isolated cortical down state. *Science* **324**, 1084.

Damasio, A. (1999) *The Feeling of What Happens*. Heinemann, London.

Elmkuist, J.K., Flier, J.S. (2004) The fat-brain axis enters a new dimension. *Science* **304**, 63–64.

Goto, Y., Grace, A.A. (2008) Limbic and cortical information processing in the nucleus accumbens. *Trends Neurosci.* **31**, 552–558.

Russo, S.J., Dietz, D.M., Dumitriu, D., Morrison, J.H., Malenka, R.G., Nestler, E.J. (2010) The addicted synapse: mechanisms of synaptic and structural plasticity in the nucleus accumbens. *Trends Neurosci.* **33**, 267–276.

Schulkin, J., Morgan, M.A., Rosen, J.B. (2005) A neuroendocrine mechanism for sustaining fear. *Trends Neurosci.* **28**, 629–635.

Section N

Ahmed, O.J., Mehta, M.R. (2009) The hippocampal rate code: anatomy, physiology and theory. *Trends Neurosci.* **32**, 329–338.

Benito, E., Baco, A. (2010) CREBs control of intrinsic and synaptic plasticity: implications for CREB-dependent memory models. *Trends Neurosci.* **33**, 230–240.

Colgin, L.L., Denninger, T., Fyhn, M., Hafting, T., Bonnevie, T., Jensen, O., Moser, M.B., Moser, E.I. (2009) Frequency of γ oscillations routes flow of information in the hippocampus. *Nature* **462**, 353.

Han, J-H., Kushner, S.A., Yiu, A.P. *et al.* (2009) Selective erasure of a fear memory. *Science* **323**, 1492.

Harrison, S.A., Tong, F. (2009) Decoding reveals the contents of visual working memory in early visual areas. *Nature* **458**, 632.

Harvey, C.D., Collman, F., Dombeck, D.A., Tank, D.W. (2009) Intracellular dynamics of hippocampal place cells during virtual navigation. *Nature* **461**, 941–946.

Kampa, B.M., Letzkus, J.J., Stuart, G.J. (2007) Dendritic mechanisms controlling spike timing-dependent plasticity. *Trends Neurosci.* **30**, 456–463.

Lee, J.L.C., Everitt, B.J., Thomas, K.L. (2004) Independent cellular processes for hippocampal memory consolidation and reconsolidation. *Science* **304**, 839–843.

Massey, M.V., Bashir, Z.I. (2007) Long-term depression: multiple forms and implications for brain function. *Trends Neurosci.* **30**, 176–184.

Monfils, M-H., Cowansage, K.K., Klan, E., Ledoux, J.E. (2009) Extinction – re-consolidation boundaries: key to persistent attenuation of fear memories. *Science* **324**, 951.

Rudoy, J.D. (2009) Strengthening individual memories by reactivating them during sleep. *Science* **326**, 1079.

Sherman, S.M. (2006). The neural substrate of cognition. *Trends Neurosci.* **29**, 295 *et seq.* (Special issue.)

Wang, D.O., Kim, S.M., Zhao, Y., Hwang, H., Miura, S.K., Sossin, W.S., Martin, K.C. (2009) Synapse and stimulus-specific local translation during long-term neural plasticity. *Science* **324**, 1536.

Section O

Barros, L.T., Porras, O.H., Bittner, C.X. (2005) Why glucose in the brain matters for PET. *Trends Neurosci.* **28**, 117–119.

Bartels, A., Logothetis, N.K., Moutoussi, K. (2008) fMRI and its interpretations: an illustration of directional sensitivity in area V5/MT. *Trends Neurosci.* **31**, 444–453.

Knöpfel, T., Diez-Garcia, J., Akemann, W. (2006) Optical probing of neuronal circuit dynamics: genetically encoded versus classical fluorescent sensors. *Trends Neurosci.* **29**, 150–166.

Lee, S-J.R., Escobedo-Lozoya, Y., Szatmari, E.M., Yasuda, R. (2009) Activation of CaMKII in a single dendritic spine during long-term potentiation. *Nature* **458**, 299.

Nature Insight: Neurotechniques (2009). *Nature* **461**, 899–939.

Special Section: Neuroscience methods (2009). *Science* **326**, 385–403.

Wyart, C., Del Bene, F., Warp, E., Scott, E.K., Trauner, D., Baier, H., Isacoff, E.Y. (2009) Optogenetic dissection of a behavioural module in the vertebrate spinal cord. *Nature* **461**, 407.

Abbreviations

NB   A few terms share the same standard abbreviation. In each case the one intended will be obvious from the context.

AII	angiotensin II	COMT	catechol-*O*-methyltransferase
ACE	angiotensin converting enzyme	COX2	cyclooxygenase 2
ACh	acetylcholine	CPG	central pattern generator
AChE	acetylcholinesterase	CR	conditioned response
ACTH	adrenocorticotrophic hormone	CRE	cAMP response element
ADH	antidiuretic hormone	CREB	cAMP response element binding protein
ADP	adenosine diphosphate		
AgRP	agouti-related peptide	CREM	cAMP response element modulator
AMP	adenosine monophosphate		
AMPA	α-amino-3-hydroxy-5-methyl-4-isoxazole proprionic acid	CRH	corticotrophin releasing hormone
ANS	autonomic nervous system	CS	conditioned stimulus
AP	action potential	CSF	cerebrospinal fluid
ATN	anterior thalamic nuclei	CVA	cerebrovascular accident
ATP	adenosine 5′-triphosphate	CVLM	caudal ventrolateral medulla
AVP	arginine vasopressin	CVO	circumventricular organ
BAT	brown adipose tissue	DAG	diacylglycerol
BDNF	brain derived neurotrophic factor	DβH	dopamine-β-hydroxylase
		DCML	dorsal column–medial lemniscal system
BMR	basal metabolic rate		
BOLD	blood oxygen level detection	DCN	dorsal column nuclei
BZ	benzodiazepine-binding site	2-DG	2-deoxyglucose
CaM	calmodulin	DHC	dorsal horn cell
CaMKII	calcium–calmodulin-dependent protein kinase II	DLPN	dorsolateral pontine nucleus
		DMH	dorsomedial hypothalamic nucleus
cAMP	cyclic adenosine monophosphate		
		DNX	dorsal vagal nucleus
CART	cocaine- and amphetamine-related transcript	L-DOPA	L-3,4-dihydroxyphenylalanine
		DOPAC	dihydroxyphenyl acetic acid
CAT	computer assisted tomography	DOPEG	3,4-dihydroxy phenylglycol
cbf	cerebral blood flow	DRG	dorsal root ganglion
CC	cingulate cortex	DRG	dorsal respiratory group
CCK	cholecystokinin	DYN	dynorphin
CF	characteristic frequency	EEG	electroencephalography
cGMP	3′,5′-cyclic guanosine monophosphate	ENK	enkephalin
		EP	evoked potential
ChAT	choline acetyltransferase	epp	endplate potential
CNGC	cyclic-nucleotide-gated channel	epsp	excitatory postsynaptic potential
CNS	central nervous system	ERK	extracellular signal-related kinase
CoA	coenzyme A		

ERP	event-related potential		ILD	interaural level difference
FEF	frontal eye field		ION	inferior olivary nucleus
FF	fast fatiguing		IP_3	inositol 1,4,5-trisphosphate
fMRI	functional magnetic resonance imaging		ipsp	inhibitory postsynaptic potential
FR	fatigue resistant		IT	inferotemporal cortex
FRA	flexor reflex afferents		ITD	interaural time difference
FSH	follicle stimulating hormone		JGA	juxtoglomerular apparatus
G_i	inhibitory G protein		K_v	voltage-dependent potassium channel
G_q	G protein coupled to phospholipase		KFN	Kölliker–Fuse nucleus
G_s	stimulatory G protein		LC	locus coeruleus
G_t	transducin		LDCV	large dense-core vesicle
GABA	γ-aminobutyrate		LdT	lateral dorsal tegmental nucleus
GAD	glutamic acid decarboxylase		LFP	local field potential
GC	guanylyl cyclase		LGIC	ligand-gated ion channel
GDP	guanosine 5′-diphosphate		LGN	lateral geniculate nucleus
GFP	green fluorescent protein		LH	luteinizing hormone
GH	growth hormone		LSO	lateral superior olivary nucleus
GHRH	growth hormone releasing hormone		LTD	long-term depression
GnRH	gonadotrophin releasing hormone		LTM	long-term memory
			LTP	long-term potentiation
GPCR	G-protein-coupled receptors		LVA	low voltage-activated
GPe	globus pallidus pars externa		M	magnocellular pathway
GPi	globus pallidus pars interna		MI	primary motor cortex
GR	glucocorticoid receptor		MII	secondary motor cortex
GTO	Golgi tendon organs		mAChR	muscarinic cholinergic receptor
GTP	guanosine 5′-triphosphate		MAO	monoamine oxidase
HD	Huntington's disease		MAP	mean arterial (blood) pressure
5-HIAA	5-hydroxyindoleacetic acid		MB	mammillary bodies
HPA	hypothalamic–pituitary–adrenal (axis)		MDMA	3,4-methylenedioxymethamphet-amine
HPG	hypothalamic–pituitary–gonadal (axis)		MEG	magnetoencephalography
			mepp	miniature endplate potential
HPT	hypothalamic–pituitary–thyroid (axis)		mf	mossy fibres
			MFB	medial forebrain bundle
5-HT	5-hydroxytryptamine (serotonin)		MGN	medial geniculate nucleus
5-HTP	5-hydroxytryptophan		MLR	mesencephalic locomotor region
HVA	high voltage-activated		MOPEG	3-methoxy,4-hydroxyphenylglycol
HVA	homovanillic acid		mpsp	miniature postsynaptic potential
IaIN	Ia inhibitory interneurons		Mr	relative molecular mass
IbIN	Ib inhibitory interneurons		MR	mineralocorticoid receptor
IC	inferior colliculus		MRI	magnetic resonance imaging
ICSS	intracranial self-stimulation		mRNA	messenger RNA
IGF-l	insulin-like growth factor 1		MSN	medium spiny neurons
iGluR	ionotropic glutamate receptor		MSO	medial superior olivary complex
			MST	medial superior temporal cortex

M/T	mitral/tufted cells
Na_v	voltage-dependent sodium channel
NA	noradrenaline (norepinephrine)
NA	nucleus ambiguus
nAc	nucleus accumbens
nAChR	nicotinic cholinergic receptor
nBM	nucleus basalis of Meynert
NMDA	N-methyl-D-aspartate
NMDAR	N-methyl-D-aspartate receptor
nmj	neuromuscular junction
NMR	nuclear magnetic resonance
NOS	nitric oxide synthase
NPY	neuropeptide Y
NREM	non-rapid eye movement sleep
NRM	nucleus raphe magnus
NRPG	nucleus reticularis paragigantocellularis
NST	nucleus of the solitary tract
OCD	obsessive–compulsive disorder
OFC	orbitofrontal cortex
OHC	outer hair cells
ORN	olfactory receptor neurons
OT	olfactory tract
OVLT	vascular organ of the lamina
P	parvocellular pathway
PAD	primary afferent depolarization
PAG	periaqueductal gray matter
PB	parvocellular–blob
PBN	parabrachial nucleus
pBOT	pre-Botzinger complex
PC	Purkinje cells
PDE	phosphodiesterase
PET	positron emission tomography
pf	parallel fibers
PFC	prefrontal cortex
PGO	pontine–geniculate–occipital spikes
PI	parvocellular–interblob
PIP_2	phosphatidyl inositol-4,5-bisphosphate
PKA	protein kinase A
PKC	protein kinase C
PLM	photoactivated localization microscopy
PM	premotor cortex
PNMT	phenyletholamine-N-methyltransferase

PNS	peripheral nervous system
POA	pre-optic area
POMC	pro-opiomelanocortin
PP	posterior parietal cortex
PPN	pedunculopontine nucleus
PPRF	paramedian pontine reticular formation
PRG	pontine respiratory group
PRL	prolactin
psp	postsynaptic potential
PVN	paraventricular nucleus
RA	rapidly adapting
REM	rapid eye movement sleep
RER	rough endoplasmic reticulum
RF	receptive field
rf	radio frequency
RHT	retinohypothalamic tract
S	slow twitch fiber
SA	slowly adapting
Sc	Schaffer collateral
SCG	superior cervical ganglion
SCN	suprachiasmatic nucleus
SER	smooth endoplasmic reticulum
SIA	stress-induced analgesia
SMA	supplementary motor area
SNpc	substantia nigra pars compacta
SNpr	substantia nigra pars reticulata
SNS	sympathetic nervous system
SOC	superior olivary complex
SON	supraoptic nucleus
SP	substance P
SPBT	spinoparabrachial tract
SPL	sound pressure level
SPW-R	sharp wave/ripple complex
SRT	spinoreticular tract
SSRl	selective serotonin reuptake inhibitors
SSV	small clear synaptic vesicle
STDP	spike-timing-dependent plasticity
STED	stimulated-emission depletion microscopy
STN	subthalamic nucleus
STT	spinothalamic tract
TB	trapezoid body
TEA	tetraethylammonium

TENS	transcutaneous electrical nerve stimulation	VLPO	ventrolateral pre-optic area
TH	tyrosine hydroxylase	V_m	resting potential
TMS	transcranial magnetic stimulation	VMAT	vesicular monoamine transporter
		VOR	vestibulo-ocular reflexes
TMN	tuberomammillary nucleus	VP	ventral pallidum
TR	thyroid hormone receptor	VPL	ventral posterior lateral nucleus (of thalamus)
TRH	thyrotrophin releasing hormone		
TRN	thalamic reticular nucleus	VPM	ventral posterior medial nucleus (of thalamus)
TRVP1	capsaicin receptor		
TSH	thyroid stimulating hormone	VRG	ventral respiratory group
TTX	tetrodotoxin	VST	ventral spinocerebellar tract
US	unconditioned stimulus	VTA	ventral tegmental area
VIP	vasoactive intestinal peptide	WDR	wide dynamic range
		WM	working memory

Index